The Sectional Anatomy Learning System:

CONCEPTS

The Sectional Anatomy Learning System:

CONCEPTS

Edith J. Applegate, M.S.
Professor of Science and Mathematics
Kettering College of Medical Arts
Kettering, Ohio

Illustrated by
David B. Ekkens

W.B. SAUNDERS COMPANY
A Division of Harcourt Brace & Company
Philadelphia London Toronto Montreal Sydney Tokyo

W. B. SAUNDERS COMPANY
A Division of Harcourt Brace & Company

The Curtis Center
Independence Square West
Philadelphia, Pennsylvania 19106

Library of Congress Cataloging-in-Publication Data

Applegate, Edith J.

Sectional anatomy learning system: concepts/Edith J. Applegate; illustrated by David B. Ekkens.
p. cm.
Includes bibliographical references and index.

ISBN 0-7216-3237-8

1. Anatomy, Surgical and topographical. I. Title.
[DNLM: 1. Anatomy, Regional. QS 4 A648s]
QM531.A65 1991

611—dc20

DNLM/DLC 90-8987

Editor: Lisa A. Biello
Developmental Editor: Shirley Kuhn
Designer: Paul Fry
Production Manager: Linda R. Turner
Manuscript Editor: Joan Powers
Illustration Coordinator: Lisa Lambert
Indexer: Dennis Dolan
Cover Designer: Jim Gerhart

THE SECTIONAL ANATOMY LEARNING SYSTEM: CONCEPTS ISBN 0-7216-3237-8

 SET ISBN 0-7216-3269-6

Printed in the United States of America

Last digit is the print number: 9 8 7 6 5 4

Dedicated to the

three favorite men

in my life:

Stan, Dave, and Doug

PREFACE

THE SECTIONAL ANATOMY LEARNING SYSTEM was created to provide teachers and students in the diagnostic imaging sciences with a tool for understanding anatomy in three dimensions. A thorough comprehension of anatomy from a sectional perspective greatly enhances an understanding of gross anatomy. The student should be able to observe a transverse, sagittal, or coronal section and mentally reconstruct the three dimensional relationships of that area. Conversely, given a gross dissection of a region, the student should be able to visualize the appearance and the relationships in the planar sections.

THE SECTIONAL ANATOMY LEARNING SYSTEM: CONCEPTS is designed for second-year radiography students and for those involved with the technologies of ultrasonography, computed tomography, and magnetic resonance imaging. It is also valuable to practicing clinicians who want to improve their skills and expertise through continuing education. Basic anatomy and physiology — at least one semester and preferably two — is necessary for background information and terminology.

This text is designed to be used in conjunction with its companion workbook, *THE SECTIONAL ANATOMY LEARNING SYSTEM: APPLICATIONS.* The common focus of both books is "anatomic relationships." The emphasis is on the relationship of one organ to another, one blood vessel to another, a blood vessel to a nerve, or one body part to another. These relationships are essential for sectional anatomy to rise above the ordinary and to become a part of understanding rather than a rote memorization process.

APPLICATIONS, the workbook, is a collection of drawings of transverse, sagittal, and coronal sections for students to label, and coloring exercises are included to help the students see the relationships of important structures and follow them through several levels. Each line drawing is accompanied by questions for students to answer.

Both *CONCEPTS,* the textbook, and *APPLICATIONS,* the workbook, are divided into five chapters, each dealing with one region of the body — the head, the thorax, the abdomen, the male and female pelves, and the extremities and articulations. Each chapter may be used as a self-contained unit for a series of mini-courses, or the books may be used in their entirety for a complete semester course. The instructor has the freedom to pick and choose topics, as the allotted class time permits.

Each chapter begins with an outline and a list of learning objectives that represent student goals. The first part of each chapter is an overview of the anatomy of the region being discussed. This provides a common starting point for all the students in the class. It reviews and refreshes the terminology appropriate to the region; thus, when studying the sectional anatomy, the student can concentrate on the anatomic relationships of structures, rather than becoming bogged down in the anatomy and terminology.

The sectional anatomy of the region is presented after the review of gross anatomy and terminology. Sectional anatomy is discussed first in the transverse plane and then in the sagittal and coronal planes. Each level of section that is presented is illustrated by a labeled line drawing.

Each chapter concludes with a series of review questions that students may use for self-testing. In general, there is at least one question pertaining to each of the objectives that are listed at the beginning of the chapter. Instructors may wish to assign them for homework or use them as test questions.

Besides the workbook, *APPLICATIONS,* other ancillary materials included in this sectional anatomy package are a slide set and an instructor's manual.

The 35-mm slide set contains photographs of CT and MRI scans and actual cadaver specimens that are sectioned in the transverse plane. These correspond to the line drawings of transverse sections in *APPLICATIONS.* There is also a corresponding slide (unlabeled) for each line drawing in *APPLICATIONS,* which can be used, at the instructor's discretion, for teaching or for testing.

And last—but certainly not least—there is an instructor's manual. This is a multifaceted teaching aid for the instructor, containing suggestions for teaching methods, content for various length courses, and techniques for testing. There is a description of each slide with comments about the important structures and the relationships evident in the slide. The instructor's manual contains the correct responses for all of the labeling exercises in the *APPLICATIONS* and for all of the questions in both the concepts textbook and the applications workbook. In addition, there is a bank of questions and answers that are different from those in either *CONCEPTS* or *APPLICATIONS.* The instructor has the option to use these questions for assignments, quizzes, tests, or in any way that will enhance the student's understanding and comprehension of sectional anatomy.

I have used these materials in a variety of ways over the last ten years—in the collegiate courses for radiography, ultrasonography, computed tomography, and magnetic resonance imaging, I use all these materials. When conducting in-service programs for technologists in a hospital department, I usually concentrate on one region and use only one chapter. Because each chapter is self-contained, this makes a flexible teaching mechanism for a continuing education series. It is my intent and hope that *THE SECTIONAL LEARNING SYSTEM* will serve as a comprehensive and effective teaching and learning package.

There is always more to learn and many different teaching methods. You can help change and improve this package with your comments. I welcome any suggestions you have to offer.

EDITH J. APPLEGATE

Acknowledgments

Writing a textbook requires the expertise, cooperation, collaboration, and encouragement of many people. *THE SECTIONAL ANATOMY LEARNING SYSTEMS* is no exception.

Three provosts of Kettering College of Medical Arts, Kettering, Ohio, have had a role in the evolution of this publication. In the spring of 1980, Winton Beaven asked me to develop a course in cross-sectional anatomy for the new ultrasonography curriculum that was to begin in September of that year. That was the beginning of this project. Early in 1989, Robert W. Williams encouraged me (maybe "nagged" is a better word) to put the material that I had developed into a publishable form. That was the second step. And now, in 1991, Peter Bath rejoices with me as *THE SECTIONAL ANATOMY LEARNING SYSTEM* comes to fruition.

I owe a great debt of gratitude to David B. Ekkens for his work in producing the line drawing illustrations for *THE SECTIONAL ANATOMY LEARNING SYSTEM: CONCEPTS.* Little did he realize what was involved when he joined the Biology Department at Kettering College of Medical Arts in 1989. (Somehow this project wasn't in his job description!)

Max Grady and his students in the Special Procedures program at Kettering College of Medical Arts provided the computed tomography (CT) and magnetic resonance imaging (MRI) illustrations. Ted Miller, M.D., from the Radiology Department of Kettering Medical Center, Kettering, Ohio, provided valuable assistance with the CT and MRI correlations.

Glenda Smith, with tremendous spirit and cooperation, spent countless hours performing the word processing and working on the reading list and glossary. She took care of many miscellaneous matters, freeing me to think and write.

My husband and my two sons have been a constant source of love, encouragement, reassurance, strength, and patience. They have understood my frustrations when things didn't go "just right." My students, with their probing quest for knowledge, have been my inspiration. Numerous colleagues and other friends have prodded me with their interest.

To all I say a big *Thank you!*

CONTENTS

CHAPTER 1

CHAPTER 2

CHAPTER 3

CHAPTER 4

CHAPTER 5

CHAPTER 1

THE HEAD AND NECK

GENERAL ANATOMY OF THE HEAD AND NECK
- **Skull**
  - *Cranium*
  - *Facial Skeleton*
  - *Paranasal Sinuses*
  - *Foramina of the Skull*
  - *Additional Bones Associated with the Head*
- **Brain**
  - *Regions of the Brain*
  - *Ventricles*
  - *Meninges*
  - *Arterial Blood Supply*
  - *Venous Drainage*
  - *Cranial Nerves*
- **Orbital Cavity and Contents**
  - *Cavity Walls*
  - *Bulbus Oculi*
  - *Vascular Supply to Orbital Contents*
  - *Protective Features of the Eye*
- **Skeletal Components of the Neck**
  - *General Structure of Vertebrae*
  - *Features of Cervical Vertebrae*
  - *Intervertebral Discs*
- **Viscera of the Neck**
  - *Salivary Glands*
  - *Pharynx*
  - *Larynx*
  - *Esophagus*
  - *Trachea*
  - *Thyroid Gland*
- **Major Blood Vessels of the Neck Region**
  - *Internal Jugular Veins*
  - *Common Carotid Arteries*
  - *Internal Carotid Arteries*
  - *External Carotid Arteries*
  - *Vertebral Arteries*
- **Muscles of the Head and Neck**
  - *Muscles of Facial Expression*
  - *Muscles of Mastication*
  - *Anterior Triangle of the Neck*
  - *Posterior Triangle of the Neck*
- **Major Nerves of the Neck**
  - *Sympathetic Trunks*
  - *Vagus Nerve*
  - *Cervical Plexus*
  - *Phrenic Nerve*

SECTIONAL ANATOMY OF THE HEAD AND NECK
- **Transverse Sections**
  - *Section Through the Lateral Ventricles*
  - *Section Through the Basal Ganglia*
  - *Section Through the Superior Cistern*
  - *Section Through the Midbrain*
  - *Section Through the Pons*
  - *Section Through the Medulla Oblongata*
  - *Section Through the Neck at Level C-1*
  - *Section Through the Neck at Level C-3*
  - *Section Through the Larynx*
- **Sagittal Sections**
  - *Section Through the Temporomandibular Joint*
  - *Section Through the Orbit*
  - *Midsagittal Section*
- **Coronal Sections**
  - *Section Through the Third Ventricle and the Brainstem*
  - *Section Through the Orbit and the Nasal Cavity*

OBJECTIVES

Upon completion of this chapter, the student should be able to do the following:

- Name the bones of the cranium and the face.
- Identify the four paranasal sinuses.
- Identify the five lobes of the cerebrum.
- Describe the relationships of the basal ganglia.
- Describe the location and structure of the diencephalon.
- Locate the components of the brainstem.
- Compare the cerebrum and cerebellum with respect to size, appearance, location, and structure.

- Trace the flow of cerebrospinal fluid through the ventricles in the brain.
- Describe the three layers of meninges.
- Identify six subarachnoid cisterns by describing their location and significance.
- Describe the arterial blood supply to the brain.
- Identify the major venous sinuses that return blood from the brain to the internal jugular vein.
- Name the twelve cranial nerves, state the foramen that serves as a passageway for each, and describe the functions of each nerve.
- Describe the structure of the eye, including the bulbus oculi, musculature, vascular supply, and protective features.
- Identify the features of "typical" vertebrae and compare these with the features unique to cervical vertebrae.
- Compare the location and relationships of the three salivary glands.
- Identify the regions of the pharynx by describing the location and features of each region.
- Describe the features of the larynx.
- Discuss the relationships of the esophagus and trachea as they descend through the neck.
- Discuss the relationships of the internal jugular vein with other vessels and anatomic structures as it descends from the jugular foramen to the brachiocephalic vein.
- Describe the pathways and the relationships of the common, external, and internal carotid arteries.
- State the origin and pathway of the vertebral arteries.
- Name five muscles of facial expression, describe the location of each, and state the insertion and innervation of this group.
- Name four muscles of mastication, describe the location of each muscle, and then state the insertion and innervation of this muscle group.
- Identify the margins of the anterior and posterior triangles of the neck, and name the principal components in each triangle.
- Name one cranial nerve, one spinal nerve, and one nerve plexus located in the neck.
- Identify the regions of the brain, nerves, blood vessels, muscles, and viscera of the head and neck in transverse, sagittal, and coronal sections.

GENERAL ANATOMY OF THE HEAD AND NECK

The principal bony structure of the head is the skull, which is especially adapted to house and to protect the brain and the pituitary gland, the two organs that integrate body activities. The head is the location of the specialized sensory organs of vision, hearing, equilibrium, taste, and smell, which provide input concerning our surroundings. The digestive and respiratory systems begin as openings in the head and continue as passageways in the neck. The vertebrae and the muscles of the neck provide support for the head but also permit a certain degree of movement. The neck also functions as a passageway for the spinal cord, blood vessels, and nerves. The intricate structure of the head and neck is an appropriate complement for the functional complexity of this region.

Skull

The bony framework of the head is called the skull. This is the most complex osseous structure of the body, and it consists of twenty-two bones connected by immovable joints called sutures. For descriptive purposes, these bones are divided into the cranium and the facial skeleton, although there is no distinct line of demarcation between the two parts. Some of the bones surround a large cranial cavity that contains the brain. The superior surface of this region is covered by the scalp. Some skull bones adjacent to the nasal cavity contain air-filled spaces called paranasal sinuses. Many of the bones in the skull have holes or openings called foramina that serve as passageways for nerves and blood vessels. In addition to the bones of the skull, there are seven other bones associated with the head. These are the auditory ossicles and the hyoid bone.

CRANIUM

BONES OF THE CRANIUM. The eight bones of the cranium form the cranial cavity, which houses the brain. The single **frontal bone** forms the forehead and the superior part of the orbit of the eye. It contains frontal paranasal sinuses, which communicate with the nasal cavity. Two **parietal bones** form most of the top of the cranium. Two **temporal bones** form a portion of the sides and the base, or the floor, of the cranium. Each temporal bone has a thin, flat, squamous portion that forms the inferior lateral part of the cranium. The posterior portion of the temporal bone is the **mastoid process.** The **external auditory canal,** the tympanic membrane, the middle ear, and the inner ear are located in the petrous portion, which extends medially to form part of the base of the cranium. The single **ethmoid bone** is located between the eyes and forms most of the medial wall of each orbit. The superior surface of the ethmoid bone forms part of the base of the cranial cavity and the roof of the nasal cavities. A thin, perpendicular plate of the ethmoid bone extends inferiorly to form part of the **nasal septum.** The superior and middle conchae (turbinates) in the nasal cavity are part of the ethmoid bone. The single **sphenoid bone** lies at the base of the skull, anterior to the temporal bones. Often described as bat-shaped, the sphenoid bone has "wings" that form the anterior lateral portion of the cranium as well as the lateral walls of the orbits. The sella turcica is found in the center of the bone and is the location of the pituitary gland. The single **occipital bone** forms the posterior portion and a part of the base of the cranium. It has a large hole, the foramen magnum, for passage of the spinal cord and the vertebral vessels. The bones of the cranium are illustrated in Figure 1–1.

CALVARIA AND SCALP. The domelike superior portion of the cranium is the **calvaria,** or skullcap. It is composed of the superior portions of the frontal, parietal, and occipital bones. The calvaria is covered by the scalp, which extends from the eyebrows to the superior nuchal line on the occipital bone.

Structurally, the scalp has five layers. The outer layer, the **skin,** covers the second layer, which is composed of highly vascularized **subcutaneous connective tissue.** The third layer, or epicranium, consists of two thin muscles that are connected by a broad, flat tendon, or aponeurosis. This musculoaponeurotic sheet has the frontalis muscle at the anterior end and the occipitalis muscle at the posterior end. The strong aponeurosis that connects the two muscles is known as the **epicranial aponeurosis,** or **galea aponeurotica.** These three layers of the scalp are bound tightly together and move as a unit. Collectively, they are often referred to as the scalp proper. A fourth layer that consists of **loose connective tissue** separates the scalp proper from the fifth layer, the **periosteum,** or pericranium. The loose connective tissue layer permits mobility of the scalp proper, but it is also considered a potentially dangerous area because it allows scalp infections to spread easily. Lacerations of the scalp typically bleed profusely because of the extensive vascularization of the subcutaneous connective tissue. The acronym SCALP serves as a useful mnemonic tool for remembering the five layers of the scalp.

S = Skin
C = Connective tissue
A = Aponeurosis
L = Loose connective tissue
P = Periosteum

CRANIAL CAVITY. The bones of the cranium surround a large cavity that contains the brain. The floor of the cranial cavity is subdivided into anterior, middle,

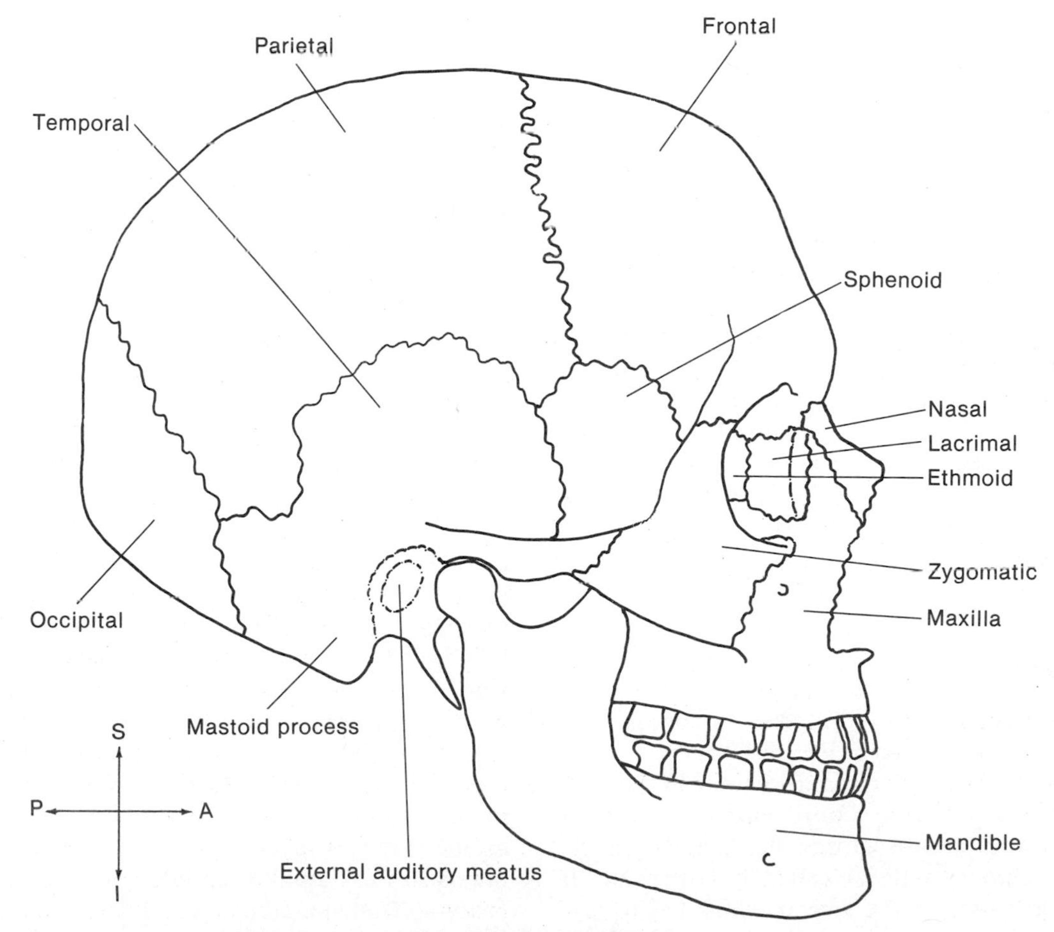

FIGURE 1-1. Bones of the skull (cranium and face).

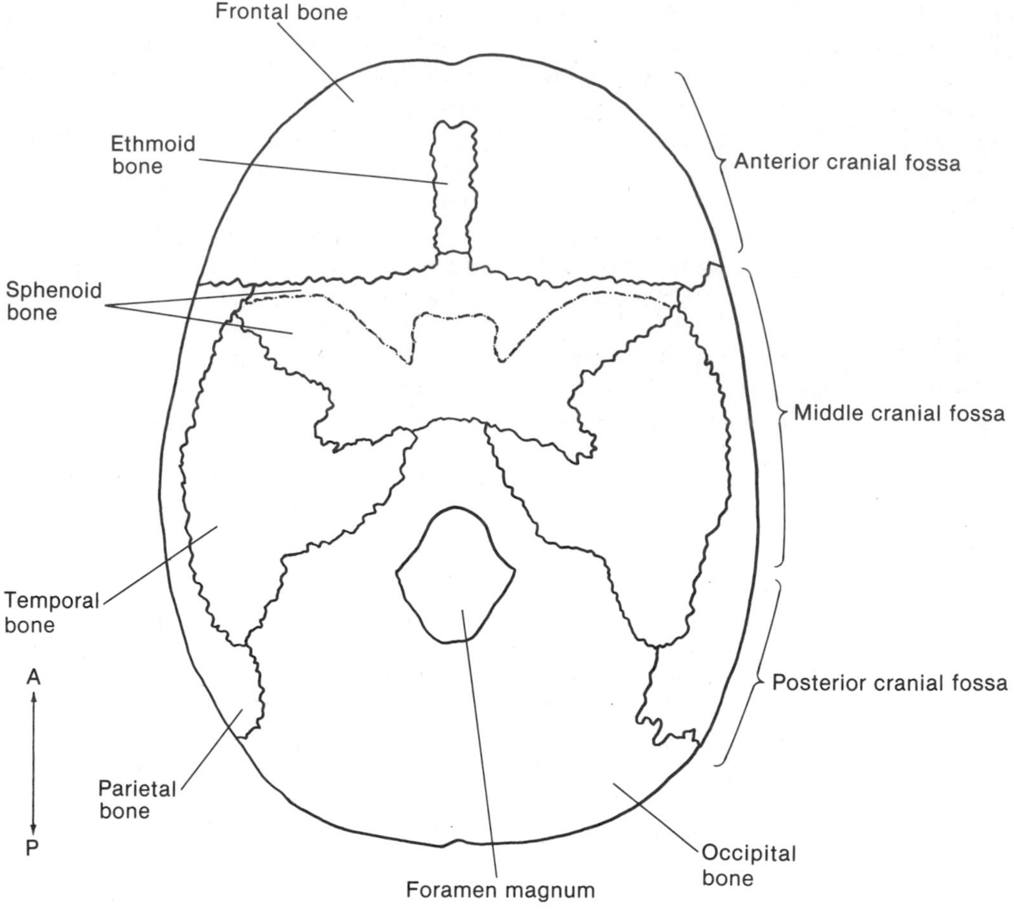

FIGURE 1–2. Cranial fossae (floor of cranial cavity).

and posterior cranial fossae, as illustrated in Figure 1–2. The **anterior cranial fossa** is formed by portions of the ethmoid, sphenoid, and frontal bones. This fossa houses the frontal lobes of the brain. The floor of the **middle cranial fossa** is composed of the body and the greater wings of the sphenoid and the squamous and petrous portions of the temporal bones. This region contains the temporal lobes of the brain. The **posterior cranial fossa** comprises the remainder of the cranial cavity and is formed by parts of the sphenoid, temporal, and occipital bones. This region contains the cerebellum, the pons, and the medulla oblongata. The inferiormost portion of the posterior cranial fossa shows the large **foramen magnum** through which the spinal cord passes.

FACIAL SKELETON

The facial portion of the skull consists of 14 bones. Some of these bones are illustrated in Figure 1–1.

The two **maxillae** (singular, **maxilla**) unite in the midline to form the upper jaw. A horizontal piece of each maxilla, the **palatine process,** forms the anterior portion of the roof of the mouth, which is called the hard palate. If the palatine processes of the two maxillae fail to join during prenatal development, a cleft palate results. Each maxilla contains a maxillary sinus, which is a large air space that communicates with the nasal cavity. Two **palatine bones** form the posterior portion of the hard palate. Two **zygomatic bones,** one on each side, form the prominence of the cheek and the lateral margin of the orbit. A posteriorly extending process of the zygomatic bone unites with the temporal bone to form the zygomatic arch. Two small, rectangular **nasal bones** join in the midline to form the bridge of the nose. Fractures of these bones are common facial injuries. The anterior part of the medial wall of each orbital cavity consists of a small, thin **lacrimal bone.** Each bone has a groove that helps to form the nasolacrimal canal, which permits the tears of the eye to drain into the nasal cavity. The single **vomer** is a thin bone that is shaped like the blade of a plow. It forms the inferior portion of the nasal septum. The two **inferior nasal conchae** are scroll-like bones that project horizontally from the lateral walls of the nasal cavities. The **superior** and **middle conchae** are part of the ethmoid bone, but the **inferior nasal conchae** are separate bones. They are covered by mucous membrane to warm and moisten the air that enters the nasal cavities. The single **mandible** forms the lower jaw. The posterior ends of the mandible extend vertically to form the **rami** (singular, **ramus**). Each ramus has a knoblike condyle and a pointed coronoid process. The mandibular condyle articulates with the temporal bone to form the temporomandibular joint. This is the only movable joint in the skull.

PARANASAL SINUSES

The **paranasal sinuses** are air-filled spaces in some of the bones adjacent to the nasal cavity. There are four sets of paranasal sinuses, and they are named according to the bone in which they are located: **frontal, ethmoidal, sphenoidal,** and **maxillary.** Usually developing after birth as outgrowths of the nasal cavity, these sinuses retain their original openings so that their secretions drain into the nasal cavity. The mucous membrane that lines the sinuses is continuous with the mucosa of the nasal cavity, but it is thinner and less vascular than the nasal mucosa.

FRONTAL SINUSES. The **frontal sinuses** are located in the frontal bone near the midline. They develop in the child and are usually visible on radiographs by the time the child is 7 years of age; however, the sinuses continue to enlarge throughout adolescence. The frontal sinuses drain into the **middle meatus** of the nasal cavity, between the middle and inferior nasal conchae, by way of a frontonasal duct.

ETHMOIDAL SINUSES. The **ethmoidal sinuses** consist of numerous air spaces in the ethmoid bone between the orbit of the eye and the upper part of the nasal cavity. Although a few ethmoidal air cells are present in the neonate, they are not readily visible on radiographs until the infant is 2 years old and do not enlarge significantly until later in the childhood years – usually around the ages of 6 to 8. The ethmoidal sinuses have numerous openings into the **superior meatus** between the superior and middle conchae and into the **middle meatus** between the middle and inferior conchae.

SPHENOIDAL SINUSES. The **sphenoidal sinuses** are located in the body of the sphenoid bone just posterior to the ethmoid sinuses and the nasal cavity. The sinuses occupy most of the volume of the sphenoid body so that only a thin plate of bone separates the sinuses from the pituitary gland, the optic nerve, the optic chiasma, the internal carotid artery, and the cavernous sinus. Tiny sphenoidal sinuses may be present in the newborn but their development is more likely to occur in a child that is around 2 years of age, with additional growth occurring during late childhood and adolescence. The sphenoidal sinuses drain into the **sphenoethmoidal recess** above the superior nasal concha.

MAXILLARY SINUSES. Small **maxillary sinuses,** which are located in the bodies of the maxillae, are present in the newborn and grow slowly until the child reaches puberty. The accelerated development of the maxillary sinuses during the child's adolescence contributes to the apparent change in facial features that typically occurs during this period. When fully developed, the maxillary sinuses are the largest of the paranasal sinuses. The maxillary sinus drains into the nasal cavity by way of a relatively long **hiatus semilunaris,** which opens into the **middle meatus.** A couple of factors hinder the drainage of this sinus: the hiatus traverses a superior direction when the body is erect, thus necessitating drainage "against gravity"; and the opening from the sinus into the hiatus is in a superior location. The drainage problem is further complicated by the fact that the maxillary sinus is the most inferiorly located of the paranasal sinuses, and the communicating channels allow the other sinuses to drain into the maxillary sinus. These are the reasons why the maxillary sinus is the sinus most often involved in infections. The frontal, ethmoidal, and sphenoidal sinuses are innervated by branches of the ophthalmic division of the trigeminal nerve. The maxillary sinus is innervated by branches of the maxillary division of the fifth cranial nerve, which will be discussed later in this chapter.

FORAMINA OF THE SKULL

Foramina are openings in bones that serve as passageways for nerves and blood vessels. Because the vessels that transport blood and the nerves that carry impulses must pass through the sutured, helmetlike bones of the skull on their way to and from the brain, there are numerous foramina in these bones. See Table 1 – 1 for a summary of the foramina of the skull.

ADDITIONAL BONES ASSOCIATED WITH THE HEAD

There are 6 auditory ossicles (3 pairs) and a single hyoid bone that are associated with the head in addition to the 8 bones of the cranium and the 14 bones of the facial skeleton. There are three chambers in the ear, called the inner ear, the middle ear, and the external ear. The ear ossicles,

TABLE 1 – 1. Major Foramina of the Skull

Foramen	Location	Structures Transmitted
Carotid canal	Temporal bone	Internal carotid artery; sympathetic nerves
Hypoglossal canal	Occipital bone	Hypoglossal nerve
Infraorbital	Maxilla	Maxillary branch of trigeminal nerve; infraorbital nerve and artery
Internal auditory meatus	Temporal bone	Vestibulocochlear nerve
Jugular	Temporal bone	Internal jugular vein; vagus, glossopharyngeal, and spinal accessory nerves
Magnum	Occipital bone	Medulla oblongata/spinal cord; accessory nerves; vertebral arteries
Nasolacrimal canal	Lacrimal bone	Nasolacrimal (tear) duct
Olfactory	Ethmoid bone	Olfactory nerves
Optic canal	Sphenoid bone	Optic nerve; central artery and vein of retina
Ovale	Sphenoid bone	Mandibular branch of trigeminal nerve
Rotundum	Sphenoid bone	Maxillary branch of trigeminal nerve
Stylomastoid	Temporal bone	Facial nerve

TABLE 1-2. Summary of Bones Associated with Head

Bone	Number	Total
Bones of the Cranium		8
Frontal	1	
Parietal	2	
Occipital	1	
Temporal	2	
Sphenoid	1	
Ethmoid	1	
Bones of the Face		14
Maxillae	2	
Nasal	2	
Zygomatic	2	
Lacrimal	2	
Mandible	1	
Vomer	1	
Inferior nasal conchae	2	
Palatine	2	
Middle Ear Bones		6
Malleus	2	
Incus	2	
Stapes	2	
Hyoid Bone		1

the **malleus,** the **incus,** and the **stapes,** are located within the middle ear chamber in the petrous portion of the temporal bone. The ossicles transmit and amplify sound waves through the middle ear. The single **hyoid bone** is located in the neck just superior to the larynx. The hyoid is unique because it does not attach directly to any other bone; instead, it is suspended by ligaments. Muscles associated with the hyoid bone are described later in this chapter in the discussion of the anterior triangle of the neck. Table 1-2 gives a summary of the bones associated with the head.

Brain

The predominant structure in the cranial cavity is the brain—a rather unimpressive looking mass of tissue that weighs approximately 3 lb. It is composed of organized regions of white matter and gray matter. The **white matter** consists of nerve fibers that are covered with a white, fatty substance called myelin. The **gray matter** consists of nerve cell bodies and unmyelinated fibers. Some of the gray matter is grouped together to form regions called basal ganglia. Generally, the gray areas are regions of **synapse**—electrical communication between neurons. Spaces called **ventricles** are located within the brain and are surrounded by brain tissue. The brain is separated from the cranial bones by layers of connective tissue called **meninges** that help protect the surface of the brain. Further protection is provided by cerebrospinal fluid, which circulates through the ventricles and around the brain. Although the brain accounts for only about 2 percent of the total body weight, it is very metabolically active; consequently, it receives 15 to 20 percent of the cardiac output through its arterial blood supply. After the blood circulates through the capillaries to provide oxygen for the brain tissue, it is returned to the heart by the veins. Another feature of the brain involves the 12 pairs of cranial nerves that emerge from the inferior surface. These nerves provide pathways for incoming sensory impulses, which are processed and interpreted by the brain, and for outgoing motor impulses, which travel from the brain to a muscle or gland and effectuate an action.

REGIONS OF THE BRAIN

For descriptive purposes the brain may be divided into the **cerebrum, diencephalon, brainstem,** and **cerebellum.**

CEREBRUM. The largest portion of the brain is the **cerebrum,** which consists of two cerebral hemispheres connected by a mass of white matter called the **corpus callosum.** The anterior end of the corpus callosum is called the **genu,** and the posterior end is called the **splenium.** The deep cleft between the two cerebral hemispheres is the **longitudinal fissure.** The surface of the cerebrum exhibits numerous convolutions that greatly increase the surface area of the cerebral cortex. The ridges are called **gyri** (singular, **gyrus**), whereas the furrows between them are designated as **sulci** (singular, **sulcus**).

Superficially, the cerebrum is divided into lobes. A **central sulcus** separates the **frontal lobe** from the **parietal lobe.** Posteriorly, the **parieto-occipital sulcus** separates the parietal lobe from the **occipital lobe.** Laterally, the **temporal lobe** is situated below the **lateral fissure.** A fifth lobe that is called the **insula,** or island of Reil, is located deep within the lateral fissure. The lobes and fissures are illustrated in Figure 1-3.

The surface layer of the cerebrum is composed of **gray matter,** which consists of nerve cell bodies and unmyelinated fibers. This is called the cerebral cortex and is 2 to 4 mm thick. Lying beneath the cerebral cortex is the lighter colored **white matter,** which is made up of myelinated nerve fibers.

Scattered throughout the white matter are distinct regions of gray matter called **basal ganglia.** Two of the larger basal ganglia are the **caudate nucleus** and the **lentiform** or **lenticular nucleus.** In sectional anatomy, the caudate nucleus is usually visualized in association with the lateral ventricle. The lentiform nucleus is rather centrally located in each cerebral hemisphere. It is subdivided into the lateral, or external, **putamen** and the medial, or internal, **globus pallidus.** The **claustrum,** another of the basal ganglia, is a thin layer of gray matter that is located just lateral to the lentiform nucleus and deep to the cortex of the insula. A band of white matter that is situated medial to the lentiform nucleus is called the **internal capsule.** The white matter that is present between the lentiform nucleus and the claustrum is the **external capsule.** The claustrum is separated from the insula by the **extreme capsule.** Because of their appear-

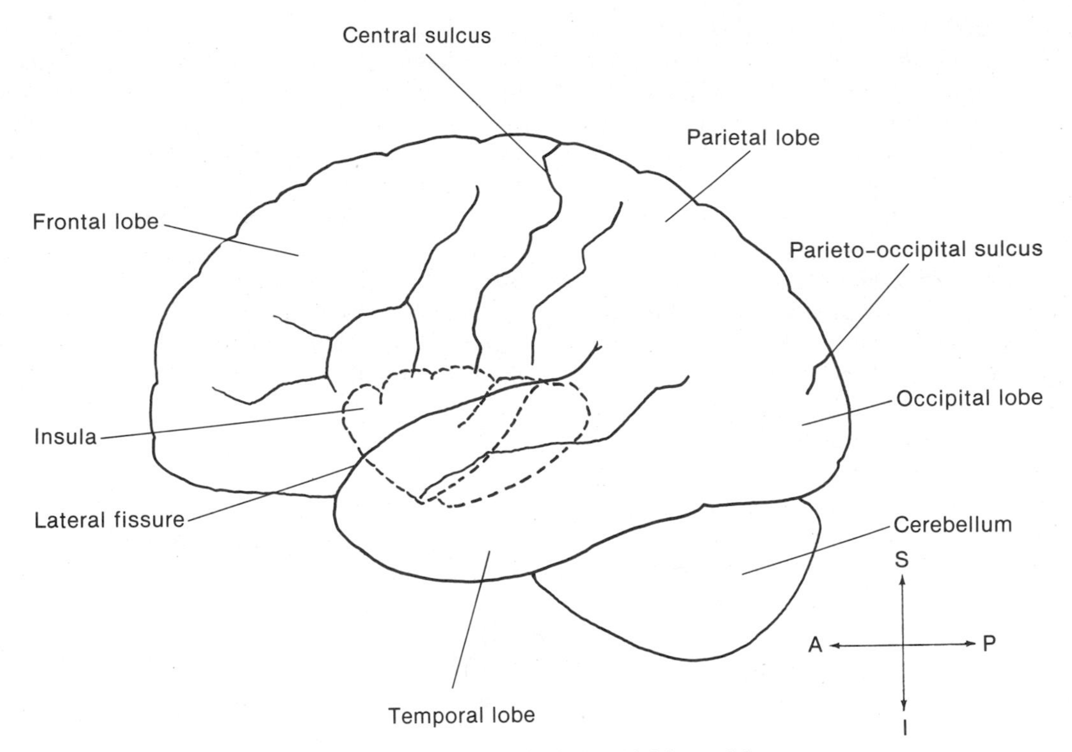

FIGURE 1-3. Surface of the brain with lobes and fissures.

ance, the caudate nucleus, the internal capsule, and the lentiform nucleus are sometimes referred to as the **corpus striatum.** Figure 1-4 illustrates the arrangement of the basal ganglia in a transverse section.

DIENCEPHALON. The diencephalon is centrally located and is nearly hidden from view by the large cerebral hemispheres. It surrounds the midline third ventricle and consists of the **epithalamus, thalamus,** and **hypothala-**

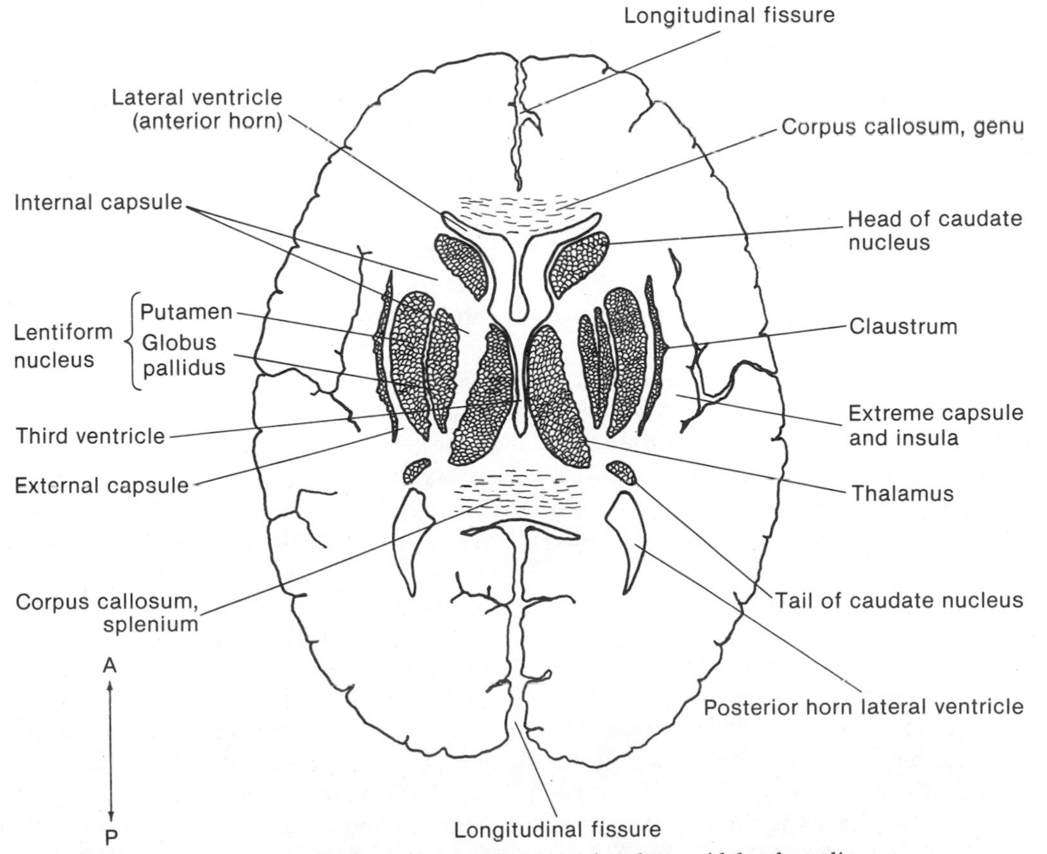

FIGURE 1-4. Transverse section of cerebrum with basal ganglia.

mus. The largest portion, the thalamus, is a mass of gray matter that lies on either side of and that forms the lateral walls of the third ventricle. This is a major relay station of the afferent, or sensory, pathway, which carries impulses to the cerebral cortex. The epithalamus forms the roof of the third ventricle. A midline projection of the epithalamus forms the **pineal gland.** The hypothalamus forms the floor of the third ventricle. On the inferior aspect of the hypothalamus the following structures are located: the **infundibulum,** or the **pituitary stalk;** the **optic chiasma,** where the optic nerves cross over and then emerge as optic tracts; and the **mammillary bodies,** which are two spherical masses of gray matter surrounded by a layer of white matter. The mammillary bodies function in some swallowing reflexes.

BRAINSTEM. The brainstem is subdivided into the midbrain, the pons, and the medulla oblongata. The smallest division is the **midbrain,** which is located between the diencephalon and the pons. The midbrain surrounds the **cerebral aqueduct,** a long, slender channel for cerebrospinal fluid. Four rounded protuberances, the **corpora quadrigemina,** are visible on the dorsal aspect of the midbrain. The upper pair, the **superior colliculi,** functions in the visual pathway, whereas the lower pair, the **inferior colliculi,** functions in the auditory pathway. Just above the corpora quadrigemina is a small glandular structure projecting from the diencephalon. This is the **pineal body,** or **gland.** On the ventral aspect of the midbrain, there are two ropelike bundles called **cerebral peduncles.** These are composed of motor fibers that extend from the cerebral cortex to the spinal cord. A narrow band of darkly pigmented cells crosses each cerebral peduncle. This is the **substantia nigra.** The dark color is caused by melanin in the cells. The substantia nigra seems to be involved in the production of dopamine in the brain, and it also functions in muscle tone reflexes.

The **pons** appears as a prominent band of fibers located between the midbrain and the medulla oblongata. Most of the fibers in the pons connect the two halves of the cerebellum, but some extend from the cerebellum to other parts of the brain.

The **medulla oblongata** looks somewhat conical and extends from the pons to the foramen magnum, where it is continuous with the spinal cord. A **median fissure** is located on the anterior surface of the medulla, and on either side of this fissure, there is a small swelling called a **pyramid.** The **cerebral aqueduct** widens about halfway along the medulla to form the fourth ventricle.

CEREBELLUM. Situated posterior to the pons and the medulla oblongata, the **cerebellum** occupies the posterior cranial fossa. It consists of two **cerebellar hemispheres** connected by a central **vermis,** which resembles a coiled-up worm. The term *vermis* is derived from the Latin word *verm,* which means worm. The surface of the cerebellum is covered by a layer of gray matter, the **cerebellar cortex.** Deep to the cortex is the white matter. Because the gray and white matter are laminated or foliated in appearance, the arrangement is sometimes called **arbor vitae. Cerebellar peduncles** connect the cerebellum with other portions of the brain. There are three pairs of cerebellar peduncles. The **superior cerebellar peduncles** connect the cerebellum to the midbrain. Fibers of the **middle cerebellar peduncles** connect the cerebellum and the pons. The **inferior cerebellar peduncles** consist of fibers that pass between the cerebellum and the medulla oblongata. The cerebellum plays an important role in controlling muscle tone and in coordinating muscular activity of the body. Some of the structures and regions of the brain are illustrated in Figure 1–5.

VENTRICLES

Ventricles are fluid-filled cavities within the brain. These ventricles include the lateral ventricles and the third and fourth ventricles. The ventricles and their communicating channels are illustrated in Figure 1–6.

LATERAL VENTRICLES. There is a large **lateral ventricle** within each cerebral hemisphere. The major portion of each lateral ventricle is located in the parietal lobe. These ventricles extend into the frontal lobes as the **anterior horns,** into the occipital lobes as the **posterior horns,** and into the temporal lobes as the **inferior horns.** The lateral ventricles are separated from each other medially by a thin vertical partition called the **septum pellucidum.** Each lateral ventricle communicates with the third ventricle by a small opening called the **interventricular foramen,** or the **foramen of Monro.**

THIRD VENTRICLE. The **third ventricle** is a narrow midline chamber enclosed by the diencephalon. The lateral walls of the third ventricle are formed by the right and left masses of the thalamus. The epithalamus and hypothalamus form the ventricle's roof and floor, respectively. A small band of white fibers called the **intermediate mass** passes through the ventricle between the right and the left thalami. The third ventricle communicates with the fourth ventricle by means of a relatively long **cerebral aqueduct,** also called the **aqueduct of Sylvius,** which passes through the midbrain.

FOURTH VENTRICLE. The **fourth ventricle** lies internal to the pons and the medulla oblongata at the level of the cerebellum. There are two openings called the **foramina of Luschka** in the lateral walls of the fourth ventricle. In the medial aspect of the dorsal wall there is a single opening called the **foramen of Magendie.** The ventricles communicate with the subarachnoid space through these three openings. The fourth ventricle is continuous with the narrow central canal that extends throughout the length of the spinal cord.

CHOROID PLEXUS. Specialized vascular structures, comprising the **choroid plexus,** are located in the lateral, third, and fourth ventricles. The choroid plexus produces the **cerebrospinal fluid** by filtration and secretion. Originating in the ventricles, the fluid circulates outwardly through the foramina in the fourth ventricle and into the

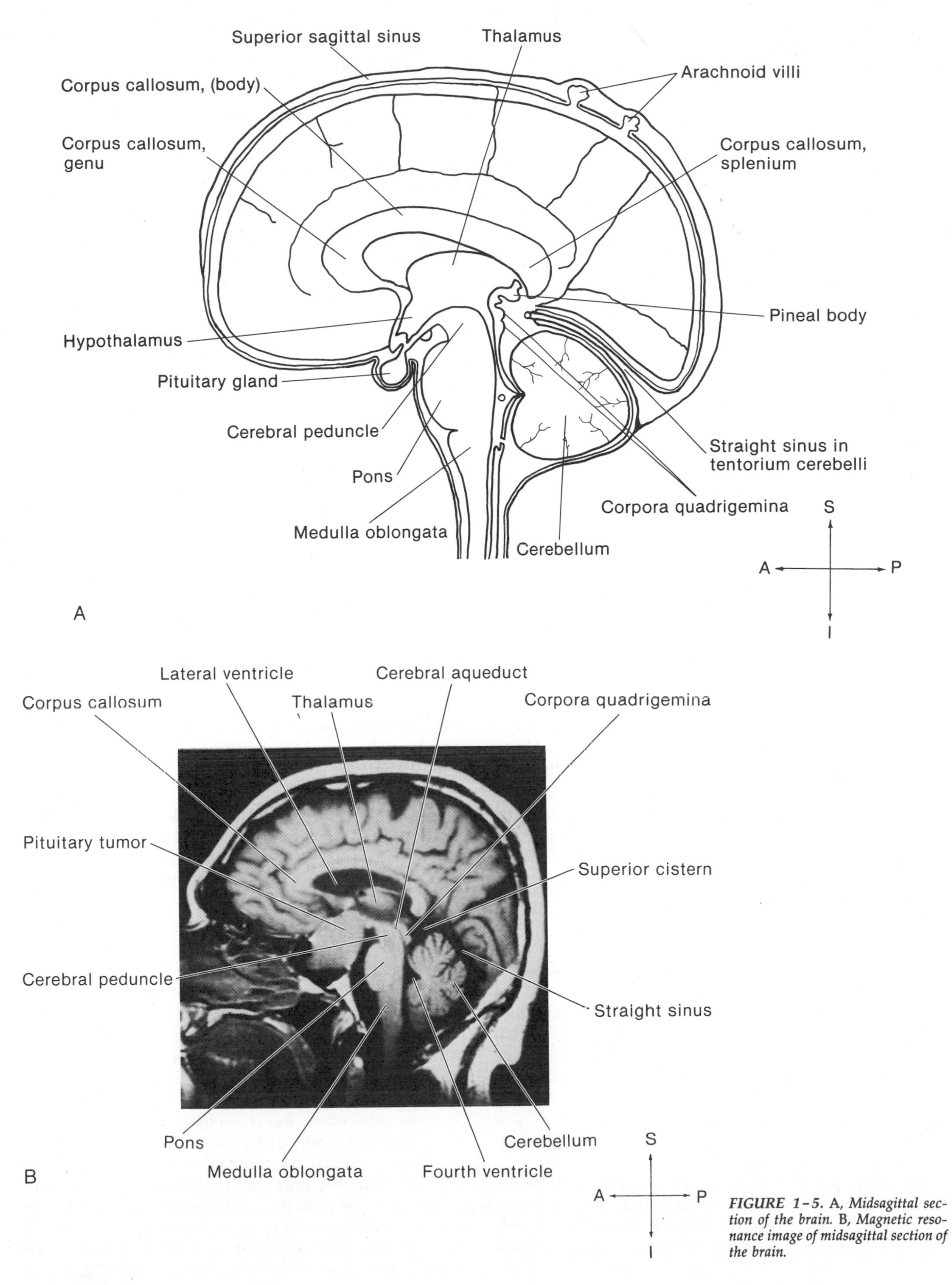

FIGURE 1-5. A, *Midsagittal section of the brain.* B, *Magnetic resonance image of midsagittal section of the brain.*

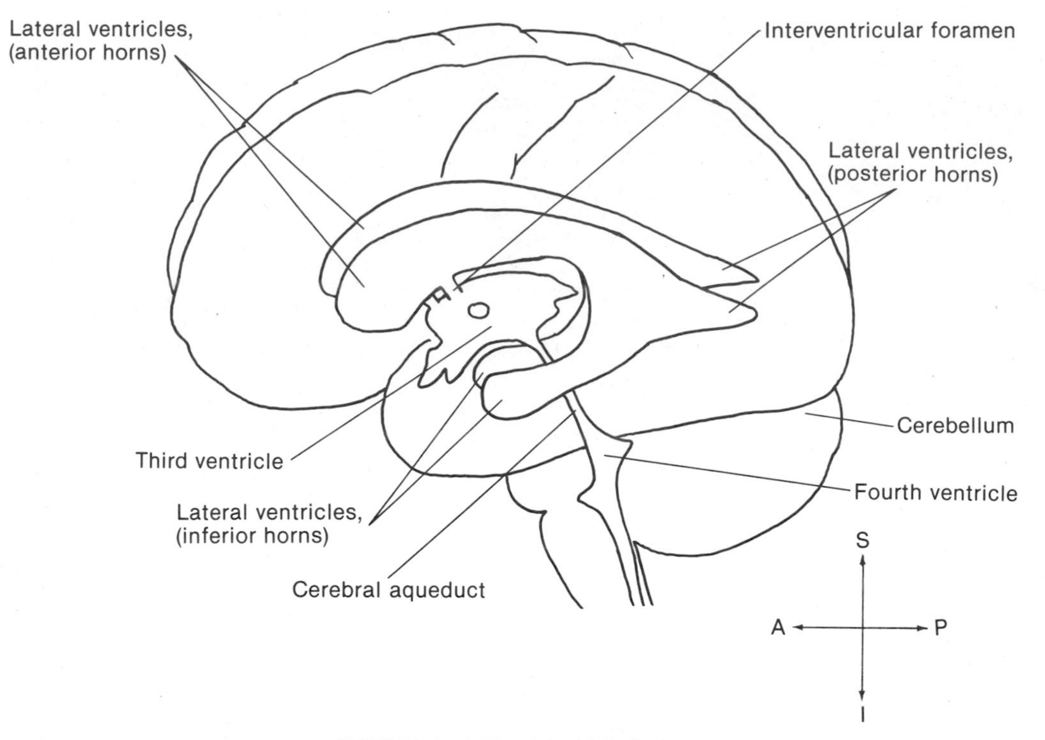

FIGURE 1-6. Ventricles of the brain.

subarachnoid space around the brain and the spinal cord. From there, the fluid is reabsorbed into the venous system and returned to the heart as part of the blood.

MENINGES

Three distinct connective tissue membranes called **meninges** cover the brain. These meninges include the dura mater, the arachnoid, and the pia mater. Cerebrospinal fluid circulates in the subarachnoid space between the arachnoid and the pia mater. In certain areas the arachnoid and the pia mater are widely separated, which creates spaces called cisterns.

DURA MATER. The outermost layer of the meninges is the **dura mater,** which is composed of tough fibrous connective tissue. This forms a strong outer covering that serves as a supportive and protective structure for the brain. Although the dura mater is sometimes described as consisting of two layers, it is important to realize that what is being called the outer layer of the dura mater is actually the endosteum (internal periosteum) of the calvaria. This outer layer is continuous with the external periosteum at the sutures and foramina. The inner, or meningeal, layer is the true dura mater. This meningeal layer is continuous with the spinal dura mater at the foramen magnum, and it also provides tubular sheaths for the cranial nerves as they pass through the foramina in the floor of the cranial fossa. The endosteum and true meningeal dura mater are closely adherent, except where there are venous sinuses.

The meningeal, or true, dura mater forms four inwardly projecting folds that partially divide the cranial cavities into compartments. These four extensions of the dura mater include the **falx cerebri,** which is located between the cerebral hemispheres; the **falx cerebelli,** which is found between the cerebellar hemispheres; the **tentorium cerebelli,** which is situated between the cerebrum and cerebellum; and the **diaphragma sellae,** which forms a bridge over the sella turcica and covers the hypophysis.

ARACHNOID. The middle layer of the meninges is an extremely thin and delicate **arachnoid.** This layer is separated from the dura mater by a small subdural space that contains just enough fluid to keep the adjacent surfaces moist. The arachnoid is separated from the innermost pia mater by the **subarachnoid space,** which contains the cerebrospinal fluid and the larger blood vessels of the brain. Samples of cerebrospinal fluid may be withdrawn from the subarachnoid space and may be examined for evidence of infections or of subarachnoid bleeding. This is usually done in the lumbar region of the vertebral column to minimize danger of damage to the brain or spinal cord. From the inner surface of the arachnoid, minute trabeculae extend across the subarachnoid space to become continuous with the pia mater. This presents a cobweblike appearance and is the basis of the name arachnoid. In the vicinity of the venous sinuses, there are numerous outgrowths, or diverticula, of the arachnoid that penetrate the dura and project into the venous sinuses. The interior of these **arachnoid villi** is continuous with the subarachnoid space and contains cerebrospinal fluid. This permits reabsorption of the fluid into the venous system. Arachnoid villi are illustrated in Figure 1-7.

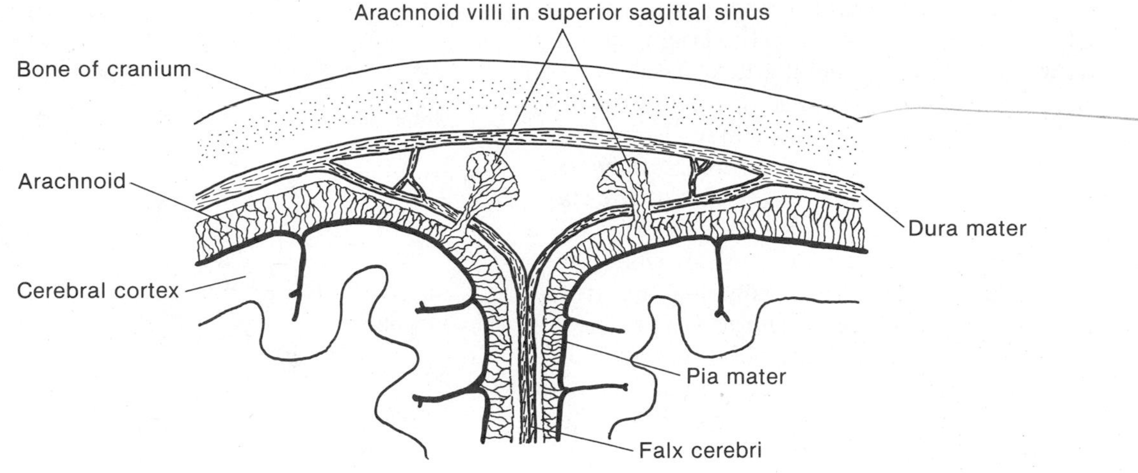

FIGURE 1-7. Meninges and arachnoid villi.

PIA MATER. The **pia mater** is the innermost layer of the meninges. It is a thin, highly vascular layer that intimately adheres to the cortical tissue of the brain surface and closely follows the contours of the brain. The arachnoid and pia mater, together, are frequently referred to as the **leptomeninges.** These two layers of meninges are in close contact at the crests of the gyri, but as the pia follows the dips of the sulci and as the arachnoid forms a bridge over the top of the gyri, they become separated, and triangular subarachnoid spaces are formed.

SUBARACHNOID CISTERNS. In addition to the triangular subarachnoid spaces between the gyri described above, there are certain other areas around the base of the brain where the arachnoid and pia mater are widely separated. This creates spaces called **cisterns,** which contain relatively large amounts of cerebrospinal fluid. Some of the cisterns discussed in this section are illustrated in Figure 1-8. The **cerebellomedullary cistern,** or cisterna magna, is formed by the arachnoid as it bridges the interval between the medulla oblongata and the inferior surface of the cerebellum. The foramen of Magendie (median aperture) from the fourth ventricle opens into this cistern. For cases in which the performance of a lumbar puncture to obtain samples of cerebrospinal fluid would be especially difficult or dangerous, the fluid may be obtained from the cerebellomedullary cistern by a cisternal puncture. The **pontine cistern** is a space on the ventral surface of the pons. This cistern contains the basilar artery and receives cerebrospinal fluid from the fourth ventricle through the foramina of Luschka (lateral apertures). As

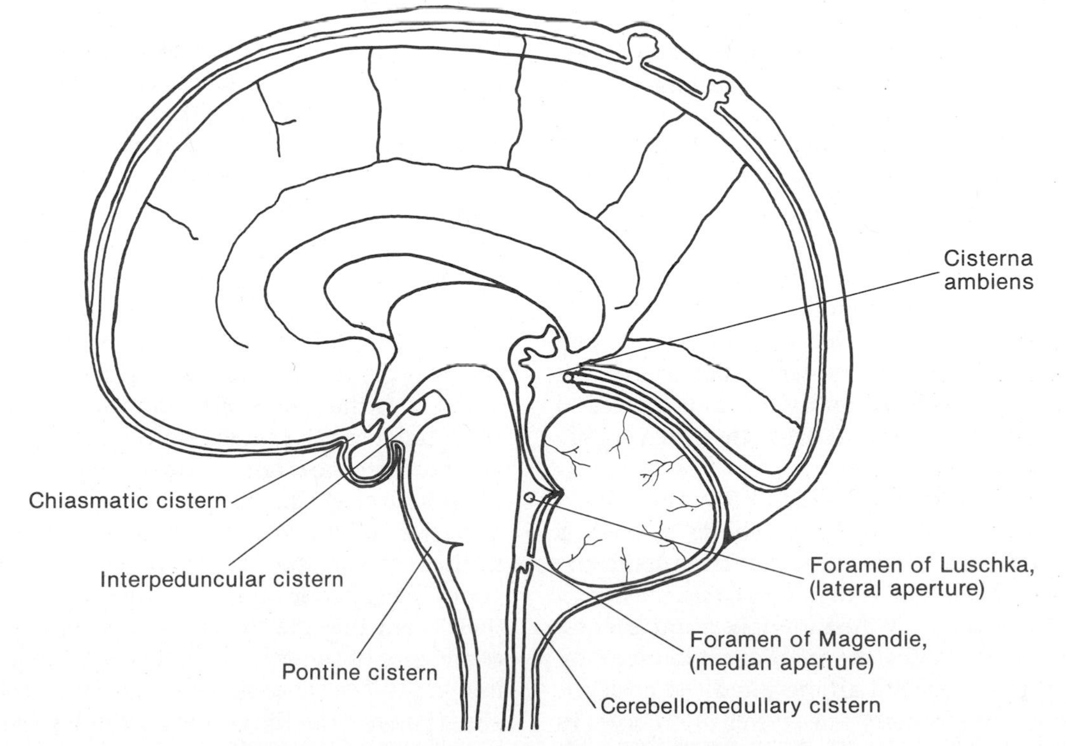

FIGURE 1-8. Midsagittal section of the brain showing cisterns.

the arachnoid bridges the gap from the temporal lobe to the frontal lobe, it forms the **cistern of the lateral sulcus,** which contains the middle cerebral artery. Between the two temporal lobes, the arachnoid is separated from the cerebral peduncles by an **interpeduncular cistern** that contains the circle of Willis. Anteriorly and superiorly, the interpeduncular cistern continues as the **chiasmatic cistern.** The **cisterna ambiens** occupies the interval between the splenium of the corpus callosum and the superior surface of the cerebellum. It contains the great cerebral vein and the pineal gland. The cisterna ambiens is also called the cistern of the great cerebral vein, superior cistern, or quadrigeminal cistern. The pineal gland, which is located in this cistern, usually becomes calcified after adolescence and can be visualized on normal radiographs. This characteristic makes it an important landmark in neuroradiography and neurosurgery.

ARTERIAL BLOOD SUPPLY

The blood is supplied to the brain by two pairs of arteries, the **internal carotid** and the **vertebral arteries.** Also discussed here is the **circulus arteriosus cerebri.**

INTERNAL CAROTID ARTERY. The cerebral portion of the internal carotid artery extends to the medial end of the lateral cerebral fissure, where it divides into the anterior cerebral and middle cerebral arteries. The right and left **anterior cerebral arteries** pass forward and medially toward the longitudinal fissure, where they are connected by a small **anterior communicating artery.** The two arteries then run parallel to each other in the longitudinal fissure and give off numerous branches to supply much of the frontal and parietal lobes. The **middle cerebral artery** passes through the lateral fissure to spread out over the lateral surface of the brain. A third branch of the internal carotid artery, the **posterior communicating artery,** runs posteriorly to anastomose with the posterior cerebral artery.

VERTEBRAL ARTERIES. The right and left vertebral arteries, which are branches of the subclavian arteries, course superiorly through the transverse foramina of the cervical vertebrae beginning at C-6. As they pass through the foramen magnum, they pierce the dura mater to enter the cerebellomedullary cistern of the subarachnoid space. The right and left vertebral arteries join to form the single **basilar artery,** which passes over the surface of the pons. The basilar artery then divides to form the two **posterior cerebral arteries,** which supply the occipital lobes.

CIRCULUS ARTERIOSUS CEREBRI. The typical configuration at the base of the brain shows the vessels anastomosing to form a "circle" called the **circulus arteriosus cerebri,** or the **circle of Willis.** This circle, illustrated in Figure 1–9, is formed by the internal carotid arteries, the anterior cerebral arteries, the anterior communicating artery, the posterior cerebral arteries, and the posterior communicating arteries. Berry aneurysms often occur in the vessels of the circle of Willis. Figure 1–9 shows the

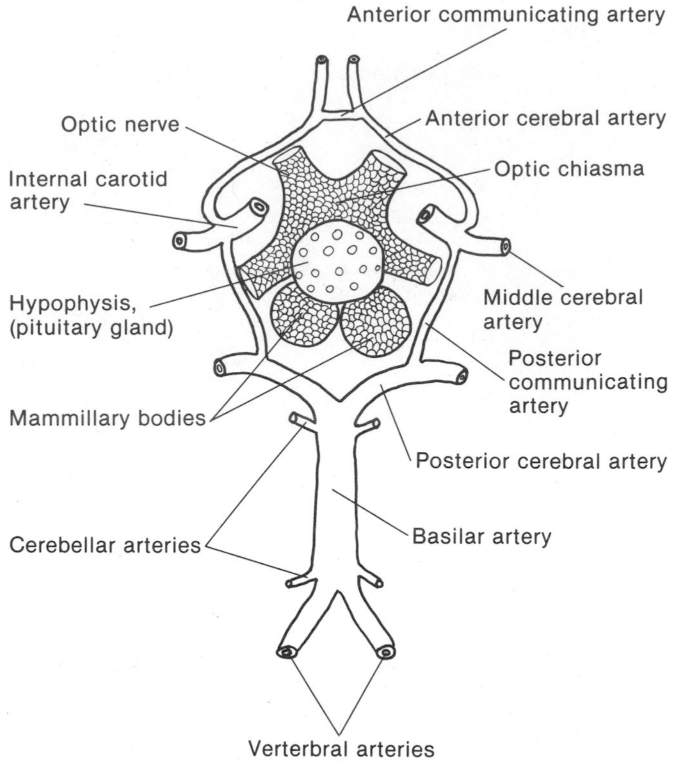

FIGURE 1–9. Circulus arteriosus cerebri.

circulus arteriosus cerebri, or circle of Willis, which is located in the interpeduncular cistern and encloses the optic chiasma, the infundibulum, and the mammillary bodies.

VENOUS DRAINAGE

Venous channels that drain blood from the brain and the meninges are called **sinuses.** They are generally located between the endosteum of the calvaria and the meningeal dura or between two layers of dura mater. Unlike other veins, venous sinuses contain no valves. The **superior sagittal sinus** appears triangular in cross section; it occupies the entire length of the superior portion of the falx cerebri and increases in size as it passes posteriorly. At the internal occipital protuberance, it usually continues as the right lateral sinus. The smaller **inferior sagittal sinus** occupies the free inferior edge of the falx cerebri. At the junction of the falx cerebri and the tentorium cerebelli, the inferior sagittal sinus receives the great cerebral vein and becomes the **straight sinus,** which courses along the tentorium cerebelli. At the internal occipital protuberance, the straight sinus usually continues as the left lateral sinus. Therefore, the **lateral sinuses** are continuations of either the superior sagittal or the straight sinuses, but at their origin, they may form a common space known as the confluence of sinuses. The lateral sinuses are subdivided into transverse and sigmoid portions. The **transverse sinus** passes from the occipital protuberance (confluence of sinuses) to the junction of the petrous and mastoid

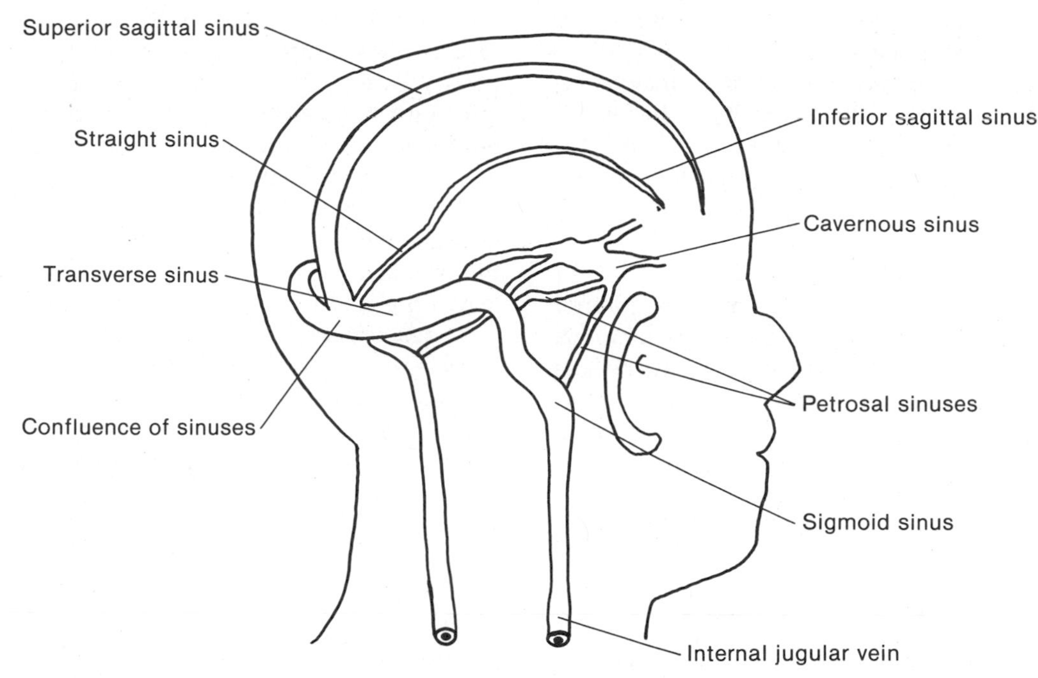

FIGURE 1-10. Venous sinuses.

portions of the temporal bone. The **sigmoid sinus,** which is a continuation of the transverse sinus, follows an S-shaped path that loops over the petrous and mastoid portions to the jugular foramen, where it becomes the **internal jugular vein.** The venous sinuses are illustrated in Figure 1-10. In addition to these main sinuses, there are numerous smaller ones that drain specific portions of the brain and then empty into these larger sinuses.

A rather large **cavernous sinus** is located on each side of the body and sella turcica of the sphenoid bone (see Fig. 1-10). These sinuses receive venous blood from the ophthalmic and middle cerebral veins and are drained by small petrosal sinuses that empty into the sigmoid sinus or internal jugular vein. The internal carotid artery enters the cavernous sinus through the foramen lacerum, makes a hairpin turn at the end of the sinus, and then exits the sinus to enter the subarachnoid space. The abducens nerve is closely related to the internal carotid artery as it traverses the cavernous sinus. Associated with the lateral wall of the cavernous sinus, superior to inferior, are the oculomotor (III), trochlear (IV), and the ophthalmic and maxillary divisions of the trigeminal (V) nerves. The relationships of the vessels and nerves in the cavernous sinus are illustrated in Figure 1-11. Injuries in the region of the

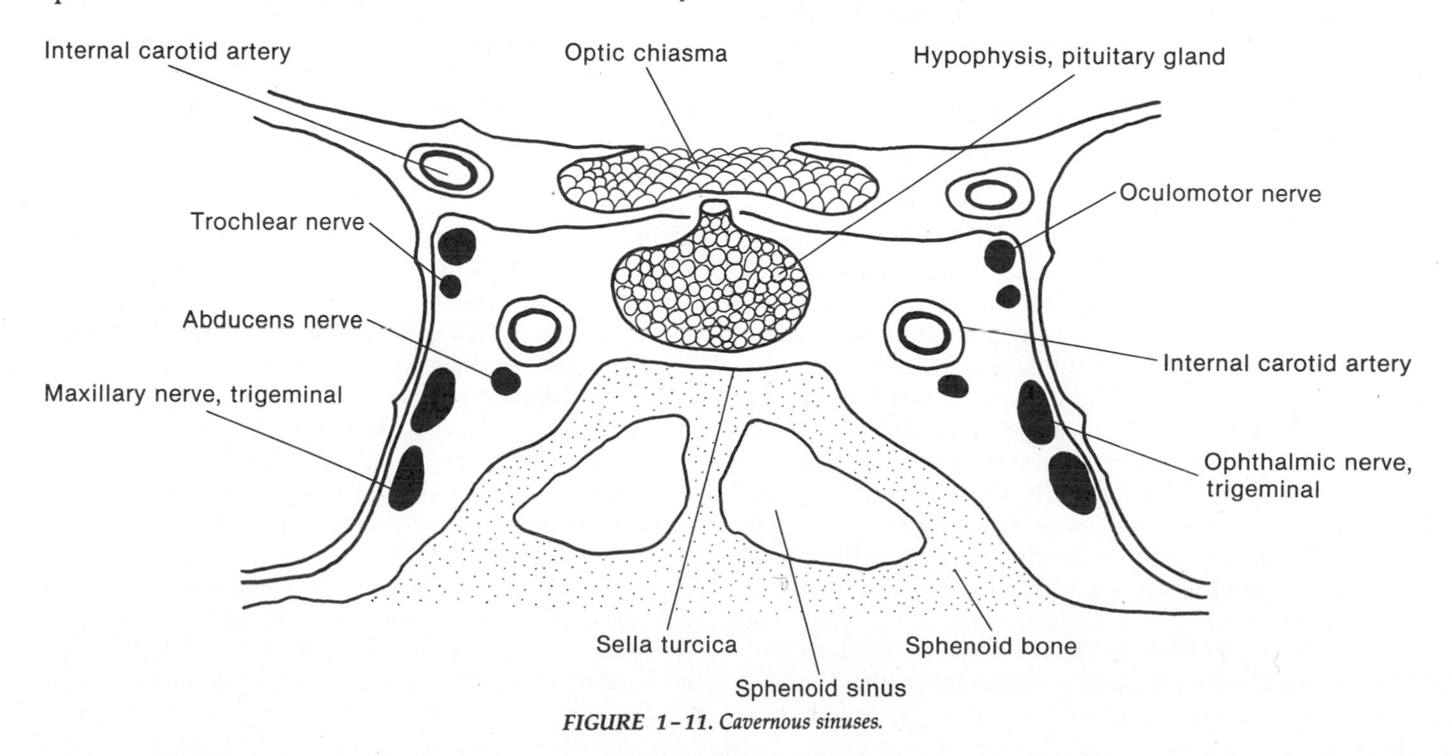

FIGURE 1-11. Cavernous sinuses.

cavernous sinus exhibit a variety of signs because there is an intimate relationship between the vessels and the nerves within the sinus. Some of these signs may be detected in the orbit of the eye. The proximity of the cavernous sinus to the sphenoidal paranasal sinus contributes to the development of meningitis as a sequela to sinusitis.

CRANIAL NERVES

Twelve pairs of cranial nerves emerge from the inferior surface of the brain. These nerves pass through foramina of the skull to innervate structures in the head and neck as well as viscera in the body. The cranial nerves are designated both by name and by Roman numerals, according to the order in which they appear on the inferior surface of the brain.

CRANIAL NERVE I (OLFACTORY). With the function of providing the body with the sense of smell, the olfactory nerves begin in the mucous membrane of the olfactory region of the nasal cavity. They continue through the olfactory foramina in the cribriform plate of the ethmoid bone and enter the olfactory bulb on the inferior surface of the brain. From the olfactory bulb, the olfactory tract proceeds posteriorly to a region just anterior to the optic chiasma.

CRANIAL NERVE II (OPTIC). Mediating visual function, the optic nerve originates in the nerve cells of the retina. From here, the two optic nerves—one for each eye—are directed posteriorly and medially through the optic foramen of the sphenoid bone into the cranial cavity, where they meet at the optic chiasma on the inferior surface of the hypothalamus. In the optic chiasma, the fibers from the medial part of the retina of each eye cross to the opposite side, whereas the fibers from the lateral portions of the retinas remain on the same side. After this partial decussation, or crossing over, in the chiasma, the fibers continue as the optic tract to the visual area of the occipital lobe.

CRANIAL NERVE III (OCULOMOTOR). Cranial nerve III is a motor nerve to the superior, inferior, and medial rectus muscles and to the inferior oblique muscles of the eye. These muscles are responsible for a variety of eye movements. In addition to these four extrinsic muscles, the nerve also has fibers that innervate the levator palpebrae superioris muscle, which raises the upper eyelid to open the eye. Some parasympathetic fibers supply the sphincter muscle of the iris and the ciliary muscle, which changes the shape of the lens in accommodation. In passing from the cranial cavity to the orbital cavity, it passes through the superior orbital fissure.

CRANIAL NERVE IV (TROCHLEAR). The trochlear, a motor nerve, supplies the superior oblique muscle of the eye. In passing from the cranial cavity to the orbital cavity, the trochlear nerve passes through the superior orbital fissure.

CRANIAL NERVE V (TRIGEMINAL). The largest of the cranial nerves is the trigeminal, which contains motor fibers for the muscles of mastication and sensory fibers from the head. After emerging from the lateral side of the pons, the nerve divides into three branches. The ophthalmic branch passes through the superior orbital fissure to receive sensory impulses from the conjunctiva and cornea of the eye, the upper eyelid, the forehead, the nose, and the scalp. The maxillary branch first passes through the foramen rotundum and then curves around to enter the orbital cavity through the inferior orbital fissure. It leaves the cavity through the infraorbital foramen to receive sensory impulses from the skin of the cheek, from the lateral sides of the nose, the upper lip, and the teeth. The mandibular branch emerges through the foramen ovale. It receives sensory impulses from the skin over the mandible, the temporal region, the tongue, the floor of the mouth, the lower teeth and gingivae, and the buccal surface of the cheek. It also has motor fibers that stimulate the muscles of mastication.

CRANIAL NERVE VI (ABDUCENS). The abducens nerve emerges from the inferior surface of the brain at the junction of the pons and the medulla. It enters the orbital cavity through the superior orbital fissure to supply motor impulses to the lateral rectus muscle of the eye.

CRANIAL NERVE VII (FACIAL). The seventh cranial nerve, the facial nerve, contains both sensory and motor fibers. After leaving the skull through the stylomastoid foramen, it passes through the substance of the parotid gland, where it branches to supply motor impulses to the muscles of facial expression. The sensory component receives impulses from the taste buds on the anterior two thirds of the tongue. The nerve also stimulates the sublingual and submaxillary salivary glands and the lacrimal glands associated with the eye.

CRANIAL NERVE VIII (VESTIBULOCOCHLEAR). The eighth cranial nerve is a special sensory nerve with two distinct components. The vestibular branch functions in equilibrium by receiving sensory impulses from the semicircular canals, the utricle, and the saccule. These structures of the inner ear detect the position and the movement of the head. The cochlear branch functions in hearing by receiving impulses from the organ of Corti in the cochlear duct, also in the inner ear. Both branches enter the cranial cavity through the internal auditory meatus. The vestibulocochlear nerve is sometimes referred to as the acoustic, or auditory, nerve.

CRANIAL NERVE IX (GLOSSOPHARYNGEAL). As the name implies, the glossopharyngeal nerve is chiefly distributed to the tongue and pharynx. It is a mixed nerve, with sensory and motor functions. This nerve supplies motor impulses to muscles that aid in swallowing and to the parotid salivary gland. The sensory component may be divided into three functional groups. Some of the fibers convey the special sensation of taste from the posterior one third of the tongue. Others transmit the general sensations of pain, temperature, and touch from the pharynx and middle ear. The third group is concerned with the regulation of respiration and blood pressure by receiving impulses from the chemoreceptors and pressure receptors associated with the carotid arteries in the neck. The ninth cranial nerve exits the cranial cavity through the jugular foramen.

CRANIAL NERVE X (VAGUS). The word "vagus" is derived from the Latin word meaning "wandering" – an appropriate name for the tenth cranial nerve, which has the most extensive distribution of all the nerves. It leaves the cranial cavity through the jugular foramen. The fibers of this mixed nerve may be divided into four groups: (1) somatic motor fibers supply the skeletal muscles of the pharynx and larynx; (2) visceral motor fibers carry impulses to the thoracic and abdominal viscera; (3) somatic sensory fibers convey impulses concerned with pain, temperature, and touch from the external ear; and (4) visceral sensory fibers function in the regulation of heart rate, blood pressure, and respiration by transmitting impulses from the stretch receptors in the heart, aorta, superior vena cava, and lungs. This component also receives sensory impulses from the abdominal viscera.

CRANIAL NERVE XI (ACCESSORY). The function of the eleventh cranial nerve is entirely motor-oriented. The accessory nerve leaves the cranial cavity through the jugular foramen. Motor fibers from this nerve stimulate the trapezius and the sternocleidomastoid muscles to contract.

CRANIAL NERVE XII (HYPOGLOSSAL). The numerous roots of the twelfth cranial nerve emerge through the hypoglossal canal in the occipital bone. After passing through the canal, the roots unite to form the hypoglossal nerve. This nerve, which functions as a motor nerve, stimulates contraction of the tongue muscles.

SUMMARY OF THE CRANIAL NERVES. The cranial nerves are summarized in Table 1–3. Most cranial nerves have both sensory and motor components. Three of the nerves (I, II, VII) are associated with the special senses of smell, vision, hearing, and equilibrium, and consist of sensory fibers only. Five other nerves (III, IV, VI, XI, XII) are primarily motor in function but do have some sensory fibers for proprioception. The remaining four nerves (V, VII, IX, X) consist of significant amounts of both sensory and motor fibers.

Orbital Cavity and Contents

CAVITY WALLS

The orbital cavity is a pyramidal structure with an apex, base, and four triangular walls. The **optic foramen** is at the apex in the posterior part of the orbit. The base, which is the anterior part that opens onto the face, is formed by the zygomatic, maxilla, and frontal bones. The nearly parallel medial walls of the two orbits have portions of the ethmoid and sphenoid sinuses situated between them. Each medial wall is formed by the lacrimal bone and the fragile orbital plates of the ethmoid and palatine bones. The superior wall, or the roof, of the cavity is formed by the orbital plate of the frontal bone, whereas the maxilla and a small portion of the zygomatic compose the inferior wall, or the floor. The lateral walls of the two orbits are positioned at angles of 90 degrees to each other and, if extended, would intersect in the region of the pituitary gland. The sturdy lateral wall is formed by the

TABLE 1–3. Summary of Cranial Nerves

Number	Name	Associated Foramen	Type	Function
I	Olfactory	Olfactory foramina in cribriform plate of ethmoid bone	Sensory	Sense of smell
II	Optic	Optic foramen of sphenoid bone	Sensory	Vision
III	Oculomotor	Superior orbital fissure of sphenoid bone	Motor	Movement of eye and eyelid
IV	Trochlear	Superior orbital fissure of sphenoid bone	Motor	Movement of the eye
V	Trigeminal		Mixed	
	Ophthalmic branch	Superior orbital fissure of sphenoid bone	Sensory	Cornea, skin of nose, forehead, scalp
	Maxillary branch	Foramen rotundum of sphenoid bone	Sensory	Cheek, nose, upper lip, and teeth
	Mandibular branch	Foramen ovale of sphenoid bone	Mixed	Skin over mandible, tongue, lower lip, and teeth; contraction of muscles of mastication
VI	Abducens	Superior orbital fissure of sphenoid bone	Motor	Eye movement
VII	Facial	Stylomastoid foramen of temporal bone	Mixed	Contraction of muscles of facial expression; lacrimal and submaxillary gland secretion; taste from anterior two thirds of tongue
VIII	Vestibulocochlear	Internal auditory meatus of temporal bone	Sensory	Hearing and equilibrium
IX	Glossopharyngeal	Jugular foramen of temporal bone	Mixed	Taste from posterior one third of tongue; contraction of muscles used in swallowing; parotid gland secretion
X	Vagus	Jugular foramen of temporal bone	Mixed	Contraction of muscles of pharynx and larynx; gastric motility; general visceral sensation; alters heart rate, respiration, blood pressure
XI	Accessory	Jugular foramen of temporal bone	Motor	Contraction of trapezius and sternocleidomastoid muscles
XII	Hypoglossal	Hypoglossal canal of occipital bone	Motor	Contraction of muscles of tongue

zygomatic bone and the greater wing of the sphenoid bone. A depression for the lacrimal gland is located in the superior portion of the lateral wall.

BULBUS OCULI

The primary structure located in the orbital cavity is, of course, the **bulbus oculi,** or **eyeball.** It is somewhat spherical, with a diameter of approximately 2 to 3 cm, and it has an anterior bulge. The eyeball is surrounded by orbital fat within the orbital cavity.

The wall of the bulbus oculi is made up of three concentric coats, or **tunics.** The **external,** or **fibrous, tunic** is the supporting layer. It consists of the white opaque **sclera,** which covers the posterior five sixths of the eyeball, and the transparent **cornea,** which covers the anterior one sixth of the eyeball.

The **middle,** or **vascular, tunic** consists of the **choroid,** the **ciliary body,** and the **iris.** The **choroid** is a highly vascular layer, which contains a brown pigment, and is located between the sclera and the retina. It is the largest part of the middle tunic and lines most of the sclera, although it is only loosely connected to the fibrous coat and can easily be stripped away. The choroid is, however, firmly attached to the retina. Anteriorly, the choroid is continuous with the **ciliary body. Suspensory ligaments** connect the ciliary body to the lens of the eye. Internally, the ciliary body contains numerous fingerlike ciliary processes that secrete the aqueous humor. Externally, the ciliary body contains ciliary muscle. When this muscle contracts, the suspensory ligaments relax and the lens bulges to allow focusing for close vision. The **iris** is the conspicuous, colored portion of the eye. It is a doughnut-shaped diaphragm with a central aperture called the **pupil.** The iris muscles continually contract and relax to change the size of the pupil, which regulates the amount of light entering the eye.

The **innermost,** or **nervous, tunic** is the **retina,** which has several layers. The outer layer of the retina is deeply pigmented and firmly attached to the choroid. The inner retinal layer contains the **rods and cones,** which are the light receptor cells. Other layers consist of bipolar neurons and ganglion cells. The axons of the ganglion cells

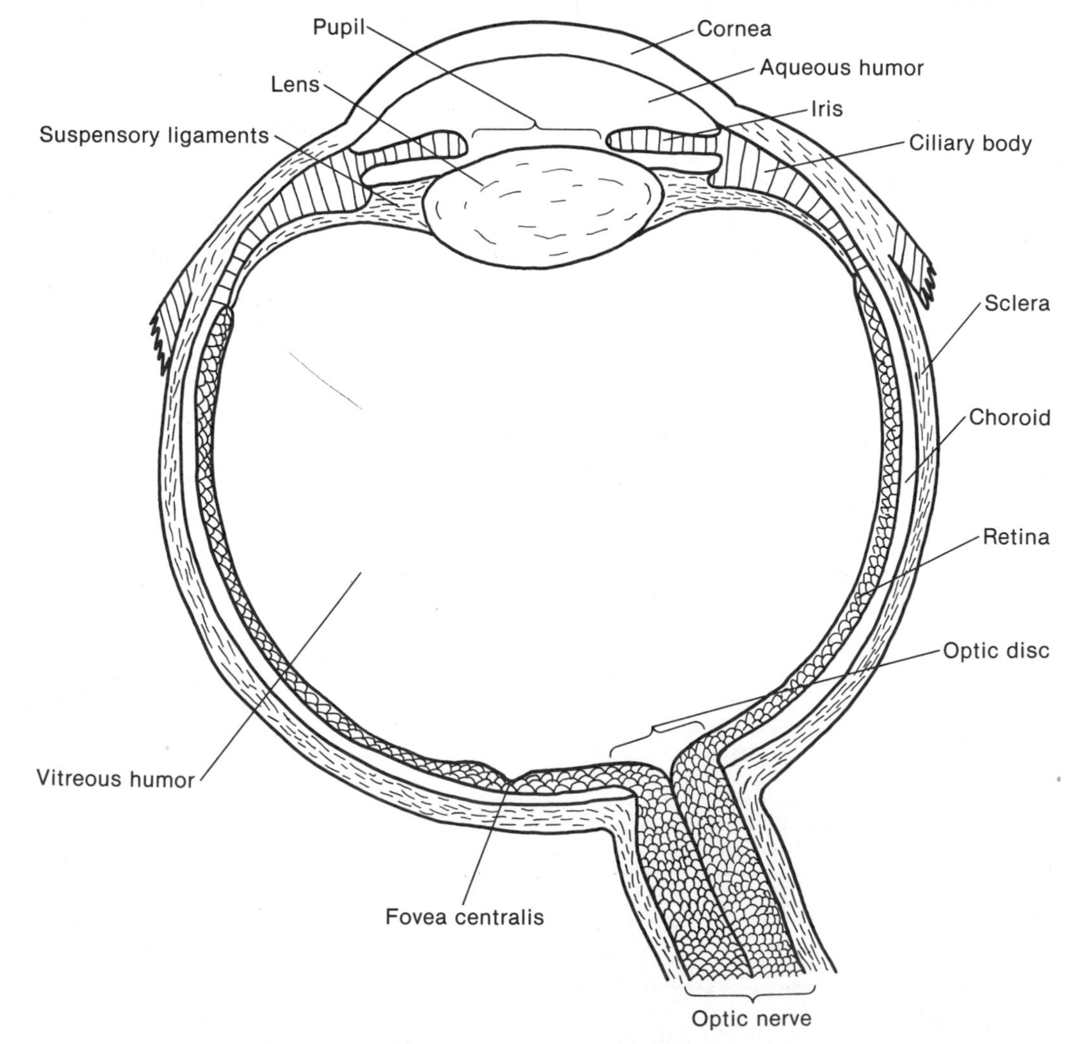

FIGURE 1-12. Structure of the bulbus oculi.

TABLE 1-4. Extrinsic Muscles Associated with Eye Movement

Muscle	Function	Innervation
Superior rectus	Rotates eye upward and laterally	Oculomotor (cranial nerve III)
Inferior rectus	Rotates eye downward and medially	Oculomotor (cranial nerve III)
Lateral rectus	Rotates eye laterally	Abducens (cranial nerve VI)
Medial rectus	Rotates eye medially	Oculomotor (cranial nerve III)
Superior oblique	Rotates eye downward and laterally	Trochlear (cranial nerve IV)
Inferior oblique	Rotates eye upward and laterally	Oculomotor (cranial nerve III)

converge to form the optic nerve, which penetrates the tunics at the optic disc and passes through the apex of the orbital cavity to reach the brain.

A transparent, biconvex **lens** is located just posterior to the iris. The curvature of the lens surface is continually changing by action of the ciliary muscles and the suspensory ligaments. The space anterior to the lens is the anterior cavity, which is filled with **aqueous humor.** The posterior cavity, between the lens and the retina, is filled with a colorless, transparent, gel-like **vitreous humor.** Unlike the aqueous humor, which is continually being replaced, the vitreous humor is formed during embryonic development and is not exchanged. Figure 1-12 illustrates the structure of the bulbus oculi.

Six extrinsic ocular muscles that insert on the sclera are associated with movements of the eye. These are summarized in Table 1-4.

VASCULAR SUPPLY TO THE ORBITAL CONTENTS

Most of the vascular supply to the orbital contents is by way of the **ophthalmic artery,** a branch of the internal carotid artery. Ciliary branches of the ophthalmic artery provide the blood supply to the sclera, the choroid, the ciliary body, and the iris. One of the smallest but most important branches of the ophthalmic artery is the **central artery of the retina.** This vessel emerges through the optic disc, along with the optic nerve; it then branches over the surface of the retina to supply it with blood. If the central artery is blocked by a tumor, thrombus, or massive edema, the result is sudden blindness.

Venous drainage of the orbital cavity is through the **superior and inferior ophthalmic veins,** which pass through the orbit to enter the cavernous sinus located adjacent to the pituitary gland. There is a **central vein of the retina** that follows the pathway of the central artery along the optic nerve. The central vein drains into the cavernous sinus along with the ophthalmic veins. Increased intracranial pressure restricts the blood flow in the central vein as it passes through the subarachnoid space. This restricted venous drainage results in edema at the optic disc, which is directly observable with an ophthalmoscope. Papilledema, or swelling at the optic disc, is one of the earliest indications of increased intracranial pressure.

PROTECTIVE FEATURES OF THE EYE

The eyes are protected by the **eyelids,** or **palpebrae.** The upper eyelid partially covers the iris and is essentially a fold of skin that covers the muscle beneath known as the **levator palpebrae superioris.** Externally, the covering is a thin skin, whereas the lining is a highly vascular **conjunctiva.** The levator palpebrae superioris muscle, which elevates the upper eyelid to open the eye, is innervated by the oculomotor nerve.

The **lacrimal apparatus** is another protective structure associated with the eye. The **lacrimal gland,** which is located in the upper and lateral part of the orbit, produces the lacrimal fluid, or tears. The lacrimal fluid moistens the surface of the eye, lubricates the eyelids, and washes away foreign particles. It also contains an enzyme that destroys certain bacteria. After spreading across the surface of the eye, the fluid drains through the nasolacrimal duct into the inferior meatus of the nasal cavity.

Skeletal Components of the Neck

The bony skeleton of the neck consists of seven cervical vertebrae. These vertebrae are similar to all the other bones of the vertebral column, but they have some features that are unique. In order to make comparisons, it is necessary to first describe the "typical" vertebra.

GENERAL STRUCTURE OF VERTEBRAE

Vertebrae, in general, have thick, disc-shaped **bodies.** Two small bridges of bone, called **pedicles,** project posteriorly from the vertebral body. The pedicles continue as flat plates of bone, called **laminae,** that unite with each other posteriorly. The body, pedicles, and laminae form a complete circle, which encloses the **vertebral foramen,** a large opening for the passage of the spinal cord. Two **transverse processes** project laterally from the laminae, and a single **spinous process** projects posteriorly from the midline.

FEATURES OF CERVICAL VERTEBRAE

Cervical vertebrae differ from all others in that there are holes, or foramina, in the transverse processes. These **transverse foramina,** as the holes are called, are for the passage of the vertebral blood vessels. The spinous processes of cervical vertebrae are forked, or **bifid.** The spinous process of the seventh, or last, cervical vertebra may not be bifid and is usually long and pointed, which makes it easily palpable. This is a good reference point for counting the vertebrae.

Cervical vertebrae collectively form a normal curvature that is convex anteriorly. This curvature develops in the infant within a few weeks after birth, when the infant begins to hold its head erect.

The first two cervical vertebrae, which have special names (the **atlas** and the **axis**), are unique.

ATLAS. The atlas is the most superior cervical vertebra. It has no vertebral body mass but instead is shaped like an oval ring. It has smooth facets on the superior surface that articulate with the condyles on the occipital bone. Inferior facets articulate with the second cervical vertebra.

AXIS. The second cervical vertebra, called the **axis,** has a superiorly projecting process called the **dens,** or the **odontoid process.** The dens projects into the vertebral foramen of the atlas above to form a pivot around which the atlas and skull rotate.

INTERVERTEBRAL DISCS

The most important structures that join the vertebrae to form a continuous column are the **intervertebral discs.** Each disc consists of a fibrocartilaginous outer ring called the **annulus fibrosus** and an inner soft core called the **nucleus pulposus.** The size and shape of each disc correspond to the adjoining vertebrae. In the cervical region, the discs are slightly thicker anteriorly than they are posteriorly, which contributes to the normal convex cervical curvature. The discs form a tight joint between the vertebrae; however, by their design, they are able to absorb the high compression forces that are generated in walking and other daily activities. When a disc degenerates or undergoes trauma, the soft nucleus pulposus may protrude through an opening or tear in the annulus fibrosus. This is called a **herniated disc.** Symptoms depend on the location and the severity of the herniation.

Viscera of the Neck

The neck region represents the connection between the head and the trunk of the body. Although the term *viscera* usually refers to the organs of the thoracic and abdominopelvic cavities, it is used here as a collective term for the miscellaneous structures of the head and neck that cannot be classified as nerves, muscles, or blood vessels. The salivary glands are associated with the face. The pharynx, larynx, trachea, and esophagus represent passageways for food and air. The thyroid and parathyroid glands are important endocrine glands in the neck.

SALIVARY GLANDS

There are three pairs of **salivary glands** that are located in the region of the face: the parotid, the submandibular, and the sublingual. These glands are usually considered to be a part of the digestive system because they secrete a fluid that moistens food particles for taste and swallowing. They also secrete an enzyme, salivary amylase, which initiates digestion of carbohydrates. All of the salivary glands are easily seen in sectional views.

PAROTID GLAND. The largest of the glands is the **parotid gland,** which is wedged between the ramus of the mandible and the mastoid portion of the temporal bone. The parotid gland occupies the space just anterior and just inferior to the auricle of the ear, and a portion of the gland overlies the masseter muscle in the cheek. The well-defined duct of the parotid gland that is known as Stensen's duct extends across the masseter muscle, turns to penetrate the buccinator muscle, and opens into the vestibule of the mouth near the upper second molar.

SUBMANDIBULAR GLAND. The **submandibular (submaxillary) gland** is located medial to the body and angle of the mandible. The gland can be palpated as a small lump along the inferior border of the posterior half of the mandible. The secretions of this gland reach the oral cavity by means of the submandibular (Wharton's) duct, which opens near the midline beneath the tongue.

SUBLINGUAL GLAND. The third salivary gland is the **sublingual gland.** It is the smallest and the most deeply situated of the salivary glands. Located under the mucous membrane in the floor of the mouth, the two sublingual glands unite anteriorly to form a glandular mass around the lingual frenulum. These glands open into the floor of the mouth by means of a major sublingual duct (Bartholin's duct) and by numerous small sublingual ducts (of Rivinus) along the midline. Sometimes ducts from the sublingual glands may open into the submandibular duct.

PHARYNX

The **pharynx** is a muscular tube that is about 12 cm long and that extends from the base of the skull to the level of the sixth cervical vertebra, where it becomes the **esophagus.** The wall of the pharynx consists of overlapping pharyngeal constrictor muscles that are lined with mucous membrane. For descriptive purposes, it is divided into the nasal, the oral, and the laryngeal portions.

NASOPHARYNX. The nasopharynx, the region of the pharynx posterior to the nose, extends from the base of the skull to the soft palate. The anterior wall is somewhat lacking in structural elements; instead, it presents open space where the nasal cavity communicates with the pharynx through the **choanae,** or internal nares. The **eustachian,** or auditory, tubes open into the nasopharynx through the lateral walls. The openings of the eustachian tubes are recognized by the **torus tubarius,** which is a rounded protuberance of cartilage that outlines the posterior wall of the opening. Posterior to the torus tubarius, the walls of the nasopharynx extend laterally and form the **pharyngeal recess.** An aggregation of lymphoid tissue is found in the mucosa of the posterior wall of the nasopharynx. This is the **pharyngeal tonsil,** and when enlarged, it is commonly known as the "adenoids."

OROPHARYNX. The **oropharynx,** which is located posterior to the mouth, extends from the soft palate down to the tip of the epiglottis. The opening from the mouth into the pharynx is called the **fauces.** Collections of lymphoid tissue are found in the wall of the oropharynx. These are the **palatine tonsils,** commonly referred to as "the tonsils." The base of the tongue, with the associated lymphoid tissue, called the **lingual tonsils,** forms part of the anterior wall of the oropharynx. Between the tongue and the epiglottis are two valleys called **valleculae.** These present a potential hazard because foreign objects may become lodged there.

LARYNGOPHARYNX. The laryngeal portion of the pharynx, which is called the **laryngopharynx,** or **hypopharynx,** extends downward from the superior border of the epiglottis to the cricoid cartilage at the junction of the larynx and the trachea at the level of the sixth cervical vertebra (C-6). The walls extend laterally around the opening of the larynx to form the **piriform recesses.** Foreign objects that enter the pharynx may become lodged in the recesses. Inferiorly, the laryngopharynx is continuous with the esophagus.

RETROPHARYNGEAL SPACE. The **retropharyngeal space** is a potential space between the prevertebral fascia that surrounds the vertebral column and its associated muscles and the pharyngeal fascia that surrounds the pharynx. It contains loose connective tissue and permits the movement of the pharynx, larynx, trachea, and esophagus during swallowing. The retropharyngeal space is closed off by the skull superiorly, but it opens into the superior part of the thorax inferiorly. Laterally, the carotid sheath forms a barrier to the space. Infections in the region of the fascial layers may penetrate the fascia to enter the retropharyngeal space. Here, pus from the infections may form abscesses that bulge into the pharynx and may cause difficulty in speaking and swallowing. Once an infection enters the retropharyngeal space, it has a direct pathway into the mediastinum.

LARYNX

Although the larynx is an essential part of the air passageway, it is especially modified for voice production. In the male, the larynx typically extends from the level of the third cervical vertebra (C-3) to the sixth cervical vertebra (C-6). It is usually somewhat higher than this in females and in children. The skeleton of the larynx is formed by nine cartilages joined by ligaments. The major cartilages are the single **thyroid, cricoid,** and **epiglottic cartilages** and the paired **arytenoid cartilages.** In addition, there are small **corniculate** and **cuneiform** cartilages, which are paired. The thyroid cartilage is the largest, and it forms the anterior wall of the larynx. It is larger in the male than it is in the female, and it is commonly referred to as the "Adam's apple." The thyroid, cricoid, and arytenoids are hyaline cartilage. With age, the hyaline cartilages may calcify, which will make them visible on radiographs. The other cartilages (epiglottic, corniculate, and cuneiform) are elastic cartilage. The corniculate cartilages are attached to the tips of the arytenoids and are covered by a fold of tissue called the aryepiglottic fold. The cuneiform cartilages are also enclosed within the aryepiglottic fold; they are located lateral to the arytenoids. The interior of the larynx is subdivided into three portions by folds of tissue that project from the lateral laryngeal wall. The upper folds are the **vestibular folds,** which form the inferior margin of the **vestibule,** the most superior chamber of the larynx. The opening between the two vestibular folds is called the **rima vestibuli.** The vestibular folds are frequently mistaken for the vocal cords; consequently, they are sometimes referred to as the false vocal cords, although they play little or no part in voice production. The lower projections are the **vocal folds** (the true vocal cords), and the slit (the opening between them) is the **rima glottidis.** This aperture's shape changes, depending on the position of the folds during breathing and phonation. The space between the vestibular folds and the vocal folds is the **ventricle of the larynx.** Sometimes the ventricle is called the laryngeal sinus. Located in the middle, this is the smallest of the three regions. The remainder of the laryngeal cavity is the **infraglottic portion,** which extends from the vocal folds to the trachea. The term **glottis** refers to the true vocal folds and rima glottidis, collectively.

If a foreign particle, such as a bit of food, enters the larynx, the musculature goes into a spasm. This creates tension on the vocal folds and closes the rima glottidis. Air is then prevented from reaching the trachea and the lower air passageways, and the individual is in danger of asphyxiation.

ESOPHAGUS

The **esophagus** is a thick, distensible, muscular tube that extends from the pharynx at the level of the cricoid cartilage (C-6) in the neck to the stomach in the abdomen. As it descends through the neck, it is situated near the midline, between the trachea and the vertebral bodies. At the root of the neck on the right side, the esophagus is related to the **pleura** of the apex of the lung. On the left, the **thoracic duct** and **subclavian artery** are positioned between the esophagus and the pleura. These relationships are illustrated in Figure 1–13.

TRACHEA

The **trachea** begins as a continuation of the larynx in the neck at the vertebral level of C-6. It is anterior to the esophagus as it descends through the neck and enters

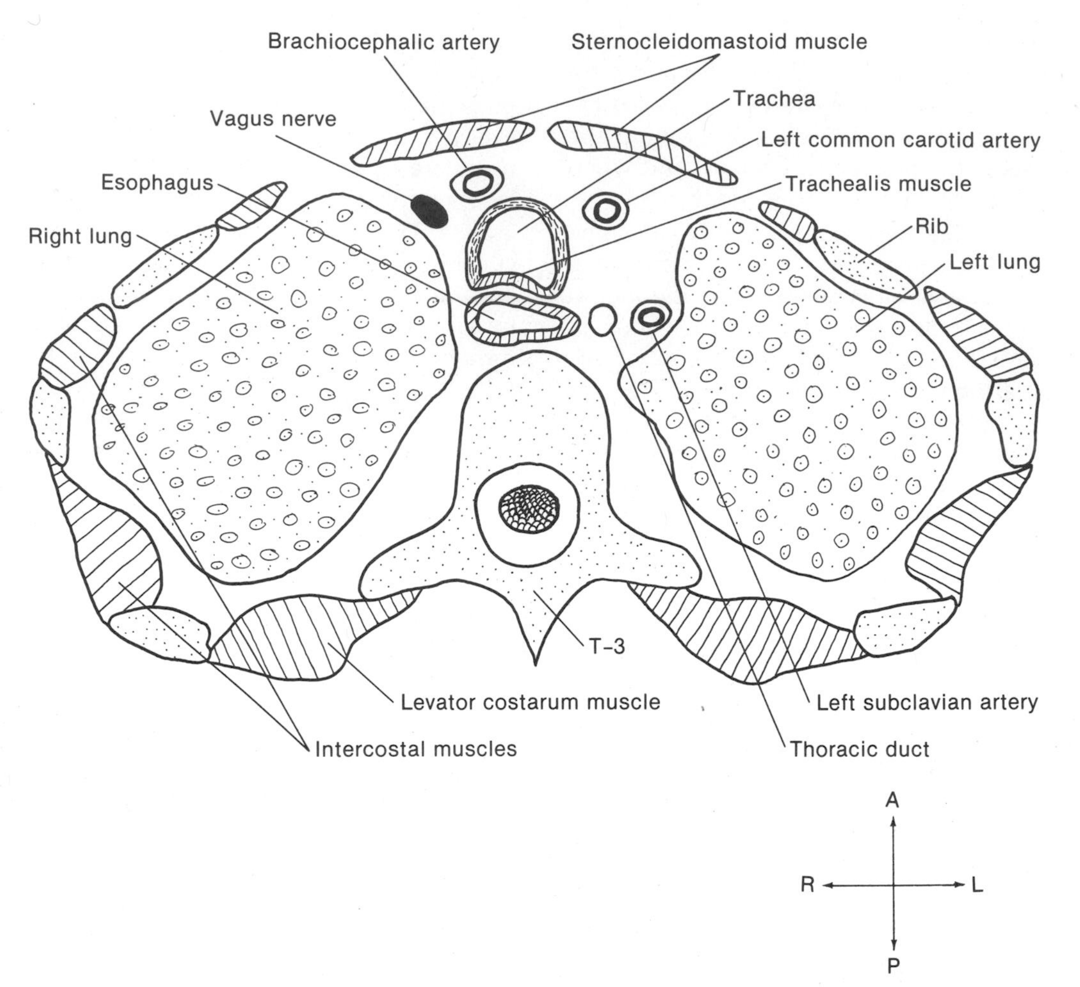

FIGURE 1-13. Relationships of esophagus at root of neck.

the **superior mediastinum** of the thorax a little to the right of the midline. It extends to vertebral level T-5. The wall of the trachea is supported by 16 to 20 incomplete cartilaginous rings. The rings of hyaline cartilage are deficient on the posterior side, where they are related to the esophagus; therefore, the posterior margin of the trachea is flattened. The tracheal airway is kept open by the rings of cartilage, whereas the soft tissue that fills the posterior gap between the tips of the rings allows for expansion of the esophagus into tracheal space during swallowing. The **common carotid arteries** and the lobes of the **thyroid gland** are lateral to the trachea in the neck. At lower levels, near the aortic arch, the **brachiocephalic artery** (trunk) is located anteriorly and to the right of the trachea.

THYROID GLAND

The **thyroid,** an important endocrine gland, consists of right and left lobes that are usually connected by an **isthmus.** The lobes are located lateral to the lower portion of the larynx and upper part of the trachea, and they may extend posteriorly enough to be related to the esophagus. The isthmus connecting the lobes passes over the second and the third tracheal rings, and the lobes may extend inferiorly to the sixth tracheal ring. Four small **parathyroid glands** are usually embedded along the posterior margin of the thyroid, but these are difficult to visualize.

Major Blood Vessels of the Neck Region

INTERNAL JUGULAR VEINS

The **internal jugular vein** begins as a continuation of the **sigmoid sinus** at the jugular foramen in the posterior cranial fossa. Usually the largest vein in the neck, it is generally larger on the right than on the left side. As the internal jugular vein descends, it passes deep to the **sternocleidomastoid muscle** and it courses anteriorly to unite with the **subclavian vein** to form the **brachiocephalic vein,** behind the sternal end of the clavicle. The internal jugular vein is located within the carotid sheath along with the common carotid artery (or internal carotid artery at higher levels) and vagus nerve. The internal jugular vein is lateral to the common carotid artery and the vagus nerve is located between and slightly posterior to these two vessels. These relationships are illustrated in Figure 1-14. At higher levels, the internal jugular vein is

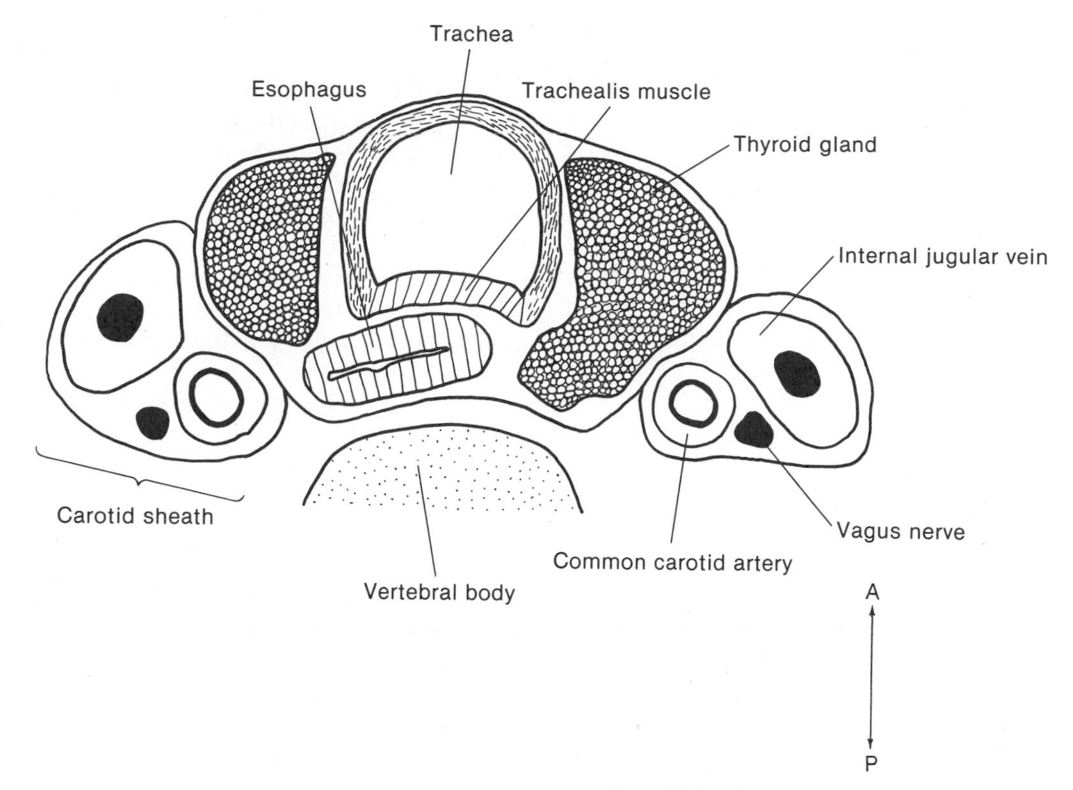

FIGURE 1-14. Contents of carotid sheath.

posterior to the internal carotid artery because the jugular foramen is posterior to the carotid canal. At lower levels, the internal jugular vein is anterior to the common carotid artery because the vein drains into the anteriorly positioned brachiocephalic vein.

COMMON CAROTID ARTERIES

On the right side, the **common carotid artery** begins posterior to the sternoclavicular joint as a branch of the **brachiocephalic artery.** On the left, it arises from the **aortic arch.** As it ascends the neck within the carotid sheath, the common carotid artery is medial to the internal jugular vein. At the level of the superior border of the thyroid cartilage, the common carotid artery divides into the **external** and **internal carotid arteries.** This is at the level of the disc between the third and the fourth cervical vertebrae. At the bifurcation, the common carotid artery and the continuing internal carotid artery are dilated to form the **carotid sinus.** The carotid sinus contains pressoreceptors for the regulation of blood pressure.

INTERNAL CAROTID ARTERIES

Arising as a direct continuation of the common carotid artery, the **internal carotid artery** ascends almost vertically within the carotid sheath to enter the carotid canal in the petrous position of the temporal bone. The right and left internal carotid arteries are two of the four major vessels that supply blood to the brain. Branches also supply the pituitary gland and the orbit. The internal jugular vein is lateral to the internal carotid artery and the vagus nerve is posterior and lateral. Refer to Figure 1-14.

EXTERNAL CAROTID ARTERIES

The **external carotid artery** arises at the bifurcation of the common carotid artery. It then courses posteriorly as it ascends to a region deep to the neck of the mandible. Here the external carotid artery terminates by dividing into the **maxillary** and the **superficial temporal** arteries. The external carotid artery and its numerous branches supply structures external to the skull. At low levels, near the bifurcation of the common carotid artery, the external carotid artery is anterior and medial to the internal carotid artery. At higher levels, the external carotid artery becomes more superficial so that it is anterior and lateral to the internal carotid artery.

VERTEBRAL ARTERIES

Each **vertebral artery** (right and left) begins as a branch of the **subclavian artery** on that side. The artery ascends through the transverse foramina of the cervical vertebrae C-6 through C-1. The artery then courses along the superior portion of the atlas and enters the **foramen magnum.** As it passes through the foramen magnum, the artery pierces the dura mater and arachnoid to enter the subarachnoid space of the **cerebellomedullary cistern.** Within the skull, at the base of the pons, the two vertebral arteries join to form a single basilar artery. The vertebral arteries are two of the four major arteries that supply the brain.

Muscles of the Head and Neck

There are numerous muscles located in the head and neck; many of them are small and difficult to separate from adjacent muscles. They are even more difficult to isolate by imaging techniques. These muscles have functional significance because they deal with facial expression, chewing food, and head movements. Only the larger and more significant muscles are presented in this text (Table 1–5).

MUSCLES OF FACIAL EXPRESSION

The muscles of facial expression are located in the subcutaneous tissue of the face. They originate in the fascia or on the underlying bone, and they insert on the skin of the face. All are innervated by cranial nerve VII, the facial nerve. Many of these muscles are small and thin and are difficult to dissect or to distinguish on cross-sections. Actions of the facial muscles are easily observed, however, as they are used to express feelings. Five of the more prominent facial muscles are mentioned here. The **frontalis muscle** is a part of the scalp. Originating from the aponeurosis, which is situated over the top of the head, the frontalis inserts on the skin of the eyebrow and forehead. When it contracts, this muscle elevates the eyebrows and produces the transverse wrinkles in the skin of the forehead. The **orbicularis oris muscle** is an important sphincter that encircles the mouth and forms the muscular bulk of the lips. This muscle's function involves closing the mouth and puckering the lips, as in whistling. It plays an important role in the enunciation of words. A similar sphincter, the **orbicularis oculi muscle,** surrounds the eye. Contraction of these muscle fibers lessens the orbital opening, as in winking and in blinking. The **buccinator muscle** is an accessory muscle in mastication and compresses the cheeks when blowing, as in playing a musical wind instrument. It inserts on the orbicularis oris at the angle of the mouth. The **platysma muscle** is a broad, flat muscle in the subcutaneous tissue of the neck. It inserts on the mandible, the skin of the neck, and the orbicularis oris muscle. When it contracts, it depresses the lower jaw and forms ridges in the skin of the neck.

MUSCLES OF MASTICATION

The four muscles of mastication provide chewing movements by acting on the temporomandibular joint to move the mandible (see Table 1–5). All insert on the mandible and are innervated by the mandibular division of the fifth cranial nerve. They are quite readily seen on transverse sections. The fan-shaped **temporalis muscle** covers the squamosal portion of the temporal bone and is a powerful muscle used to close the mouth by elevating the mandible. The **masseter,** or **chewing, muscle** is located on the lateral aspect of the ramus of the mandible. Both the **lateral** and the **medial pterygoid muscles** originate on the lateral pterygoid plate of the sphenoid bone and insert on the medial surface of the mandible. All of the muscles of mastication move the mandible to produce the movements necessary for chewing food. When these muscles are observed in transverse sections, the temporalis will be seen in the more superior sections. In lower sections, starting at the lateral surface, you will see, in sequence, the masseter, the ramus of the mandible, the lateral pterygoid, and the medial pterygoid.

ANTERIOR TRIANGLE OF THE NECK

The muscles of the neck are often described as being located within one of two triangles, which are separated by the **sternocleidomastoid muscle.** The muscles of the anterior triangle are generally considered to be the throat muscles, whereas those of the posterior triangle are the back muscles. In addition to these, there are numerous muscles associated with the vertebral column.

The anterior triangle extends from the midline of the neck to the anterior margin of the sternocleidomastoid muscle. The lower border of the mandible forms the triangle's base, and the manubrium of the sternum forms its apex.

The muscles of the anterior triangle, which are generally considered throat muscles, also help to form the floor of the oral cavity. All of these muscles are attached to the hyoid bone. They are involved with movements of the tongue and aid in swallowing. These muscles may be divided into two groups: the **suprahyoid muscles** and the **infrahyoid muscles** (see Table 1–5).

As a group, the suprahyoid muscles are located superior to the hyoid bone. These muscles function to raise the hyoid bone during swallowing or to open the jaw, when the hyoid bone is fixed. This muscle group includes the digastric, stylohyoid, mylohyoid, and geniohyoid.

The infrahyoid muscles are inferior to the hyoid bone; they pull down on the larynx and the hyoid, returning them to their normal positions after swallowing. These muscles include the sternohyoid, sternothyroid, thyrohyoid, and omohyoid.

The common carotid artery, the internal carotid artery, and the internal jugular vein, all enclosed within the carotid sheath, are located in the anterior triangle. Some of the branches of the external carotid artery are inside the sheath and others are outside. The vagus nerve, the longest of the cranial nerves, is also enclosed in the carotid sheath.

TABLE 1-5. Muscles Associated with Head and Neck

Muscle	Origin	Insertion	Action	Innervation
Muscles of Facial Expression				
Frontalis	Aponeurosis of scalp	Skin of eyebrow and forehead	Elevates eyebrows; wrinkles forehead	Facial (cranial nerve VII)
Orbicularis oris	Maxillae and mandible	Lip mucosa and skin at corner of mouth	Closes mouth and puckers lips, as in whistling	Facial (cranial nerve VII)
Orbicularis oculi	Frontal bones and maxillae around orbit	Eyelid	Closes eye, as in winking and blinking	Facial (cranial nerve VII)
Buccinator	Buccinator ridge of mandible, alveolar processes of maxillae, pterygomandibular ligament	Orbicularis oris at angle of mouth	Compresses cheeks when blowing, as when playing a musical instrument	Facial (cranial nerve VII)
Platysma	Fascia of the cervical region	Mandible, the skin of the neck, and the orbicularis oris	Depresses lower jaw, forms ridges on neck	Facial (cranial nerve VII)
Muscles of Mastication				
Temporalis	Temporal bone	Mandible	Elevates mandible to close mouth	Trigeminal (cranial nerve V), mandibular division
Masseter	Zygomatic arch	Mandible	Elevates mandible	Trigeminal (cranial nerve V), mandibular division
Lateral pterygoid	Sphenoid	Mandible	Pulls mandible forward (protracts)	Trigeminal (cranial nerve V), mandibular division
Medial pterygoid	Sphenoid	Mandible	Protracts mandible and moves mandible laterally	Trigeminal (cranial nerve V), mandibular division
Muscles of Anterior Triangle				
Sternocleidomastoid	Sternum and clavicle	Mastoid or temporal bone	Turns head side to side; flexes neck	Spinal accessory (cranial nerve XI)
Suprahyoid Muscles				
Digastric	Mandible	Hyoid	Elevates hyoid; opens mouth	Trigeminal (cranial nerve V)
Stylohyoid	Temporal bone	Hyoid	Elevates hyoid; retracts tongue	Facial (cranial nerve VII)
Mylohyoid	Mandible	Hyoid	Elevates hyoid and floor of mouth	Trigeminal (cranial nerve V)
Geniohyoid	Mandible	Hyoid	Protracts hyoid	Hypoglossal (cranial nerve XII)
Infrahyoid Muscles				
Sternohyoid	Sternum	Hyoid	Depresses hyoid	Hypoglossal (cranial nerve XII)
Sternothyroid	Sternum	Thyroid cartilage and hyoid	Depresses thyroid cartilage and hyoid	Hypoglossal (cranial nerve XII)
Thyrohyoid	Thyroid cartilage	Hyoid	Depresses hyoid	Hypoglossal (cranial nerve XII)
Omohyoid	Scapula	Hyoid and clavicle	Depresses hyoid	Hypoglossal (cranial nerve XII)
Muscles of the Posterior Triangle				
Trapezius	Occipital bone and vertebral spines	Scapula	Elevates scapula	Spinal accessory (cranial nerve XI)
Splenius capitis	Cervical and thoracic vertebrae	Occipital bone	Extends head	Cervical nerves
Levator scapulae	Cervical vertebrae	Vertebral border of scapula	Elevates scapula	Dorsal scapular
Anterior scalene	Cervical vertebrae	First rib	Elevates rib	Cervical plexus
Middle scalene	Cervical vertebrae	First rib	Elevates rib	Cervical plexus
Posterior scalene	Cervical vertebrae	Second rib	Elevates rib	Cervical plexus

POSTERIOR TRIANGLE OF THE NECK

The posterior triangle of the neck extends from the posterior margin of the **sternocleidomastoid muscle** to the **trapezius muscle,** with the apex of the triangle at the junction of these two muscles. The base is formed by the clavicle. The muscular floor of the triangle is formed by the splenius capitus; the levator scapulae; and the anterior, middle, and posterior scalenes (see Table 1–5). Significant structures located in the posterior triangle are the external carotid artery and the phrenic nerve.

Major Nerves of the Neck

SYMPATHETIC TRUNKS

The **sympathetic trunks** are strands of nerve fibers and ganglia that lie lateral to the vertebral column. They extend from the base of the skull to the coccyx. In the cervical region, the trunks are posterior to the carotid sheath and are immediately anterior to the transverse processes of the vertebrae. There are three ganglia in the cervical region of the sympathetic trunk. The superior cervical ganglion is located at the level of the axis, and because it is large, it serves as a good landmark for locating the trunk. The middle cervical ganglion is small and is located just anterior to the vertebral artery at the level of the transverse process of the sixth cervical vertebra. The inferior cervical ganglion is usually found posterior to the vertebral artery at the level of the superior border of the neck of the first rib.

VAGUS NERVE

The tenth cranial nerve is named the **vagus** because of its wide distribution. The vagus nerve leaves the skull through the jugular foramen along with the internal jugular vein. It descends through the neck in the carotid sheath in the angle between the internal jugular vein and the carotid artery—either common or internal, depending on the level.

CERVICAL PLEXUS

The **cervical plexus** is a network of nerve fibers derived from the first four cervical nerves. It is located lateral to the first four cervical vertebrae and deep to the internal jugular vein and the sternocleidomastoid muscle. Branches from this plexus innervate the skin and the muscles of the neck, and portions of the head and shoulders. The **phrenic nerve** to the diaphragm is a major branch of the cervical plexus.

PHRENIC NERVE

The **phrenic nerve,** an important branch of the cervical plexus, is the only nerve that supplies motor impulses that stimulate contraction of the diaphragm. It descends the neck vertically, on a path across the anterior surface of the anterior scalene muscle. At the root of the neck it enters the thorax between the subclavian artery and the subclavian vein.

SECTIONAL ANATOMY OF THE HEAD AND NECK

Transverse Sections

SECTION THROUGH THE LATERAL VENTRICLES

Sections superior to the lateral ventricles demonstrate the **cerebral gyri** and **sulci** with cortical gray matter and underlying white matter. The **longitudinal fissure** containing the **falx cerebri** and the **superior sagittal sinus** is readily identifiable. In cadaveric specimens, the **dura mater** and **arachnoid** may be seen; however, the **subdural space** is frequently exaggerated because of the tissue shrinkage.

Sections that are taken at a slightly more inferior level (6 to 7 cm from the top of the head) show the roof or upper portion of the **lateral ventricles** (Fig. 1–15). Just inferior to this, the lateral ventricles are separated by a thin partition, the **septum pellucidum.** The **genu** of the corpus callosum is between the anterior horns, and the **splenium** is between the posterior horns of the lateral ventricles. The body of the **caudate nucleus** is lateral to the lateral ventricles. Depending on the angle of the plane, the **inferior sagittal sinus** may be present.

SECTION THROUGH THE BASAL GANGLIA

Proceeding inferiorly from the upper portions of the lateral ventricles, numerous internal brain structures can be noted (Fig. 1–16). This plane passes just above the tentorium cerebelli. The **genu** and the **splenium** of the corpus callosum are readily identified as bands of white fibers passing from one hemisphere to another. The anterior and the posterior horns of the **lateral ventricles** are both present at this level. Just posterior and lateral to the anterior horns—in fact, forming the floor of the anterior

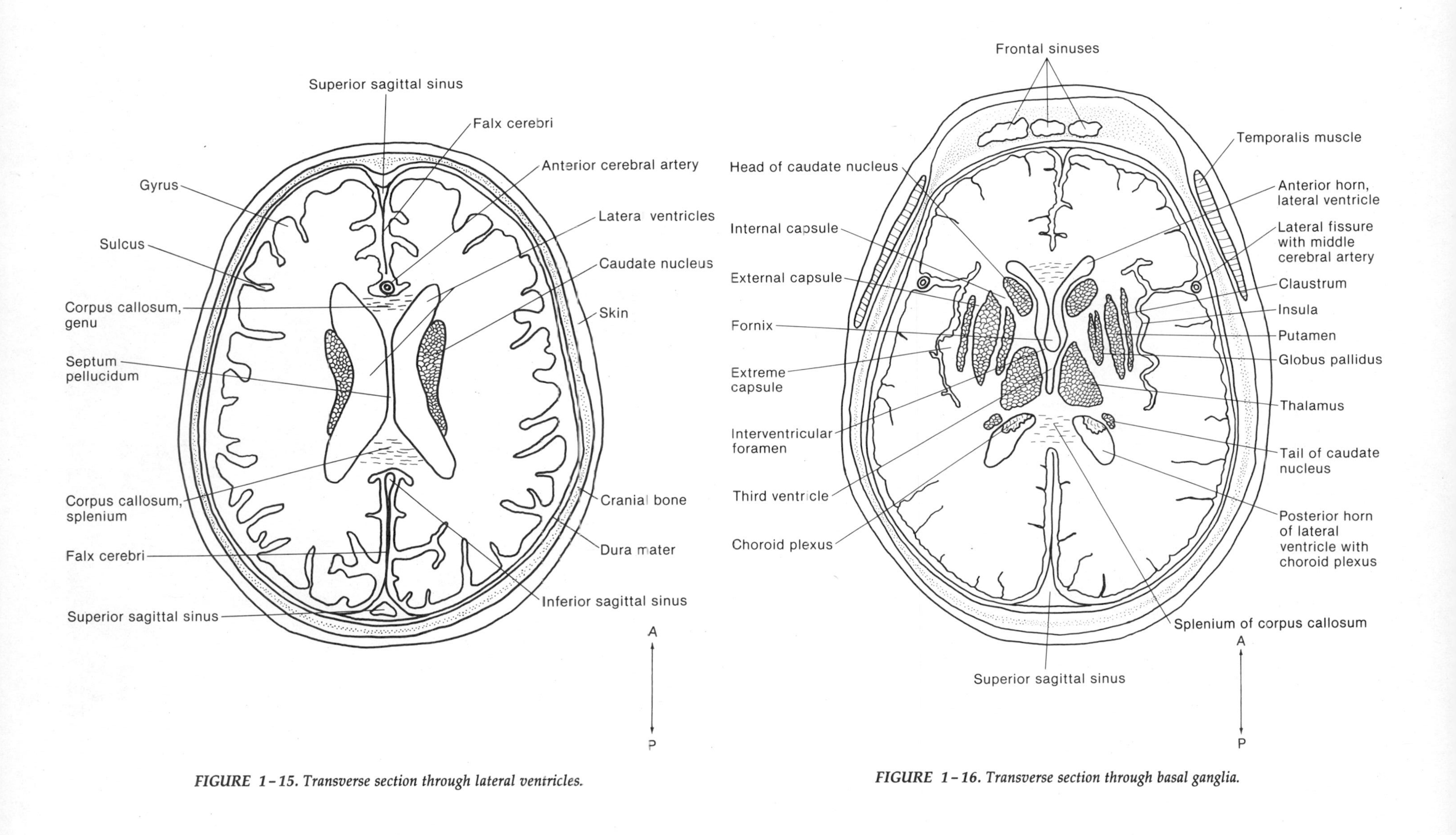

FIGURE 1-15. Transverse section through lateral ventricles.

FIGURE 1-16. Transverse section through basal ganglia.

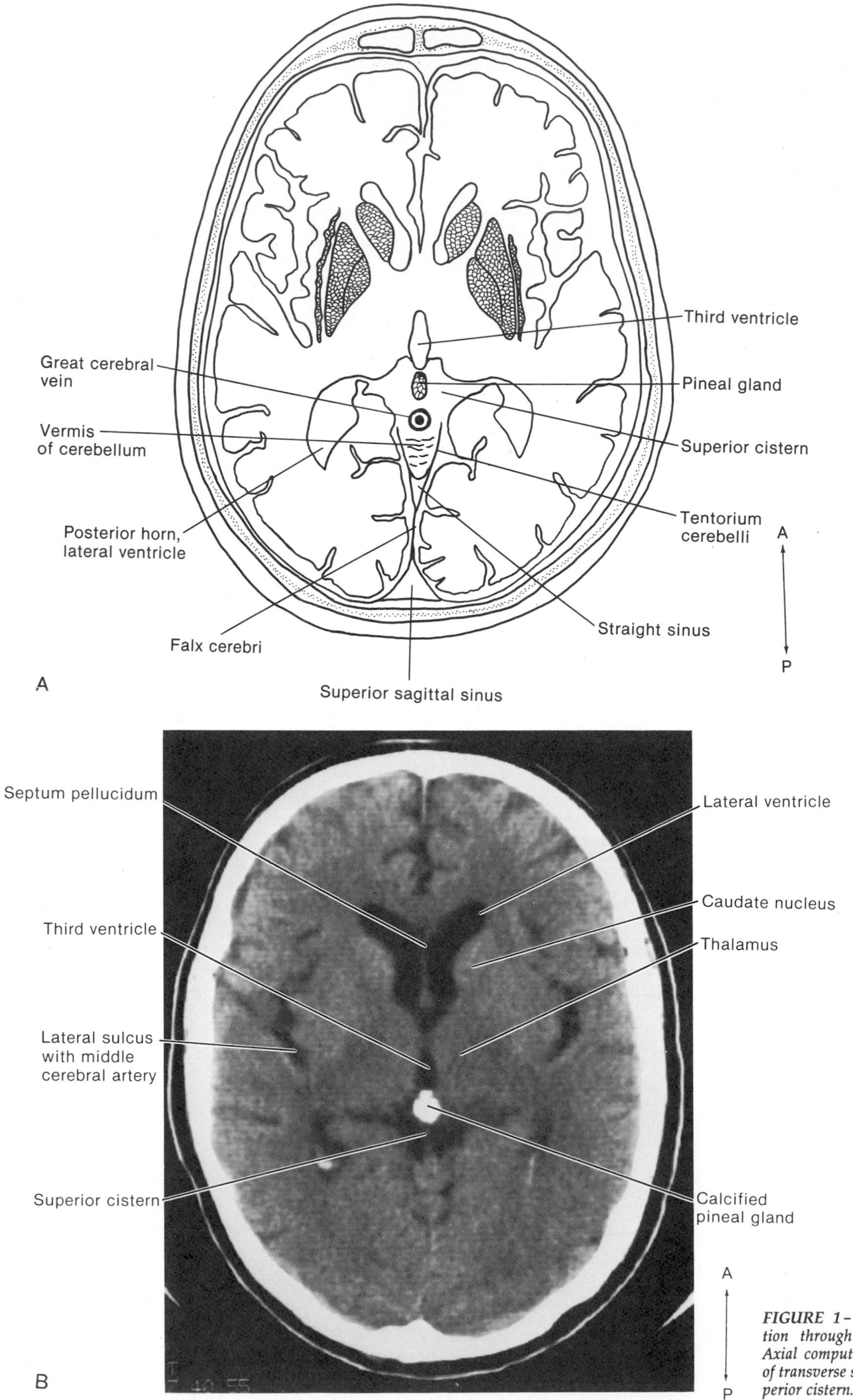

FIGURE 1-17. A, *Transverse section through superior cistern.* B, *Axial computed tomographic image of transverse section through the superior cistern.*

horns — are regions of gray matter, comprising the **head of the caudate nucleus.** The **septum pellucidum** may be seen as a thin midline partition between the two lateral ventricles. A slight enlargement of the septum pellucidum is the **fornix.** Posterior to the fornix, in a midline position, a narrow slit represents the **third ventricle.** The **interventricular foramen,** sometimes called the foramen of Monro, is an opening between each of the lateral ventricles and the single midline third ventricle. Two regions of gray matter form the lateral walls of the third ventricle. These are two **thalami** that make up a significant portion of the diencephalon. The posterior horns of the lateral ventricles are posterior to the two thalamic areas. The **choroid plexus** may be seen in the wall of the ventricles. A small area of gray matter may be seen in the roof of the posterior horns. This is the **tail of the caudate nucleus.** The posterior horns are separated by the splenium of the corpus callosum. The **insula,** or island of Reil, which is buried deep within the lateral fissure, can usually be seen at this level. Within the lateral fissure and around the insula, there are branches of the **middle cerebral artery.** Deep to the insula is a mass of gray matter that is the **lentiform nucleus.** Actually, the lentiform nucleus is made up of two parts: an outer, darker segment, called the **putamen,** and an inner, paler segment, called the **globus pallidus.** The lentiform nucleus is separated from the caudate nucleus by a band of white fibers that constitutes the **anterior limb of the internal capsule.** Between the thalamus and the lentiform nucleus, the internal capsule continues as the **posterior limb.** A thin strip of gray matter, the **claustrum,** is located between the insula and the putamen of the lentiform nucleus. A portion of the **frontal sinus** may be seen in some specimens at this level.

SECTION THROUGH THE SUPERIOR CISTERN

If a section through the basal ganglia is taken parallel to the orbitomeatal line, the posterior portion of the section will be lower than that shown in Figure 1 – 16 and will likely pass through the **superior cistern,** as shown in Figure 1 – 17 *A* and *B*. The **superior cistern** is located between the splenium of the corpus callosum, superiorly, and the tentorium cerebelli, inferiorly. The **great cerebral vein** and the body of the **pineal gland** are located in the superior cistern. The **straight sinus** will also be seen when the tentorium is present. The straight sinus runs in the junction between the falx cerebri and the tentorium cerebelli. It is continuous with the inferior sagittal sinus, and it also receives the great cerebral vein.

SECTION THROUGH THE MIDBRAIN

If the plane of a section through the midbrain is parallel to the orbitomeatal line, then it will pass through the superior part of the orbit anteriorly and cut through the cerebellum posteriorly. If the plane of the section is more oblique, say at an angle of 15 to 20 degrees to the orbitomeatal line, then anterior views will be above the orbit and more of the cerebellum will be shown posteriorly.

Figure 1 – 18 *A* and *B* illustrates the area surrounding the **midbrain.** The two anteriorly projecting **cerebral peduncles** are made particularly noticeable by the presence of a very dark substance called the **substantia nigra.** The **interpeduncular cistern** is between the cerebral peduncles. Two **mammillary bodies,** which form a portion of the floor of the **third ventricle,** project into the interpeduncular cistern. The space just anterior to the mammillary bodies is the third ventricle, and the **optic tracts** project posteriorly and laterally from the third ventricle. The aqueduct of Sylvius, or **cerebral aqueduct,** forms a small opening near the **superior colliculi of the corpora quadrigemina.** The space between the corpora quadrigemina and the cerebellum is the **superior cistern,** or the cistern of the great cerebral vein. In addition to the **great cerebral vein,** branches of the posterior cerebral arteries are located here.

In sections that are nearly parallel to the orbitomeatal line, the frontal bone exhibits a large **frontal sinus.** The **orbits,** on either side, contain some **orbital fat** or possibly some of the superior muscles, such as the **levator palpebrae superioris** or **superior rectus.** Laterally, the **temporalis muscle** is superficial to the temporal bone. Posteriorly, these sections reveal the **cerebellum** and the centrally located **vermis.** The **tentorium cerebelli** separates the cerebellum from the cerebrum. The **anterior cerebral arteries** can be seen in the longitudinal fissure, the **middle cerebral arteries** in the lateral fissure, and the **posterior cerebral arteries** in the superior cistern.

In more oblique sections through the midbrain — say at an angle of 15 or 20 degrees to the orbitomeatal line — the plane passes superior to the orbit and the frontal sinus, anteriorly, and it cuts through the cerebellum at a more inferior level. Either a **sigmoid sinus** or a **transverse sinus** will probably be present, instead of the superior sagittal sinus.

SECTION THROUGH THE PONS

Transverse sections through the pons, which are illustrated in Figure 1 – 19 *A* and *B*, show the **pontine cistern** with the single **basilar artery.** The **fourth ventricle** is posterior to the pons, with the **superior cerebellar peduncles** forming the walls of the ventricle. Large **middle cerebellar peduncles** are lateral to the superior cerebellar peduncles. Two large **trigeminal nerves** emerge laterally from the pons. In sections that are cut more or less parallel to the orbitomeatal line, the **pituitary gland** is in the sella turcica of the sphenoid bone, just anterior to the pontine cistern. In some sections, this may also cut through the body of the **sphenoid bone** with a cavernous sinus located on either side. The **internal carotid arteries** and some nerves are in the cavernous sinus (see Fig. 1 – 11). Sections at this level also show the **sphenoid sinuses** and the **ethmoid air cells.**

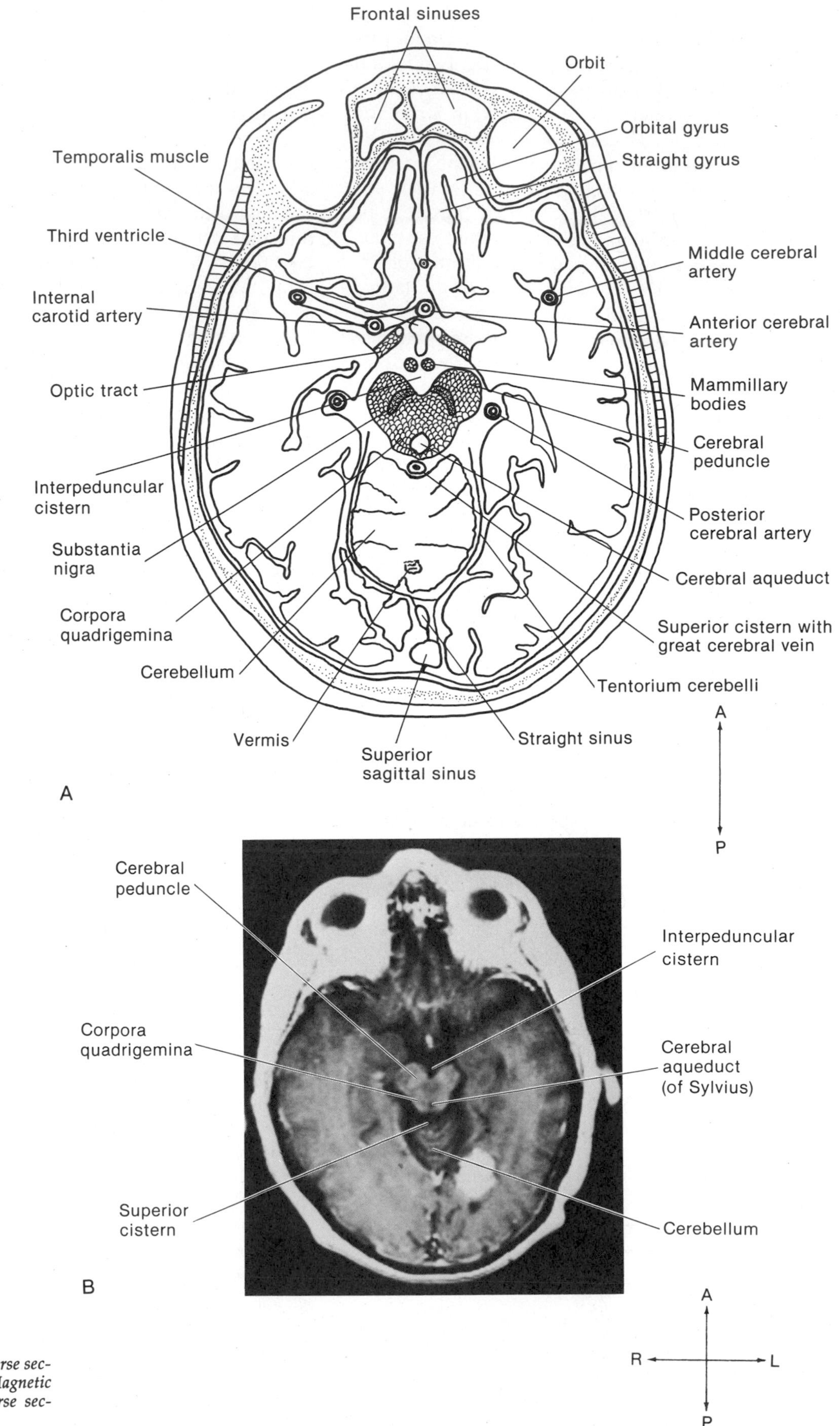

FIGURE 1-18. A, *Transverse section through midbrain.* B, *Magnetic resonance image of transverse section through the midbrain.*

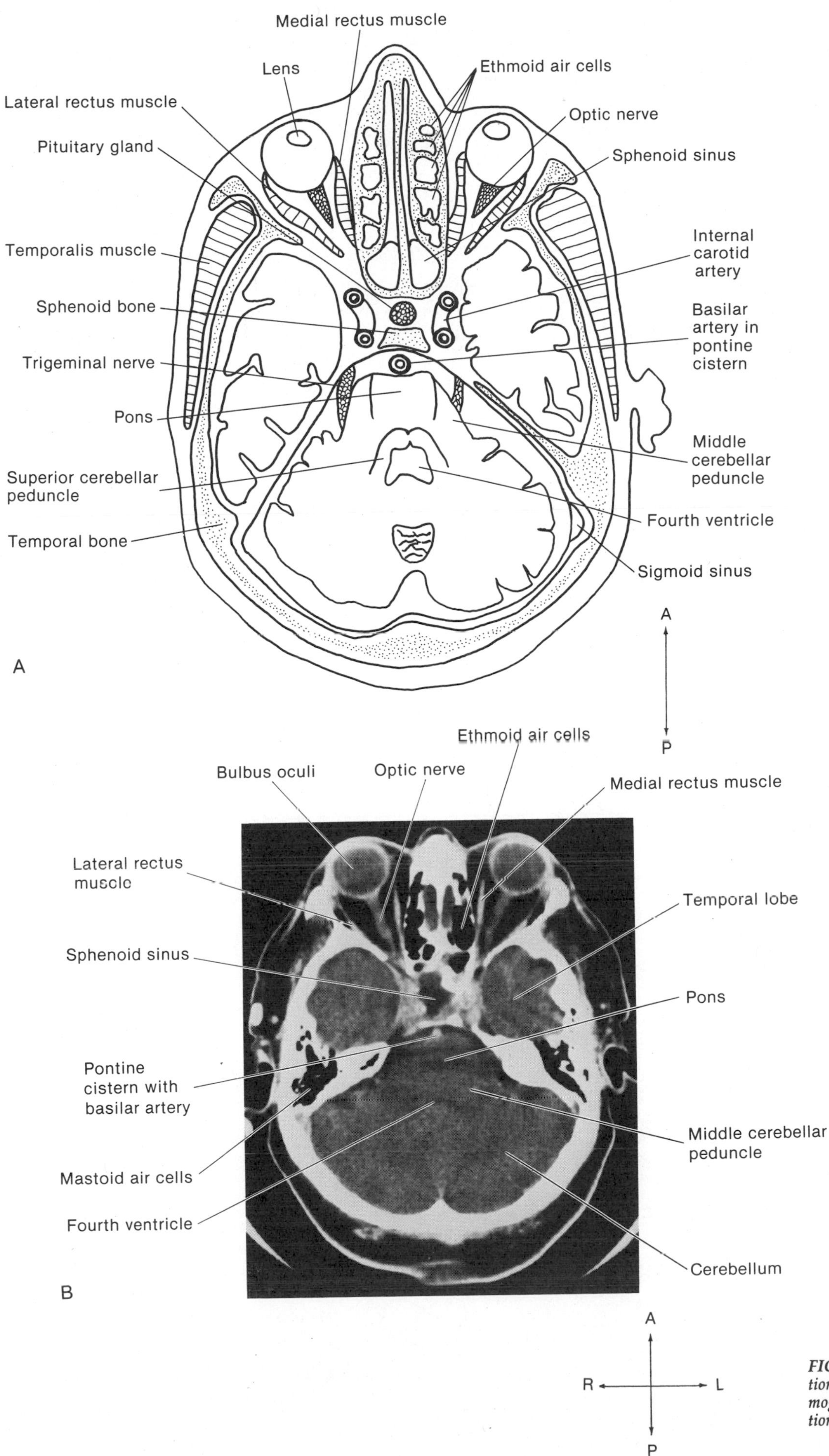

FIGURE 1-19. A, *Transverse section through pons.* B, *Computed tomographic image of transverse section through the pons.*

Figure 1–19 also shows that the **lateral walls** of the orbit are perpendicular to each other, whereas the **medial walls** are parallel. The **optic nerve** is centrally located in each orbit between the **lateral rectus muscle** and the **medial rectus muscle.**

Posteriorly, the dural venous sinuses are either **transverse** or **sigmoid sinuses,** depending on exact location. At the level of the pons, the **petrous portions of the temporal bones** project between the cerebellum and the temporal lobes of the cerebrum. The **temporalis muscle** is in the temporal fossa. In more oblique sections through the level of the pons, the anterior portion will pass above the level of the orbit and the nasal cavity and will still show the frontal lobes of the cerebrum.

SECTION THROUGH THE MEDULLA OBLONGATA

At the level of the medulla oblongata, the **right and left vertebral arteries** are present in the **cerebellomedullary cistern** just anterior to the medulla oblongata (Fig. 1–20). The venous sinuses are present as **sigmoid sinuses** by the **petrous portion of the temporal bone.** In some cases, the **jugular foramen** may be present at this level. The **internal carotid arteries** are lateral to the basilar portion of the occipital bone. In sections parallel to the orbitomeatal line, shown in Figure 1–20, the facial region will show the **nasal septum,** the **middle nasal conchae,** and the **maxillary sinuses.** The **zygomatic arch** may be seen in the cheek area and the **temporalis muscle** may be seen just medial to the arch.

SECTION THROUGH THE NECK AT LEVEL C-1

Sections through the neck at the level of the first cervical vertebra typically pass through the **hard palate** and **soft palate** (Fig. 1–21). The **pharyngeal constrictor muscle** is situated around the **oropharynx** at the posterior edge of the soft palate. The **dens,** or **odontoid process of C-2,** projects upward, posterior to the **anterior arch of C-1.** The **ramus of the mandible** appears as a thin slice of bone with the **masseter muscle** positioned lateral to it and the **medial pterygoid muscle** positioned medial to it. Another significant structure at this level is the **parotid gland.** The **sternocleidomastoid muscle** is posterior to the parotid gland, and the masseter muscle, the medial pterygoid muscle, and the ramus of the mandible are positioned anterior to this gland. Medially, the parotid gland is related to the **styloid process** of the temporal bone and the **internal jugular vein.**

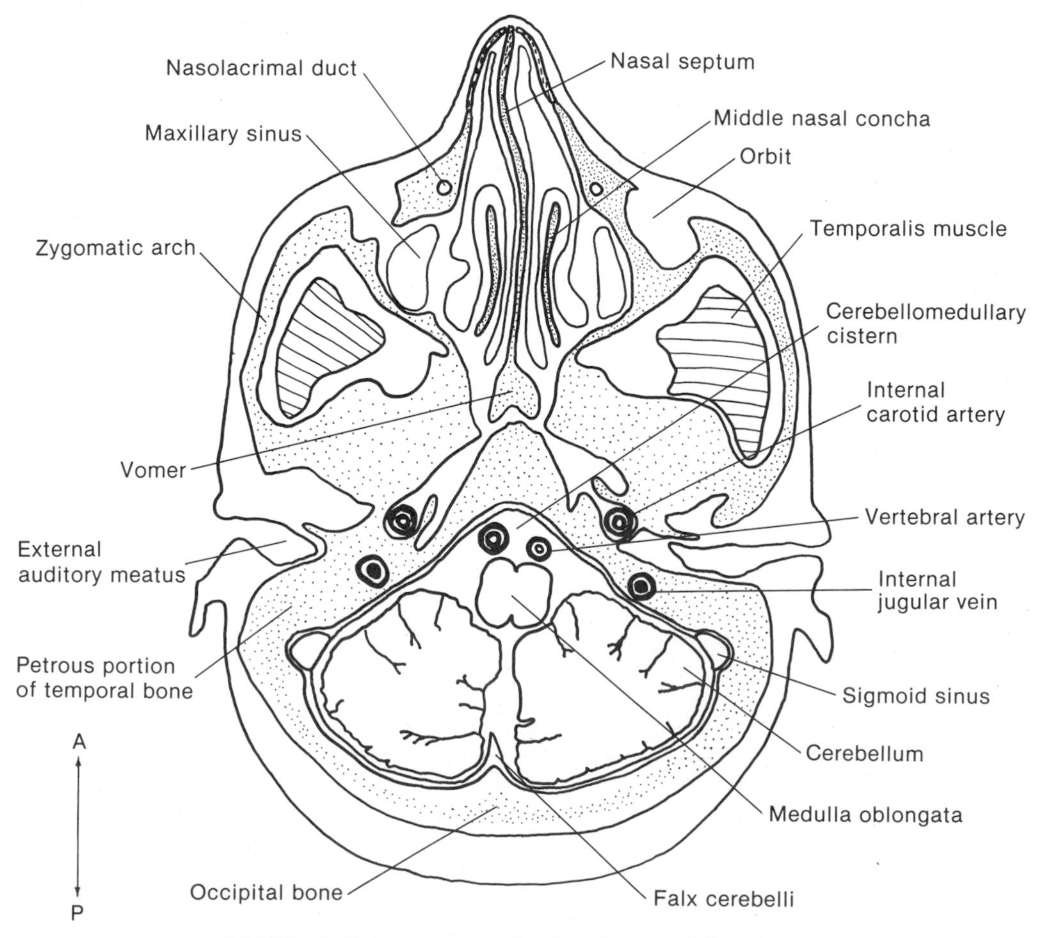

FIGURE 1–20. Transverse section through the medulla oblongata.

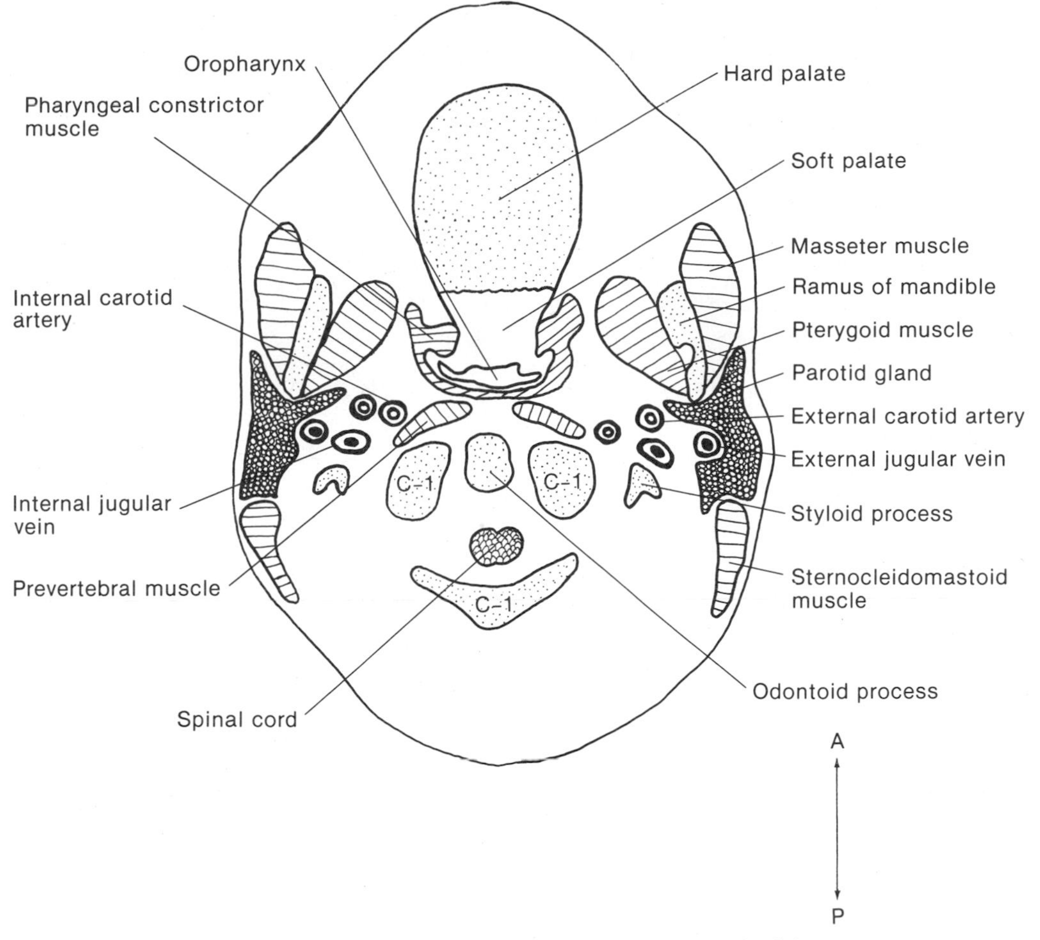

FIGURE 1-21. Transverse section through the neck at level C-1.

SECTION THROUGH THE NECK AT LEVEL C-3

Sections through the third cervical vertebra usually pass through the **mandible** and the **muscle of the tongue** (Fig. 1-22 *A* and *B*). The **submandibular gland** is medial to the mandible, and a portion of the **parotid gland** may still be present. The arrangement of the vessels in this region is of particular significance. The **external** and **internal carotid arteries** appear close together, which indicates that this is near their junction point. The external carotid artery is anterior to the internal carotid artery and the **internal jugular vein** is lateral to both of these. The **external jugular vein** is still more laterally situated and is related to the sternocleidomastoid muscle.

SECTION THROUGH THE LARYNX

The laminae of the laryngeal **thyroid cartilage** are usually evident at levels of C-5 or C-6 (Fig. 1-23). Near the upper margin of the thyroid laminae the **common carotid artery** bifurcates into the **external** and **internal carotid arteries.** At this junction the common carotid artery dilates to form the **carotid sinus,** which contains baroreceptors to monitor blood pressure. The pharynx continues through this region as the **laryngopharynx.** The muscles that form the floor of the posterior triangle of the neck are lateral to the transverse processes of the vertebrae.

Inferior to the larynx, the **trachea** is apparent with a lobe of the **thyroid gland** shown on either side or with the thyroid isthmus shown anteriorly. The posterior soft tissue of the trachea, namely the **trachealis muscle,** allows for expansion of the **esophagus** during swallowing. At this level, the **sternocleidomastoid muscles** are more anteriorly positioned.

Sagittal Sections

SECTION THROUGH THE TEMPOROMANDIBULAR JOINT

Figure 1-24 illustrates a parasagittal section through the lateral surface of the head and neck, particularly through the **temporomandibular joint.** The **condyle** of the mandible articulates in the **mandibular fossa** of the temporal bone. The **zygomatic process of the temporal bone,** using the joint as a reference, projects anteriorly, and the **external auditory meatus** is immediately posterior to the joint, followed by the **mastoid process** of the temporal bone. The superficial cheek muscle is the **masseter muscle,** which is followed posteriorly by the **parotid gland**

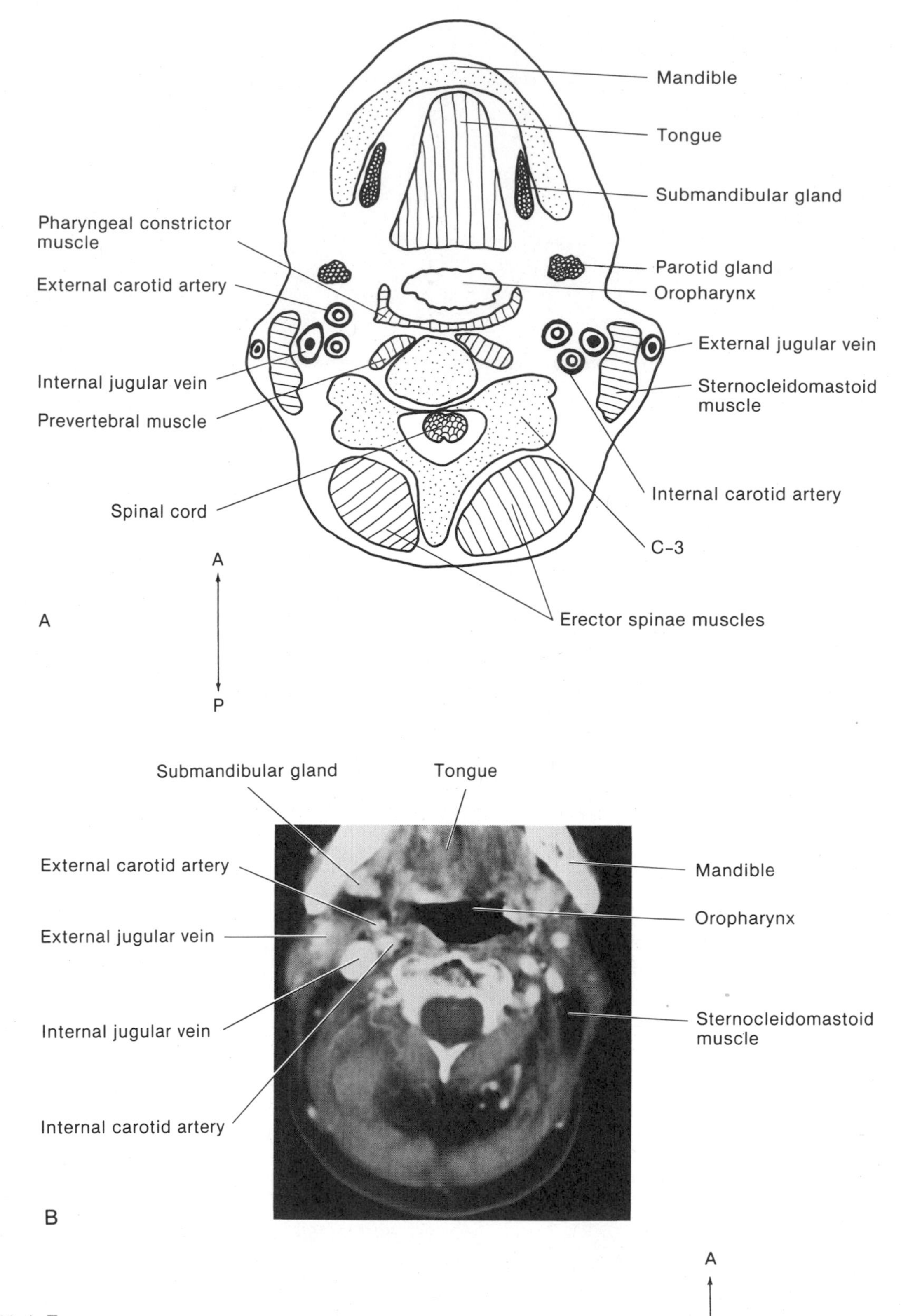

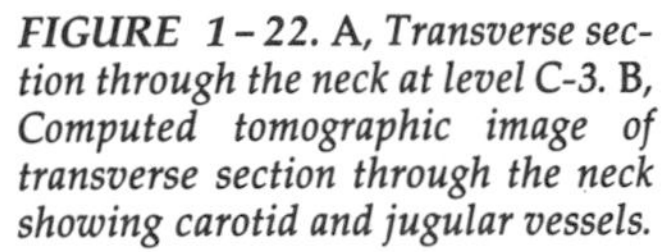

FIGURE 1-22. *A, Transverse section through the neck at level C-3. B, Computed tomographic image of transverse section through the neck showing carotid and jugular vessels.*

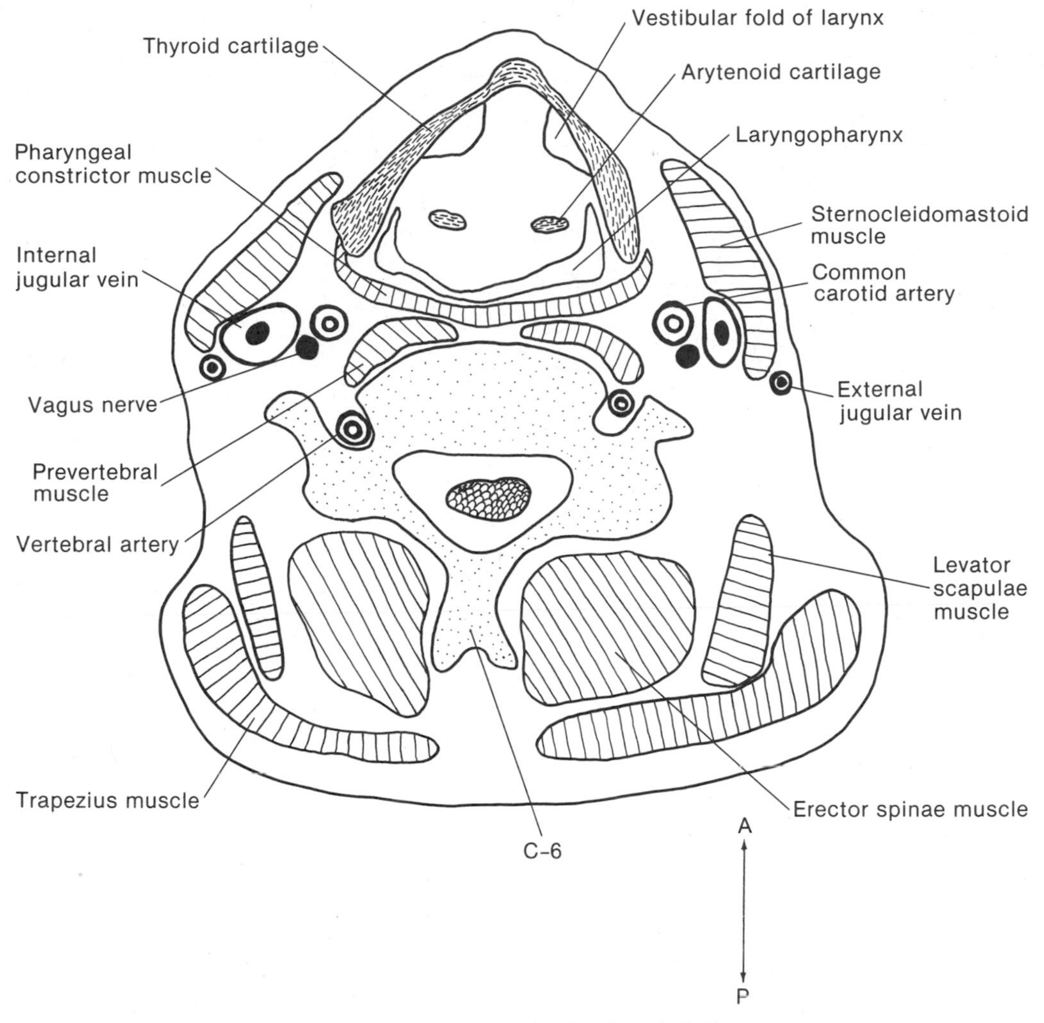

FIGURE 1-23. Transverse section through the larynx.

and the **sternocleidomastoid muscle.** The sternocleidomastoid muscle, a landmark muscle in the neck, courses obliquely along the neck to insert on the mastoid process. The **temporalis muscle** overlying the temporal bone is superior to the masseter. Only the surface of the brain is visible inside the cranial cavity.

SECTION THROUGH THE ORBIT

Sagittal sections through the orbit, as illustrated in Figure 1-25, are too lateral to show many details of the brain; however, numerous other structures of the head and neck are shown. This plane passes through the **frontal, parietal, occipital,** and **temporal lobes** of the cerebrum. The **tentorium cerebelli,** with the **transverse sinus** in its posterior margin, forms a partition between the cerebrum and cerebellum. Note that the tentorium cerebelli is anchored to the **petrous ridge** of the temporal bone and that the **sigmoid venous sinus** is located along the posterior margin of the ridge.

The **frontal bones** and the **maxillae** form the superior and inferior portions of the orbit, respectively. Within the orbit, fat surrounds the **bulbus oculi** and the **extrinsic eye muscles.** The **mandible** is a recognizable structure in the face. The **buccinator muscle,** one of the muscles of facial expression, is located in the cheek superior to the mandible, whereas the **submandibular gland** is located inferior to the mandible. The sublingual gland is not present because it is closer to the midline. A portion of the **temporalis** and **pterygoid muscles** may be evident superior and posterior, respectively, to the buccinator. In the neck, the **sternocleidomastoid muscle** is superficial anteriorly, and the **trapezius** is superficial posteriorly. The **internal jugular vein** starts at the jugular foramen in the temporal bone and descends through the neck. The **common carotid artery** is usually anterior and medial to the internal jugular vein.

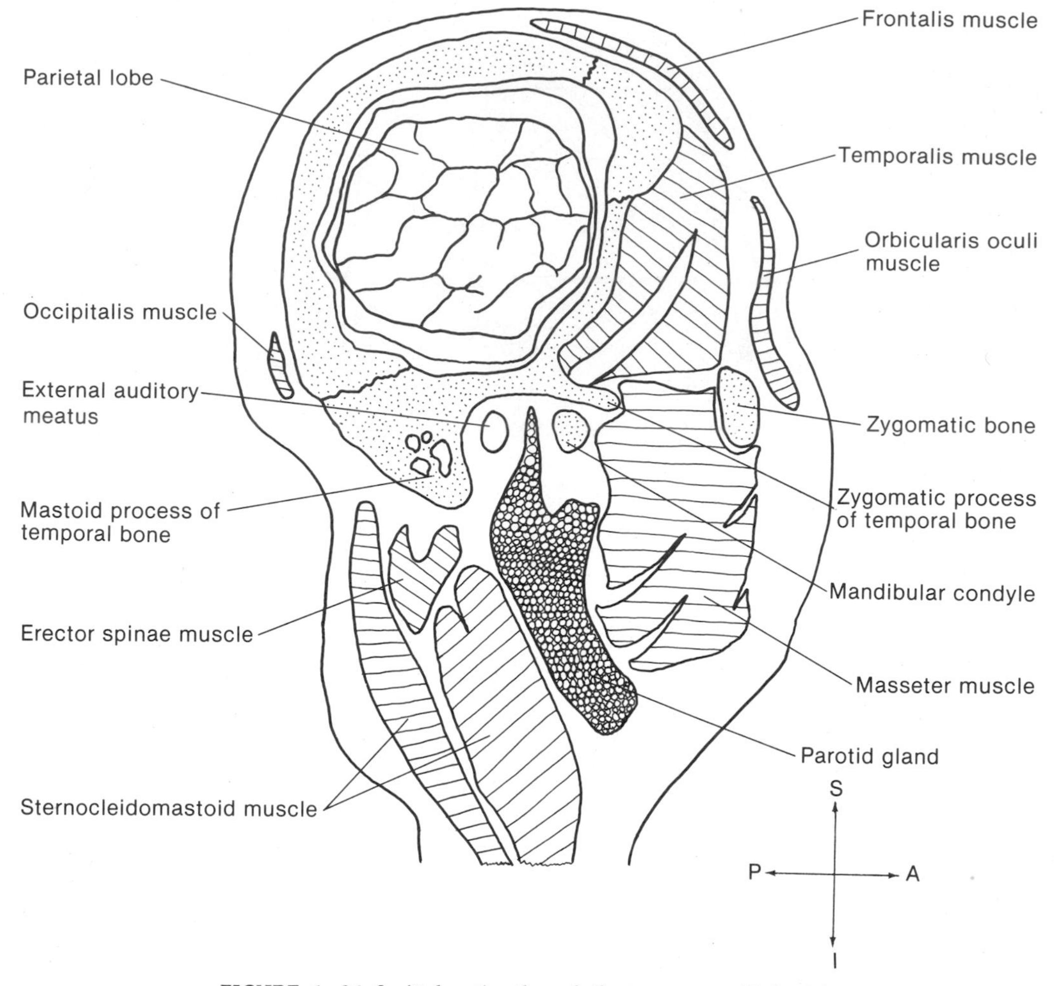

FIGURE 1-24. Sagittal section through the temporomandibular joint.

MIDSAGITTAL SECTION

Figure 1-26 illustrates a midsagittal section through the head and neck. Because many parts of the brain are midline, or nearly so, they are evident on midsagittal sections. The **cerebrum** curves around the **thalamus** to enclose it on all sides, except inferiorly. The **lateral ventricle** is superior to the thalamus. Posteriorly, the **tentorium cerebelli** forms a partition between the cerebrum and cerebellum. The brainstem, which includes the **midbrain,** the **pons,** and the **medulla oblongata,** extends inferiorly from the thalamus. A large **middle cerebellar peduncle** connects the cerebellum to the pons. The **hypophysis,** or **pituitary gland,** is located in the sella turcica of the sphenoid bone, and just anterior to this are the **sphenoid sinuses.** Midsagittal sections show the **nasal septum,** which is made up of the vomer and the perpendicular plate of the ethmoid bone.

In the region of the mouth, there is the **maxilla,** which forms the upper jaw; the **mandible,** which forms the lower jaw; and the **hard and soft palates,** which form the roof of the mouth. The **uvula** is the terminal portion of the soft palate. The largest structure present in the mouth is the **tongue.** Inferior to the tongue, located between it and the mandible, the **sublingual gland** is visible as a midline structure.

Components of the **larynx** that are quite obvious in midsagittal sections include the leaf-shaped **epiglottis,** the **vestibular folds,** and the **true vocal folds.** The space between the vestibular folds and the true vocal folds is the laryngeal **ventricle.** The **trachea,** with its cartilaginous rings, is inferior to the larynx.

The pharynx is immediately anterior to the vertebral column. The **nasopharynx,** which is located posterior to the nasal cavity, extends from the superior pharyngeal margin to the uvula. At the uvula, the nasopharynx becomes the **oropharynx,** which continues inferiorly to the epiglottis. The **laryngopharynx** continues inferiorly from the epiglottis, posterior to the larynx. At the level of the first tracheal cartilage, the laryngopharynx becomes the **esophagus.** Midsagittal sections provide nice views of the **vertebral bodies,** the **intervertebral discs,** and the **spinal cord.**

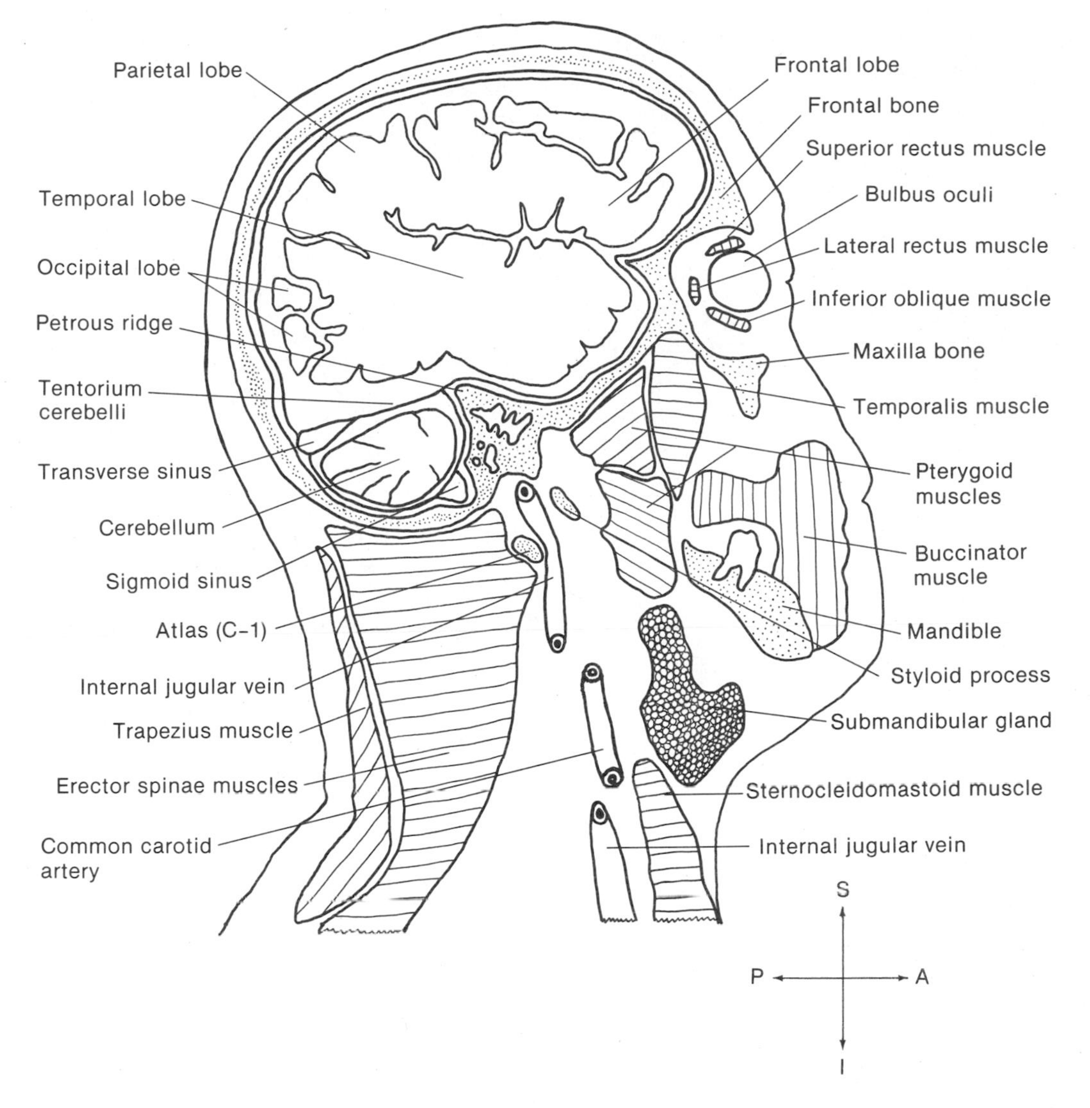

FIGURE 1-25. Sagittal section through the orbit.

Coronal Sections

SECTION THROUGH THE THIRD VENTRICLE AND THE BRAINSTEM

Figure 1-27 illustrates a coronal section through the third ventricle and the brainstem. The two cerebral hemispheres are separated by the **longitudinal fissure,** which contains the **falx cerebri** and the **anterior cerebral arteries.** The **superior sagittal sinus** is located in the superior margin of the falx cerebri. At the inferior margin of the falx cerebri, the **corpus callosum** forms a communicating band of white fibers between the two hemispheres. The two **lateral ventricles** are separated by a thin partition, the **septum pellucidum.** A region of gray matter, the **caudate nucleus,** forms the lateral portion of the floor of the lateral ventricle. Inferior to the lateral ventricles, in the midline, the **third ventricle** appears as a thin, slitlike opening, with the **thalamus** forming the wall on each side. Another region of gray matter, the **lentiform nucleus,** is lateral to the thalamus, between the thalamus and the **lateral sulcus.** The **cerebral peduncles** of the midbrain, easily identified by the small, dark band of **substantia nigra,** extend inferiorly from the thalamus and the third ventricle. The brainstem is completed by the **pons** and the **medulla oblongata.** At the **foramen magnum,** between the occipital condyles, the medulla oblongata continues as the **spinal cord.** Large **middle cerebellar peduncles** form a connection between the cerebellar hemispheres and the pons. Figure 1-27 also shows the petrous and mastoid portions of the temporal bone. The **tentorium cerebelli** extends medially from the petrous ridge to form the dural partition between the cerebrum and cerebellum. Sigmoid venous sinuses are associated with the petrous ridge.

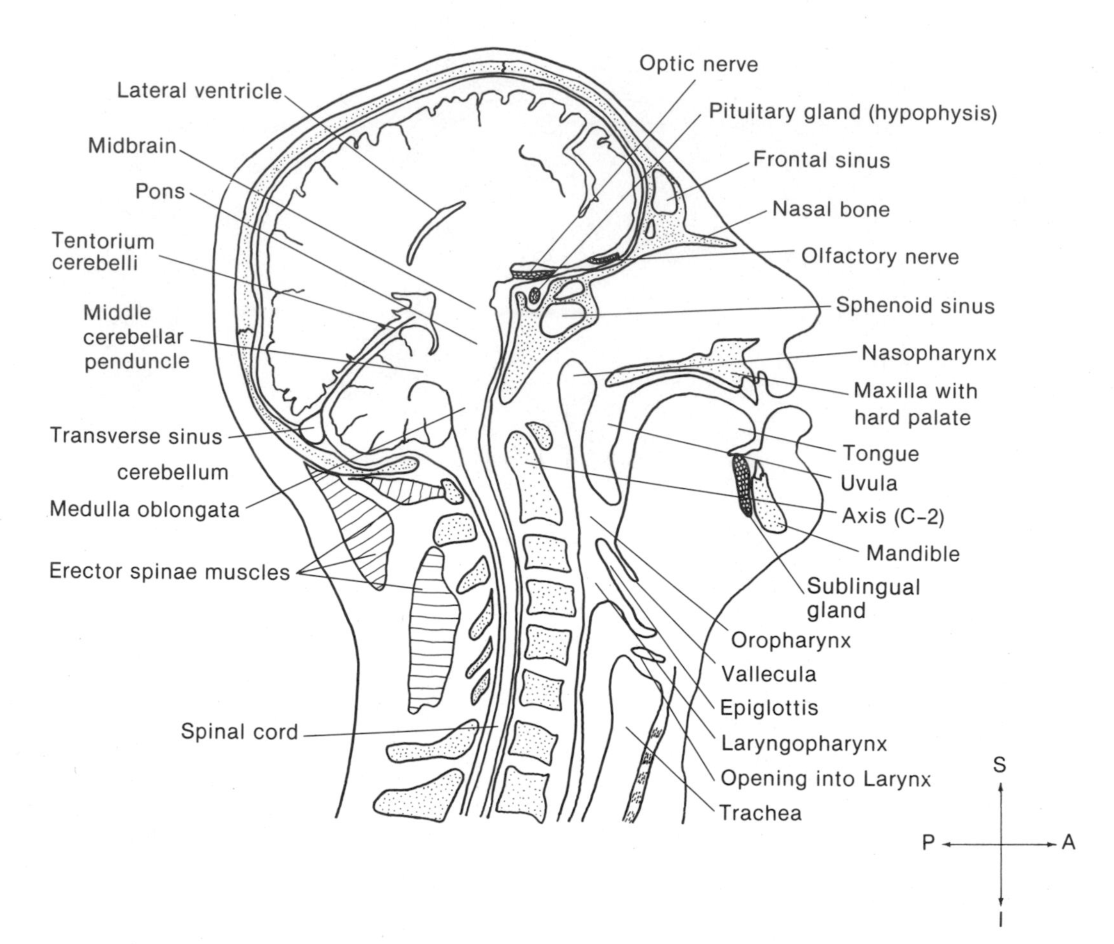

FIGURE 1-26. Midsagittal section through the head and the neck.

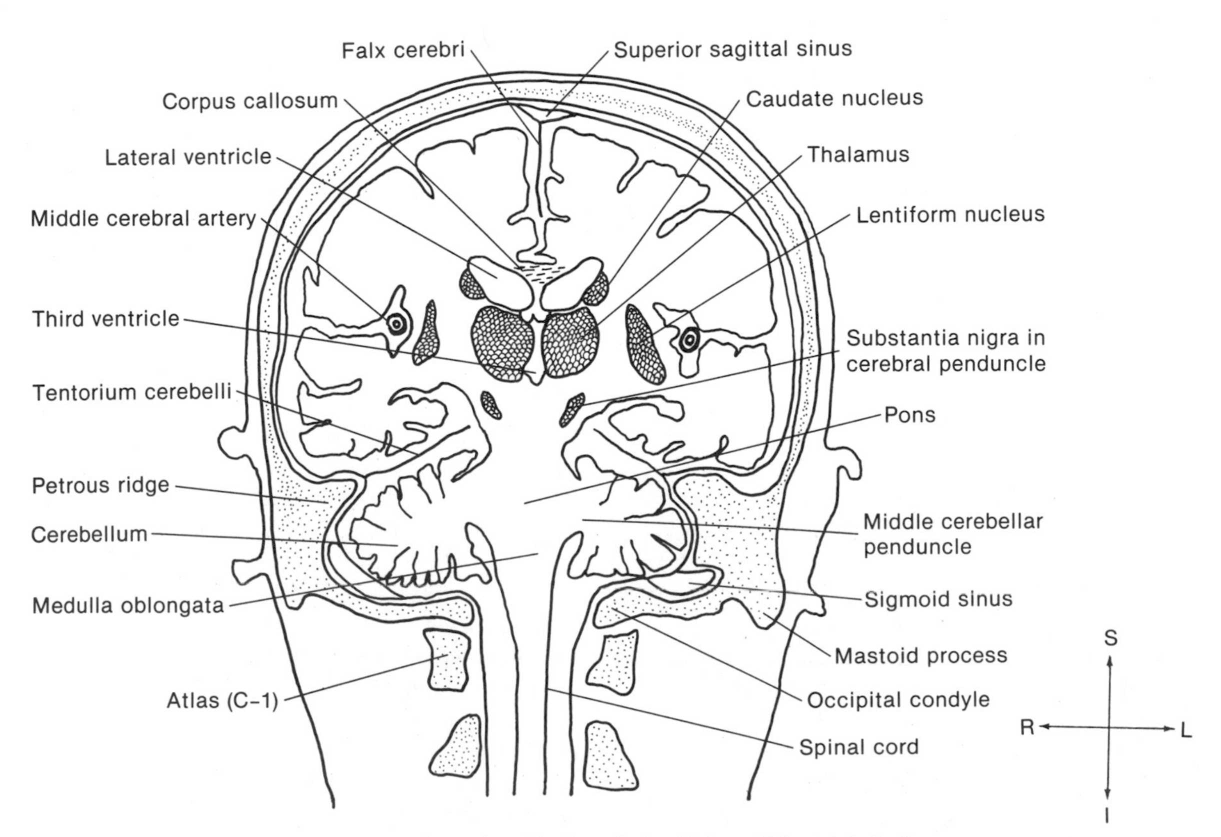

FIGURE 1-27. Coronal section through the third ventricle and the brainstem.

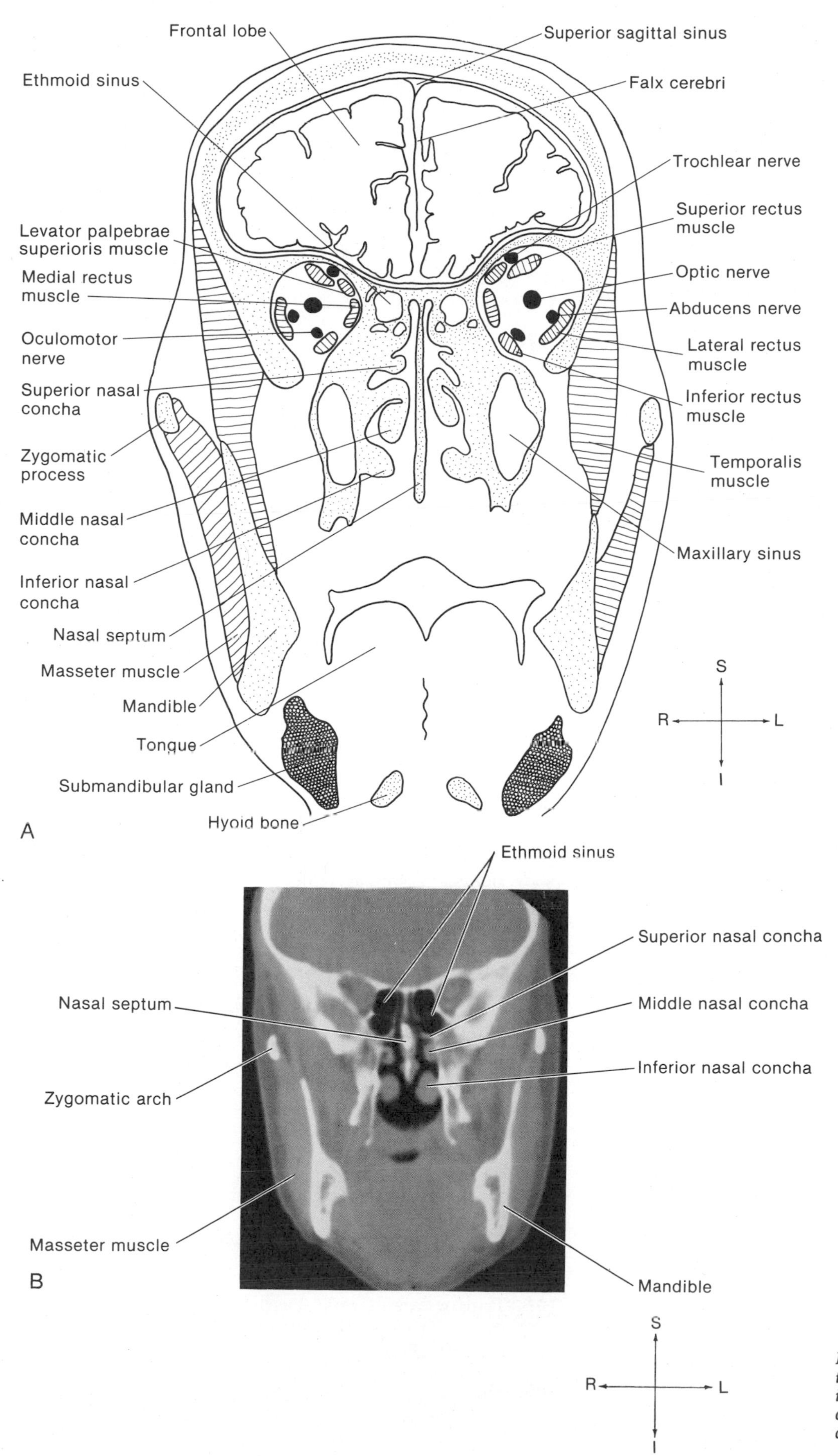

FIGURE 1-28. A, Coronal section through the orbit and the nasal cavity. B, Computed tomographic image of coronal section through the nasal cavity.

SECTION THROUGH THE ORBIT AND THE NASAL CAVITY

Figure 1–28 illustrates a coronal section through the posterior portion of the orbit and the nasal cavity. This plane intersects the frontal lobes of the cerebral hemispheres with the longitudinal fissure and falx cerebri between them. Within the orbit, the **optic nerve** appears as a central structure, with the extrinsic eye muscles and other nerves surrounding it. All are embedded in the orbital fat. The **bulbus oculi** is anterior to this plane and is not visible. **Ethmoid sinuses** are medial to the orbits. Within the nasal cavity, the central structure is the thin, **perpendicular plate of the ethmoid** that forms the nasal septum. Three **nasal conchae** project medially from the lateral walls of the cavity. The superior and middle conchae are part of the ethmoid bone, but the inferior concha is a separate bone. A portion of the large **maxillary sinus** is lateral to the nasal cavity. The palate separates the nasal cavity from the oral cavity. Bones and muscle make up the framework of the sides of the face. Extending inferiorly from the temporal bone, the **temporalis muscle** inserts on the medial side of the mandible. The **masseter muscle,** which is lateral to the mandible, originates on the zygomatic arch and the maxilla and inserts on the mandible. The **submandibular gland** is apparent near the inferior margin of the mandible.

REVIEW QUESTIONS

1. Which of the following are facial bones?
   a. zygomatic, frontal, nasal
   b. frontal, temporal, maxilla
   c. maxilla, inferior nasal conchae, zygomatic
   d. mandible, maxilla, temporal
   e. palatine, ethmoid, sphenoid

2. The largest of the paranasal sinuses is the ______________.

3. **True** or **False:** The longitudinal sulcus separates the frontal lobe from the parietal lobe.

4. Which of the following statements is **NOT** true regarding basal ganglia?
   a. The caudate nucleus is usually associated with the lateral ventricle.
   b. The lentiform nucleus is lateral to the internal capsule.
   c. The putamen is lateral to the globus pallidus.
   d. The external capsule separates the putamen from the globus pallidus.

5. The largest part of the diencephalon is the ______________.

6. The three principal parts of the brainstem are the
   a. thalamus, hypothalamus, and epithalamus.
   b. cerebral peduncles, corpora quadrigemina, and cerebral aqueduct.
   c. midbrain, forebrain, and hindbrain.
   d. midbrain, pons, and medulla oblongata.

7. The arrangement of white matter in the cerebellum is called the ______________.

8. Given the following numeric code, choose the sequence that represents the correct pathway for the flow of cerebrospinal fluid: (1) foramen of Magendie; (2) choroid plexus; (3) third ventricle; (4) fourth ventricle; (5) cerebral aqueduct; (6) lateral ventricle; (7) interventricular foramen.
   a. 6, 1, 3, 7, 4, 5, 2
   b. 2, 6, 7, 3, 5, 4, 1
   c. 1, 2, 6, 5, 3, 7, 4
   d. 6, 5, 3, 7, 4, 1, 2
   e. 6, 7, 3, 1, 4, 5, 2

9. **True** or **False:** The cerebrospinal fluid is located in the subarachnoid space between the arachnoid and the dura mater.

10. A calcified pineal body is an important landmark in neuroradiography. This structure is located in the ______________ cistern.

11. Which of the following is **NOT** true concerning the arterial supply to the brain?
    a. The two vertebral arteries join to form the basilar artery, which passes over the midbrain.
    b. The posterior cerebral arteries are branches of the basilar artery.
    c. The internal carotid arteries divide into middle cerebral arteries and anterior cerebral arteries.
    d. The middle cerebral arteries are located in the lateral sulcus.
    e. The posterior cerebral arteries supply the occipital lobe.

12. The venous sinus that follows along the tentorium cerebelli from the inferior sagittal sinus to the confluence of sinuses is the ______________.

13. The three branches of the fifth cranial nerve are the ______________, the ______________, and the ______________.

14. Which of the following statements about the eye is **NOT** true?
    a. The innermost tunic of the bulbus oculi is the retina.
    b. The lens is posterior to the iris.
    c. There are six extrinsic muscles that control eye movements.
    d. The lacrimal gland for the production of tears is located in the superior and medial margin of the orbit.
    e. The ophthalmic artery provides most of the blood supply to the eye.

15. **True** or **False:** Cervical vertebrae are unique because they are the only ones that have bifid spinous processes and transverse foramina.

16. The ______________ gland is the largest of the salivary glands, and it is closely associated with the ______________ muscle.

17. Which of the following statements about the pharynx is **NOT** true?
   a. The pharynx extends from the base of the skull to the level of the sixth cervical vertebrae.
   b. The pharyngeal tonsils are located in the oropharynx.
   c. The laryngopharynx is also called the hypopharynx.
   d. The retropharyngeal space is posterior to the pharynx but anterior to the vertebrae.
   e. The most lateral portions of the laryngopharynx are the piriform recesses.

18. The three single and largest cartilages of the larynx are the ______________, ______________, and ______________.

19. **True** or **False:** The esophagus is related to the pleura on the right, to the thoracic duct and the subclavian artery on the left, and to the trachea anteriorly.

20. Which of the following statements about the internal jugular vein is **NOT** true?
   a. As the internal jugular vein descends the neck, it is deep to the sternocleidomastoid muscle.
   b. The internal jugular vein is located within the carotid sheath.
   c. The internal jugular vein is lateral to the common carotid artery and anterior to the vagus nerve.
   d. At higher levels in the neck, the internal carotid artery is posterior to the internal jugular vein.

21. **True** or **False:** The relationships of the external and internal carotid arteries change so that at higher levels the external carotid artery is medial to the internal carotid artery but at lower levels it is lateral to the internal carotid artery.

22. The vertebral arteries ascend the neck within the ______________ of cervical vertebrae and enter the skull through the ______________.

23. Muscles of facial expression insert on the ______________ and are innervated by the ______________ nerve; muscles of mastication insert on the ______________ and are innervated by the ______________ division of the ______________ nerve.

24. Which one of the following statements about the triangles of the neck is **NOT** correct?
   a. The sternocleidomastoid muscle forms a dividing line between the anterior triangle and the posterior triangle of the neck.
   b. The mandible forms the apex of the anterior triangle.
   c. The carotid sheath is in the anterior triangle.
   d. The apex of the posterior triangle is at the junction of the sternocleidomastoid and trapezius muscles.
   e. The base of the posterior triangle is formed by the clavicle.

25. **True** or **False:** The vagus nerve is the longest nerve in the body and, among other things, provides the primary innervations of the diaphragm.

26. Identify the indicated structures on Figure *A*.

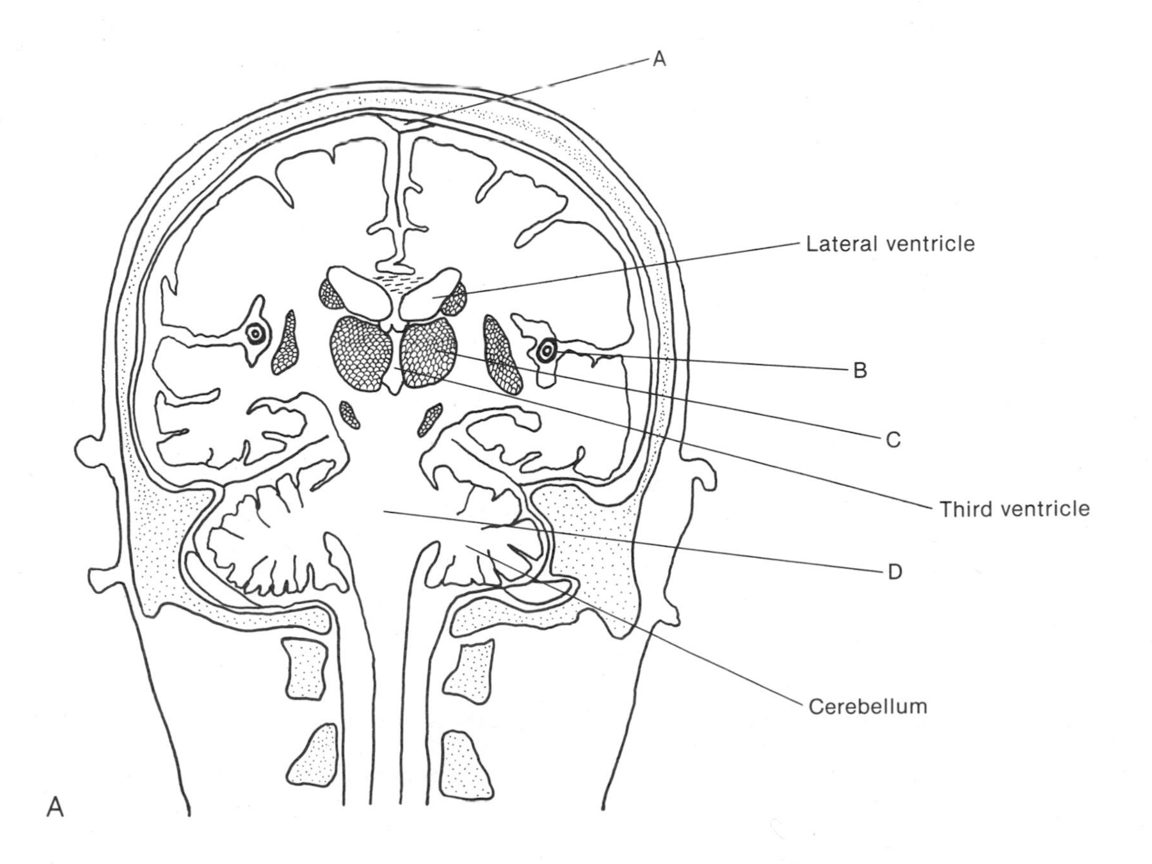

CHAPTER 2

THORAX

OBJECTIVES

Upon completion of this chapter, the student should be able to do the following:

- Identify and describe the bones that form the thoracic cage.
- State the vertebral level of the jugular notch, the sternal angle, and the xiphisternal junction.
- State the boundaries of the superior and inferior thoracic apertures.
- List three groups of muscles that form thoracic boundaries.
- Describe the pleura and pleural cavities.
- Compare the features of the right and left lungs.
- List the divisions of the mediastinum and the contents of each region.
- Describe the pericardial sac, the pericardium, and the pericardial cavity.
- Describe the structure of the heart wall.
- Define and state the location of the apex, the base, the surfaces, and the borders of the heart.
- Discuss the features and relationships of the chambers and the valves of the heart.
- Compare the right and the left coronary arteries with respect to their origins, branches, anatomic locations, and the regions that they supply.

- Trace the pathway of a stimulus through the conduction system of the heart.
- Identify the great vessels associated with the heart by describing the location and the relationships of each vessel.
- Trace the flow of blood through the heart from the right atrium to the ascending aorta.
- Discuss the location and the relationships of the thymus, the trachea, the esophagus, the azygos vein, and the hemiazygos vein.
- Identify muscles associated with the pectoral, back, and shoulder regions.
- State the origin and the location of the brachial plexus, and name five nerves that emerge from the plexus.
- Describe the structure and the hormonal control of the female breast.
- Name four groups of lymph nodes involved in lymphatic drainage of the breast.
- Identify the skeletal components, the muscles, the blood vessels, and the viscera of the thorax in transverse, sagittal, and coronal sections.

GENERAL ANATOMY OF THE THORAX

The terms **thorax** and **chest** are synonyms that refer to the region located between the neck and the abdomen. The **thoracic inlet,** formed by the first thoracic vertebra, the first pair of ribs, and the upper margin of the manubrium, separates the thorax from the root of the neck. Inferiorly, the thorax is bounded by the diaphragm, which covers the thoracic outlet. The musculoskeletal wall of the thorax provides protection for the heart and the lungs, which are contained in the thoracic cavity.

Enclosing Structures

BONES OF THE THORAX

The skeleton of the thorax is formed anteriorly by the **sternum,** posteriorly by the **12 thoracic vertebrae,** and laterally by the **ribs with their costal cartilages.** These bones form a thoracic cage that serves both as an attachment for muscles and as protection for the vital viscera that it encloses.

STERNUM. The **sternum** is an elongated, flat bone that is located in the anterior midline of the thorax. Anteriorly, it is covered only by skin, by superficial fascia, and by periosteum. It consists of three parts: the **manubrium,** the **body,** and the **xiphoid process.**

The manubrium is the most superior of the three parts. Its upper border is indented by an easily palpable midline **jugular notch,** which is sometimes called the suprasternal notch. The jugular notch is situated at the level of the disc between the second and the third thoracic vertebrae. **Clavicular notches,** which are located at the superolateral margins of the manubrium on either side of the jugular notch, form articulating surfaces for the clavicle.

Inferiorly, the manubrium is joined to the body of the sternum by fibrocartilage and ligaments. The manubrium and the body of the sternum do not articulate in a straight line; instead, their line of junction projects forward and forms the **sternal angle** (angle of Louis). This reliable landmark is generally 5 cm below the jugular notch and locates the sternal end of the second rib. This landmark also marks the level of the intervertebral disc between the fourth and fifth thoracic vertebrae. The trachea bifurcates into the two bronchi, and the aortic arch begins at the level of the sternal angle. The body, forming the bulk of the sternum, articulates with the second through the seventh costal cartilages.

The smallest and most inferior part of the sternum is the xiphoid process. It consists of hyaline cartilage in the youth but gradually ossifies during adulthood, and by the individual's fortieth year, it is usually bony in nature and is fused with the body of the sternum at the xiphisternal junction. The xiphisternal junction is usually at the level of the ninth thoracic vertebra.

RIBS. The lateral boundaries of the chest are formed by the **12 pairs of ribs.** These flat, elongated, curved, and slightly twisted bones and their respective costal cartilages extend from the thoracic vertebrae posteriorly to the sternum anteriorly. The ribs make up the major portion of the thoracic skeleton. The first seven ribs are considered **vertebrosternal ribs,** or **true ribs,** because they articulate directly with the sternum by way of their costal cartilages. The remaining five pairs are called false ribs. The costal cartilages of the eighth, ninth, and tenth ribs (the first three pairs of false ribs) are attached to the cartilages of the preceding ribs (the seventh, eighth, and ninth ribs, respectively) rather than directly to the sternum. These are called **vertebrochondral ribs.** The last two pairs of false ribs are **vertebral** or **floating ribs** because they have no anterior attachment to the sternum. The costal cartilages of the vertebrochondral ribs join so that their inferior edges form a continuous **costal margin.** As the costal margins diverge from the xiphisternal junction, they delineate the **infrasternal angle,** or the **costal arch.**

THORACIC VERTEBRAE. The posterior median skeleton of the thoracic cage is formed by the **12 thoracic vertebrae.** Some features of these vertebrae are specifically related to the thorax, including the **facets** on the transverse processes and the vertebral bodies that provide smooth surfaces for articulation with the ribs, and the long **spinous processes** along the posterior midline. When the vertebral column is flexed, the most prominent spinous process is usually that of the seventh cervical vertebra, although the first thoracic spinous process may sometimes be just as evident. If the arms are at the sides of the body, a horizontal line drawn through the tip of the third thoracic spinous process indicates the level of the base of the scapular spine. Also, the inferior angle of the scapula is at the same level as the middle of the seventh thoracic spinous process. The position of these landmarks is changed by raising the arms.

OPENINGS INTO THE THORAX

Openings into the thoracic cage include the superior and inferior thoracic apertures. The **superior thoracic aperture,** or **thoracic inlet,** is rather small and oblique in position. The most superior portion of the aperture is situated posteriorly and then slopes inferiorly in the anterior direction. It is bounded posteriorly by the first thoracic vertebra, laterally by the first pair of ribs and the costal cartilages, and anteriorly by the manubrium of the sternum. The **inferior thoracic aperture,** or **thoracic outlet,** is bounded posteriorly by the twelfth thoracic vertebra and anteriorly by the xiphisternal junction. The lateral margins of the thoracic outlet are formed by the twelfth rib and the sloping cartilage of the costal margin.

MUSCLES OF THE THORACIC WALL

Numerous muscles are attached to the skeleton of the thorax; most of these muscles move the pectoral girdle or are associated with the shoulder joint. In this section, only muscles that form a part of the thoracic boundary and that are associated with changing the intrathoracic volume during breathing are considered. An increase in the volume of the cavity decreases the pressure and permits inspiration. Conversely, a decrease in the volume increases the pressure and forces air out of the lungs during expiration. To increase intrathoracic volume, the boundaries of the cavity may increase in three different dimensions: vertical, transverse, and anteroposterior. Elastic recoil of the lungs and the weight of the thoracic wall primarily account for the decrease in each dimension during expiration.

DIAPHRAGM. The **diaphragm** covers the thoracic outlet, forming a muscular, movable partition between the thoracic and abdominal cavities. Contraction of the diaphragm enlarges the thoracic cavity in the vertical dimension during inspiration. The diaphragm is visualized to a greater extent in abdominal sections than it is in thoracic sections; thus, it will be described in greater detail in Chapter 3.

INTERCOSTALS. Three layers of intercostal muscles fill the spaces between the ribs. These are the external, internal, and innermost intercostal muscles, all of which receive motor impulses from the intercostal nerves. The **external intercostals** arise from the lower border of one rib and insert on the upper limit of the next rib below. The fibers are directed inferiorly and anteriorly. The **internal intercostals** occupy the intercostal spaces deep to the external intercostals. Also rising from the lower border of one rib and inserting on the upper limit of the next one below, the fibers are directed inferiorly and posteriorly and at right angles to the external intercostal fibers. The **innermost intercostals** appear similar to the internal intercostals, but they are separated from them by a neurovascular bundle that contains an intercostal nerve, artery, and vein.

LEVATOR COSTARUM. Twelve fan-shaped **levator costarum muscles** originate on the transverse processes of the thoracic vertebrae and insert on the rib immediately below. These are visualized on transverse sections as extending from the vertebral column to a rib. Action of the intercostal and levator costarum muscles increases both the transverse and the anteroposterior dimensions of the thoracic cavity.

Thoracic Cavity

The thoracic cavity is enclosed within the bones and muscles described in the previous paragraphs. It has three major divisions: the right and left pleural cavities, and the mediastinum. The two pleural cavities are filled with the lungs and occupy the lateral regions. The mediastinum is the central region between the two pleural cavities. It contains the heart and other structures, such as the trachea, the esophagus, and the thymus gland.

PLEURAL CAVITIES

The two pleural cavities are completely closed, separated, and lined by a serous membrane, the pleura. The **pleura** is essentially a continuous sheet in each cavity but is divided for descriptive purposes into the **visceral** and **parietal layers.** Its structure is very similar to a balloon indented by a fist. A balloon is a single sheet of material, but with a fist pushed into it, there are two layers: the outer layer and the inner layer next to your hand. The visceral pleura, or inner layer, is intimately adherent to the lung; this layer covers the entire surface of the lung and continues deeply into its fissures. The parietal pleura, or outer layer, lines the thoracic wall and is divided into four regions: the costal, the diaphragmatic, the mediastinal, and the cervical (Fig. 2–1). The **costal parietal pleura** is applied to the ribs, the costal cartilages, the intercostal muscles, and the sternum. The **diaphragmatic parietal pleura** is fused with the diaphragm and is continuous with the **mediastinal parietal pleura,** which is adjacent to the mediastinum. The **cervical parietal pleura** projects into the thoracic inlet to cover the apex of the lung. It extends up to, but not above, the neck of the first rib.

The space between the two pleural layers, parietal and visceral, is the **pleural cavity,** or the **pleural space.** In reality, this is a **potential space** that is filled with only a capillary layer of serous lubricating fluid. This fluid reduces friction, allowing the two surfaces to glide easily over each other during respiratory movements. The visceral pleura is insensitive to pain but the parietal layer is very sensitive.

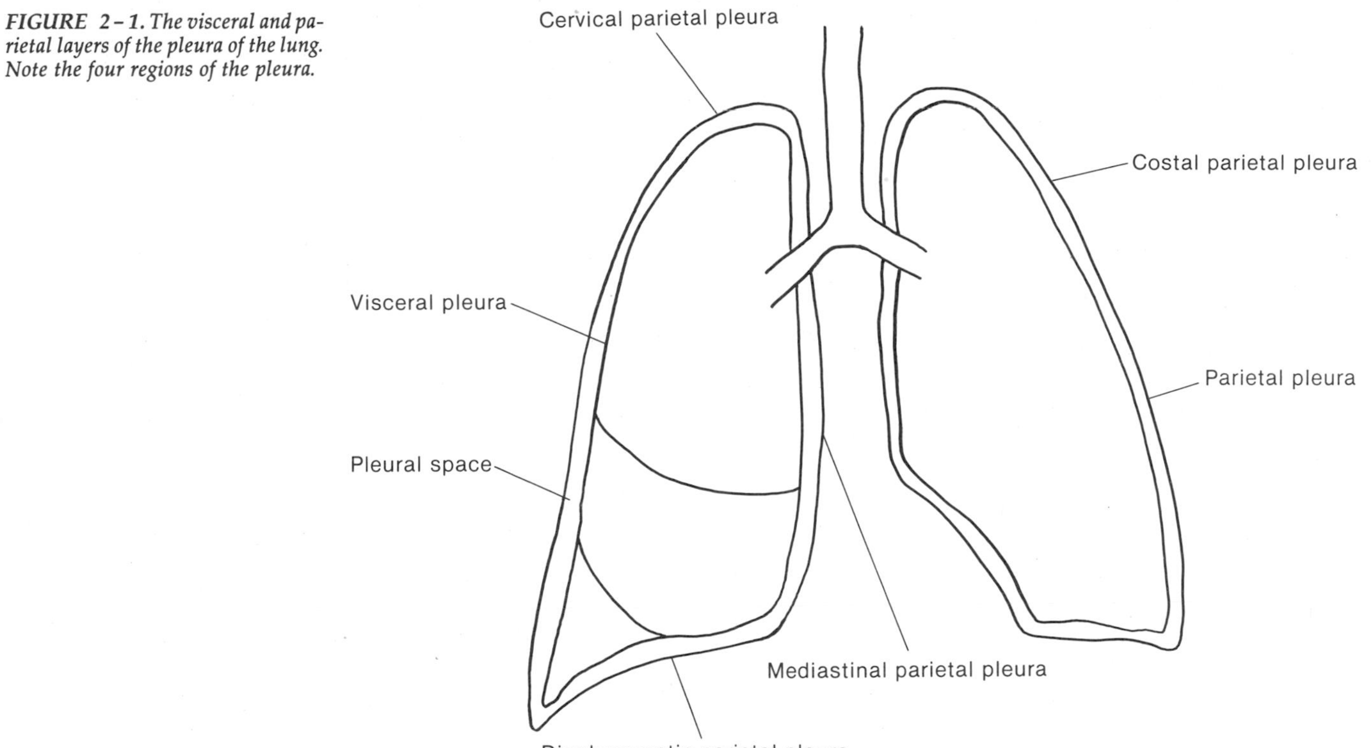

FIGURE 2-1. The visceral and parietal layers of the pleura of the lung. Note the four regions of the pleura.

LUNGS

Each lung is an elongated structure that is shaped roughly like half of a cone. The apex, or cervical dome, lies posterior to the middle third of the clavicle. It extends slightly above the first rib to project through the superior thoracic aperture. On the slightly concave medial (mediastinal) surface there is an opening called the **hilum,** at which point the bronchi, the blood vessels, the lymph vessels, and the nerves enter and leave the lung. The structures that traverse the hilum are collectively called the **root** of the lung. Each lung is freely movable within its own pleural cavity, except at the root or the hilum where it is attached. The right and left lungs are separated from each other by the mediastinum. The base of each lung is concave, conforming to the dome of the diaphragm. Although the right lung is shorter because of the volume of the liver on that side, it is also wider and has a greater volume than the left lung. The heart makes an indentation called the **cardiac notch** in the left lung. Each lung is partially transected by an **oblique fissure** that separates the lung into the **superior** and the **inferior lobes.** The right lung is further subdivided by the **horizontal fissure** to form a wedge-shaped **middle lobe.** Within the lung, each primary bronchus divides into secondary bronchi—two on the left side and three on the right. There is a secondary bronchus for each lobe of the lung. Each secondary bronchus further divides into tertiary segmental bronchi, which supply specific regions. A tertiary segmental bronchus with the specific sector of the lung that it supplies is called a **bronchopulmonary segment.**

MEDIASTINUM

The lungs and the pleura occupy the lateral portions of the thoracic cavity. All other thoracic structures are crowded into a central area called the **mediastinum.** This is divided into four regions (Fig. 2-2).

SUPERIOR MEDIASTINUM. The **superior mediastinum** is the area above the fibrous pericardium, and it is separated from the inferior mediastinum by a line that passes from the sternal angle to the intervertebral disc between the fourth and the fifth thoracic vertebrae. The superior mediastinum contains the aortic arch and its branches, the thymus, and all of the structures that pass between the neck and the thorax.

INFERIOR MEDIASTINUM. The **inferior mediastinum** is divided into the anterior, the middle, and the posterior portions of the mediastinum.

Anterior Mediastinum. The **anterior mediastinum** is a limited area that is anterior to the pericardium, posterior to the sternum, inferior to the sternal angle, and superior to the diaphragm. Superiorly, at the sternal angle, it is continuous with the superior mediastinum. It contains connective tissue with some fat and some lymph nodes.

Middle Mediastinum. The **middle mediastinum,** centrally located and limited by the fibrous pericardium, contains the heart and the roots of the ascending aorta, the pulmonary artery, the superior and inferior venae cavae, and the four pulmonary veins.

Posterior Mediastinum. The **posterior mediastinum** is posterior to the pericardium and inferior to the fourth thoracic vertebra. The diaphragm limits the posterior me-

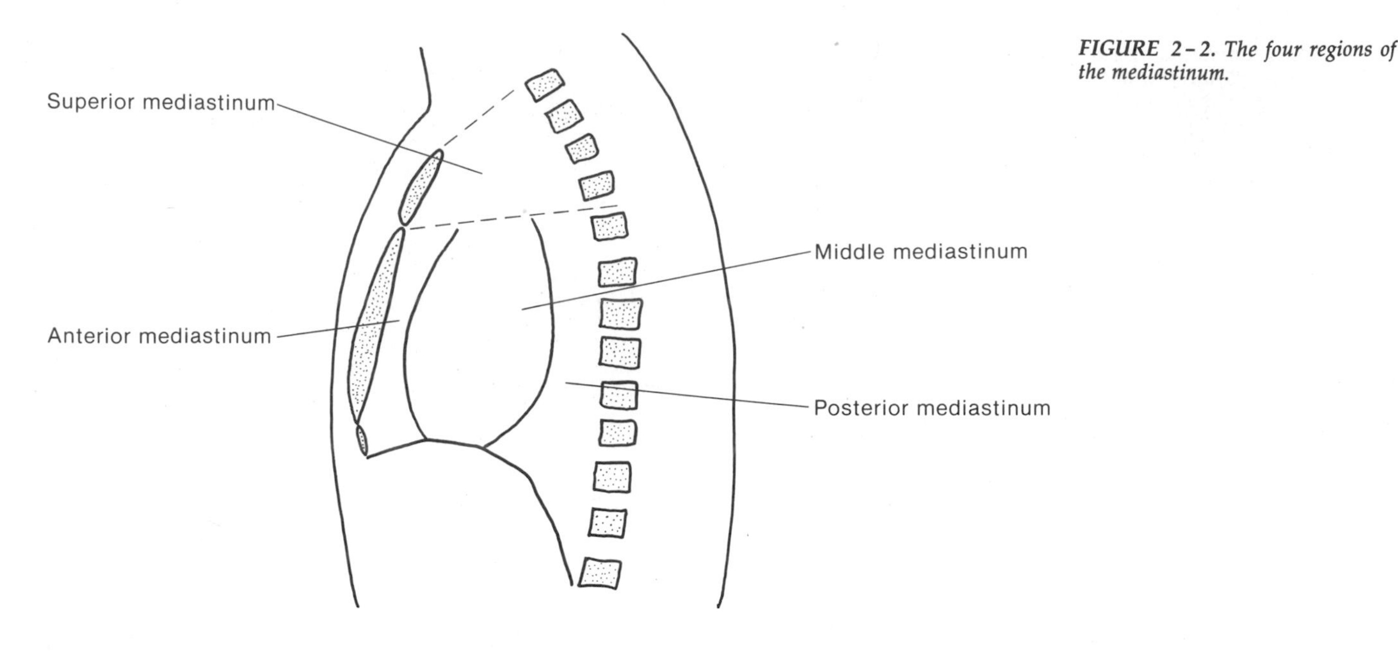

FIGURE 2-2. The four regions of the mediastinum.

diastinum inferiorly. Between the two parietal pleurae of the lungs and anterior to the vertebrae, the posterior mediastinum contains the descending thoracic aorta, the azygos and hemiazygos veins, the thoracic duct, and the esophagus (see Fig. 2-4).

Heart

The **heart** is a hollow, muscular organ that is enclosed in a fibroserous sac within the middle mediastinum. Shaped somewhat like a cone, the heart lies obliquely in the chest with two thirds of its mass to the left of the median plane and one third located to the right. About the size of a person's clenched fist, the heart weighs approximately 250 to 300 g. Superficial relationships include an apex, a base, three surfaces, and four borders.

PERICARDIUM

FIBROUS PERICARDIUM. A fibroserous sac, the **pericardium,** surrounds the heart and the proximal portions of the great vessels that enter and leave the heart. There are essentially two types of pericardium—fibrous and serous. The external, strong **fibrous pericardium** is composed of tough, fibrous connective tissue. Superiorly, at its apex, the fibrous pericardium blends with the outer layer of the great vessels. At its base, the fibrous layer fuses with the central tendon of the diaphragm so that respiratory movements influence the movement of the pericardial sac. In the anterior midline, the fibrous pericardium is attached to the posterior surface of the sternum by strong **sternopericardial ligaments.**

SEROUS PERICARDIUM. The double-layered **serous pericardium** is composed of a thin, transparent, serous membrane. The outer **parietal layer of serous pericardium,** sometimes called the parietal pericardium, forms a smooth, moist lining for the fibrous pericardium. The fibrous pericardium and the parietal layer of serous pericardium are closely adherent and difficult to separate, and together they make up the **pericardial sac.** The inner **visceral layer of the serous pericardium** covers the cardiac muscle of the heart wall. The visceral serous pericardium is often called the **epicardium** because it forms the outer layer of the heart wall.

PERICARDIAL CAVITY. The parietal and the visceral layers of the serous pericardium form a continuous closed sac around the heart in the same way that the parietal and visceral layers of the pleura surround the lungs (see earlier discussion). Between the parietal and visceral pericardia is a potential space, the **pericardial cavity,** which contains a small amount of serous fluid that is distributed as a capillary film on the opposing surfaces. The lubricating action of this fluid keeps the surfaces moist and reduces friction, which allows the layers to glide easily over each other during heart movements.

HEART WALL

The heart wall consists of three layers. The outermost layer is the **epicardium,** which is the visceral layer of serous pericardium. Heart muscle, which is called cardiac muscle, makes up the middle layer, the **myocardium.** This is the layer that contracts to produce the heart's pumping action. The myocardium makes up the bulk of the heart wall. The thickness of the myocardium in the heart wall varies from chamber to chamber. The harder a particular chamber has to work to pump blood, the thicker the wall. The atria have relatively thin walls because they are primarily "receiving" chambers rather than "pumping" chambers. Ventricles have thick walls because they forcefully eject blood from the heart. The left ventricle has the thickest wall because it pumps blood into the systemic circulation throughout the whole body.

The **endocardium,** the innermost layer, is a thin, smooth layer of simple squamous epithelium called endothelium. The endothelial lining of the heart also covers the heart valves and is continuous with the endothelial lining of the blood vessels.

SUPERFICIAL FEATURES OF THE HEART

Superficial features of the heart include an apex, a base, three surfaces, and four borders.

APEX. The **apex,** formed entirely by the left ventricle, points downward and to the left. Located in the fifth intercostal space, at the level of the eighth thoracic vertebra, the apex is the lowest part of the heart and is situated to the left of the midline. These positions vary and are dependent on the phase of respiration.

BASE. The **base** of the heart is the broad superior portion of the heart that is found opposite the apex. This means the base projects superiorly, posteriorly, and to the right. It extends between the fifth and eighth thoracic vertebrae. The two atria are the primary components of the base, and the posteriorly positioned left atrium is the predominant structure. The ascending aorta, the pulmonary trunk, and the superior vena cava emerge from the base. The base is sometimes referred to as the **posterior surface.**

SURFACES. In addition to the posterior surface, or the base, there are two other surfaces of the heart. The anterior **sternocostal surface** is created primarily by the right atrium and the right ventricle, although the left auricular appendage and the left ventricle contribute a small portion. The two ventricles, resting on the diaphragm, comprise the **diaphragmatic surface.**

BORDERS. The right atrium, in line with the superior and inferior venae cavae, forms the **right border.** The **left border** is more convex and is outlined by the left ventricle. The right ventricle along with a small contribution from the left ventricle near the apex forms the horizontal **inferior border.** The **superior border,** the place where the great vessels enter and leave the heart, is formed by both atria.

CHAMBERS AND VALVES

The heart is divided into four chambers: the right and left atria and the right and left ventricles. On the surface, the atria are separated from the ventricles by a groove that encircles the heart. This is the coronary sulcus, or the atrioventricular sulcus. Similarly, an interventricular sulcus marks the division between the right and the left ventricles.

VALVES. A system of valves keeps blood flowing through the heart in the appropriate direction. Differences in function necessitate differences in structure; thus there are two basic types of valves in the heart. Both types involve cusps, which consist of flaps of fibrous tissue that are covered with endothelium. **Semilunar valves** are found at the exit ports of the ventricles. **Atrioventricular valves** function as inflow valves at the point where the blood flows from the atria into the ventricles.

Semilunar valves consist of three cusps that balloon out from the vessel wall to prevent backflow of blood from the vessels into the ventricles. When the ventricles contract, the increased pressure forces the valve cusps flat against the vessel wall, which allows ejection of blood. After contraction, as ventricular pressure decreases, the cusps are caught in a passive backflow, and they balloon out from the walls to close the orifice. This prevents blood from flowing backward into the ventricles.

Atrioventricular valves are named for their location between the atria and the ventricles. The atria have thin walls, do not contract strongly, and generate relatively little pressure. Consequently, the atrioventricular valves must open easily to allow blood to flow from the atria into the ventricles. These valves have thin cusps that move readily in the current of flowing blood. Stringlike structures called **chordae tendineae** anchor the cusps to projections of myocardium, called **papillary muscles,** in the ventricles. When pressure in the ventricles increases as a result of contraction, these valves are forced back over the opening to prevent blood from going back into the atria. The papillary muscles contract with the ventricles, which creates a tension in the chordae tendineae and prevents the valve cusps from protruding back into the atria.

RIGHT ATRIUM. The most right-sided portion of the heart is the **right atrium.** This thin-walled chamber receives venous blood from the coronary and the systemic circulations. The right atrium has openings from the **superior** and **inferior venae cavae,** which return venous blood from systemic circulation, and an opening from the **coronary sinus,** which returns blood from the circulation that supplies blood to the heart wall.

The smooth-walled posterior region of the right atrium, where the venae cavae enter, is called the **sinus venarum.** Anteriorly, the atrial wall is roughened by muscular ridges called **pectinate muscles.** A small muscular pouch, the **auricle,** projects from the right atrium toward the left side and covers the root of the aorta. The posterior wall of the right atrium is formed by the **interatrial septum,** the partition between the right and the left atria. In the fetal circulation, there is an opening in the interatrial septum that is called the **foramen ovale,** which allows the circulating blood to pass directly from the right atrium to the left atrium and to bypass the pulmonary circulation. After birth, higher pressure on the left side of the heart pushes a flap of tissue across the opening to close it. Eventually, the foramen is sealed off, but a depression, the **fossa ovalis,** remains. The left or medial wall of the right atrium contains the right atrioventricular valve. This valve has three cusps and is called the **tricuspid valve.** The tricuspid valve has a vertical orientation and is posterior to, or slightly to the right of, the sternum at the level of the fourth intercostal space between the fourth and fifth ribs.

RIGHT VENTRICLE. The triangular **right ventricle** forms a major portion of the sternocostal or anterior surface of the heart. This chamber receives blood from the right atrium through the tricuspid valve and forces the blood into the **pulmonary trunk,** from where it travels to the lungs. A **pulmonary semilunar valve** is located at the outflow from the right ventricle into the pulmonary trunk. An average location places this valve at the third costal cartilage on the left side of the sternum. The upper anterior portion of the right ventricle that is located around the origin of the pulmonary trunk is the smooth-walled **conus arteriosus.** The remainder of the right ventricular wall is roughened by muscular ridges called **trabeculae carneae** and projections called **papillary muscles.** The stringlike **chordae tendineae** that extend from the three papillary muscles to the cusps of the right atrioventricular or tricuspid valve prevent inversion of the valve when ventricular pressure increases.

LEFT ATRIUM. The **left atrium** is the most posterior structure of the heart. It forms most of the base of the heart and consists of the atrium proper and its auricular appendage. The wall of the left atrium is slightly thicker than that of the right atrium, and its interior is smooth, except for a few pectinate muscles in the auricle. Four **pulmonary veins,** two on each side, return oxygenated blood to the heart from the lungs and enter the left atrium at its superolateral aspect. The anterior wall of the left atrium is formed by the left atrioventricular valve. This valve has two cusps and is called the **bicuspid valve,** or **mitral valve.** This is the valve that is most often affected by disease, especially rheumatic fever. The average location of the bicuspid or mitral valve is at the level of the fourth costal cartilage on the left side of the sternum.

LEFT VENTRICLE. From the left atrium, blood enters the **left ventricle** through the bicuspid or mitral valve. The left ventricle has much thicker walls than the right because it is responsible for pumping blood throughout the whole body via the systemic circulation. Most of the internal surface of the left ventricle is covered with the muscular ridges of trabeculae carnae, which are similar to those in the right ventricle. The wall is smooth in the **vestibule,** or aortic outflow region. The two papillary muscles, attached to the bicuspid valve by chordae tendineae, are usually larger than those in the right ventricle. From the left ventricle, blood enters the aorta to supply the body through the systemic circulation. Between the left ventricle and the aorta, there is an **aortic semilunar valve.** This valve is located at the level of the third intercostal space, between the third and fourth ribs, on the left side of the sternum.

INTERVENTRICULAR SEPTUM. The **interventricular septum** forms a partition between the right and left ventricles. The septum has an oblique orientation. Its position can be visualized on the surface of the heart by the anterior and posterior interventricular sulci, or grooves. The septum is mostly thick and muscular, although there is a small, oval portion that is thin and membranous. This is located just inferior to the right cusp of the aortic semilunar valve. The membranous portion of the interventricular septum is the part most frequently involved in ventricular septal defect. This defect, either alone or in combination with other defects, is present in about one half of the congenital cardiac abnormalities.

VARIATION IN FEATURES. Valve location as well as heart size and delineation varies greatly, depending on age, sex, body build, phase of respiration, and other factors. The locations and surface relationships stated here are determined for what would be the "average person" (if one existed). The valve locations are anatomic projections and do not necessarily indicate the best place for listening to the heart sounds. Figure 2–3 illustrates some of the features of the heart and the flow of blood through the heart.

BLOOD SUPPLY TO THE HEART

The heart wall is muscle; thus it requires a continual supply of oxygenated blood to function effectively. Blood is supplied to the muscular wall of the heart via the **coronary arteries** and their branches. The right and left coronary arteries originate from the aorta immediately superior to the aortic semilunar valve.

RIGHT CORONARY ARTERY. After originating from the aorta, the **right coronary artery** passes slightly forward and to the right, emerging between the pulmonary trunk and the right auricle. It then descends in the atrioventricular sulcus (coronary sulcus) to the inferior border.

During its course, the right coronary artery gives off branches to the wall of the right atrium. Near the inferior border of the heart, the right coronary artery gives off a **marginal branch** that proceeds toward the apex. The marginal branch supplies the right ventricle. After giving off the marginal branch, the right coronary artery continues in the coronary sulcus to the posterior surface, where it gives off a **posterior interventricular branch** that descends toward the apex in the posterior interventricular sulcus. This vessel gives off branches that supply the posterior wall of both ventricles and a portion of the interventricular septum.

LEFT CORONARY ARTERY. The **left coronary artery** is very short, usually only 2 to 3 cm. It emerges between the left auricular appendage and the pulmonary trunk and then divides. The **anterior interventricular (descending) branch** descends toward the apex in the anterior interventricular sulcus and then turns toward the posterior surface to anastomose with the posterior interventricular branch of the right coronary artery. This branch supplies portions of both ventricles and the interventricular septum. The **circumflex branch** of the left coronary artery follows the left atrioventricular (coronary) sulcus around the left margin of the heart to the posterior surface, where it anastomoses with the right

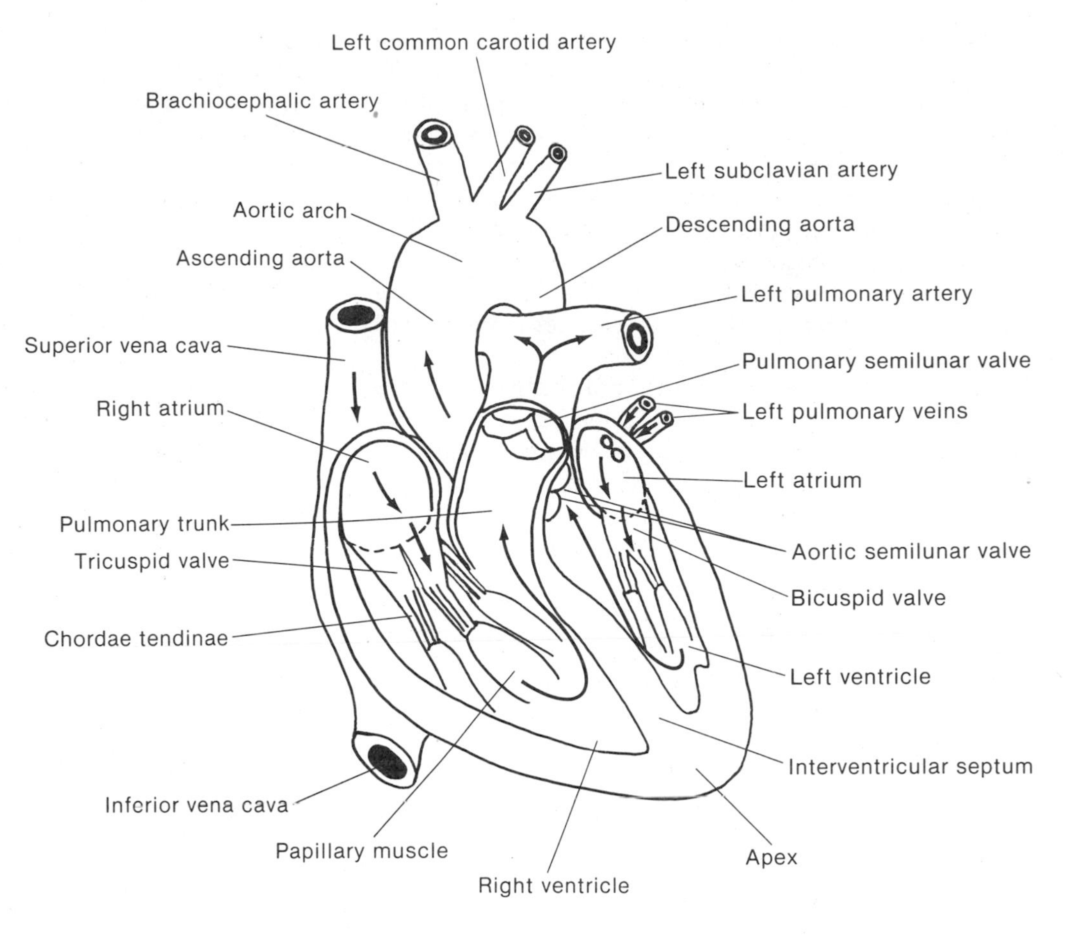

FIGURE 2-3. Chambers, valves, and other features of the heart. The sequence for the flow of blood through the heart is as follows: (1) right atrium, (2) tricuspid valve, (3) right ventricle, (4) pulmonary semilunar valve, (5) pulmonary trunk, (6) right/left pulmonary arteries, (7) lungs, (8) pulmonary veins, (9) left atrium, (10) bicuspid (mitral) valve, (11) left ventricle, (12) aortic semilunar valve, (13) ascending aorta.

coronary artery. During its course, the circumflex branch gives off vessels to the left atrium and also supplies a portion of the left ventricle. The right coronary artery and its branches supply the right atrium, portions of both ventricles, and the posterior portion of the interventricular septum. The anterior interventricular branch of the left coronary artery supplies both ventricles and a portion of the interventricular septum, whereas the circumflex branch supplies the left atrium and a portion of the left ventricle. It should be understood that variations in the branching patterns of the coronary arteries are very common.

VENOUS DRAINAGE. The primary venous drainage of the heart wall is through veins that empty into the **coronary sinus,** which then drains into the right atrium. The coronary sinus is a thin-walled, venous dilation in the coronary sulcus on the posterior surface of the heart. The main tributary of the coronary sinus is the **great cardiac vein,** which begins at the apex of the heart and ascends while located in the anterior interventricular sulcus. When it reaches the atrioventricular sulcus (coronary sulcus), the great cardiac vein passes to the left and enters the left end of the coronary sinus. Another vessel, the **middle cardiac vein,** is on the posterior surface and enters the coronary sinus on the right. It drains the posterior wall of both ventricles. Other small vessels may empty directly into chambers of the heart.

CONDUCTION SYSTEM OF THE HEART

The conduction system of the heart consists of specialized cardiac muscle fibers that initiate and coordinate the contractions of the muscular chambers. Impulses for contraction are initiated in the **sinoatrial node** (SA node), located in the wall of the right ventricle near the superior vena cava. The SA node initiates the impulses; thus, it is called the pacemaker. From the SA node, the impulses spread through the cardiac muscle cells of the atria, causing them to contract. The **atrioventricular node** (AV node), located in the interatrial septum, receives the impulses from the atrial cells and transmits them to the **atrioventricular bundle** (AV bundle). From the AV bundle and its branches, the impulses are transmitted by way of **Purkinje fibers** to the papillary muscles and then to the muscle cells of the ventricles. Even though the heart has its own rhythm, which is established by the SA node, this rhythm can be altered by stimulation from the autonomic nervous system. Sympathetic stimulation speeds up the action of the SA node, and parasympathetic (vagal) stimulation decreases the heart rate.

GREAT VESSELS OF THE HEART

AORTA. For descriptive purposes, the aorta is divided as follows: the ascending aorta in the middle mediastinum, the aortic arch in the superior mediastinum, and the descending aorta in the posterior mediastinum. The **ascending aorta** originates at the outflow orifice from the left ventricle. An **aortic semilunar valve** at the aortic orifice prevents backflow of blood from the aorta into the left ventricle. The right and left coronary arteries originate from the ascending aorta immediately superior to the aortic orifice in the region of the semilunar valve. The ascending aorta passes superiorly to the level of the sternal angle, at the level of the disc between the fourth and fifth thoracic vertebrae, where it continues as the arch of the aorta.

In the superior mediastinum, the **aortic arch** ascends to the middle of the manubrium and then arches posteriorly and to the left, passing to the left of the trachea and the esophagus. At the level of the intervertebral disc between the fourth and fifth thoracic vertebrae, the arch continues in the posterior mediastinum as the **descending thoracic aorta.**

The brachiocephalic (innominate), the left common carotid, and the left subclavian arteries arise from the arch. The **brachiocephalic artery** (trunk) is the largest and most anterior of the three vessels that arise from the aortic arch. At its origin, it is anterior to the trachea, but as it ascends, the vessel becomes more lateral. Posterior to the right sternoclavicular joint and lateral to the trachea, the brachiocephalic artery divides into the right subclavian and the right common carotid arteries.

The **left common carotid artery** is the middle branch of the aortic arch. It is located to the left of and slightly posterior to the brachiocephalic artery. In the superior mediastinum, the left common carotid artery is initially located anterior to the trachea and is then found more lateral. As it passes posterior to the left sternoclavicular joint to ascend in the neck region, it follows a course similar to that of the right common carotid artery.

The **left subclavian artery** is the third and most posterior branch from the aortic arch. As the vessel passes through the superior mediastinum, it lies very near the pleura and the left lung.

In fetal circulation, a vessel called the **ductus arteriosus** extends from the left pulmonary artery to the inferior aspect of the aortic arch. This arrangement allows fetal blood to pass directly from the pulmonary artery into the aorta and to bypass the lungs, which are not yet functioning. The fetal ductus arteriosus usually closes by the end of the third month after birth to become the **ligamentum arteriosum.**

There are numerous variations in the arch of the aorta and the origins of its branches. Some of these are asymptomatic but others are not compatible with life and must be surgically corrected.

PULMONARY TRUNK. Arising from the right ventricle, the **pulmonary trunk** ascends on the left side of the ascending aorta to the level of the aortic arch. At its origin from the right ventricle, the pulmonary trunk is anterior to the ascending aorta, but as it ascends, the trunk becomes more posterior. When it reaches the aortic arch, at the level of the sternal angle, the pulmonary trunk divides into the right and the left pulmonary arteries. The **right pulmonary artery** is longer and wider in diameter than the left. As it proceeds to the lungs, the right pulmonary artery passes posterior to the superior vena cava and the ascending aorta but anterior to the right bronchus. The **left pulmonary artery** courses horizontally to the left, anterior to the left bronchus and the descending aorta. At the hilum of the lung, the pulmonary arteries divide according to the divisions of the bronchial tree.

VENAE CAVAE. The **superior vena cava** is formed in the superior mediastinum by the union of **the right and the left brachiocephalic veins.** The left brachiocephalic vein crosses horizontally, anterior to the aorta and the pulmonary trunk to join with the vein on the right; thus formed, the superior vena cava descends on the right side of the superior mediastinum to enter the right atrium vertically from above. The lower half of the superior vena cava is enclosed within the pericardium in the middle mediastinum. It is located to the right of the aorta and anterior to the trachea and esophagus. The superior vena cava returns blood to the heart from structures above the diaphragm, except for the lungs.

The **inferior vena cava** ascends the abdominal cavity to the right of the midline. It penetrates the diaphragm and enters the middle mediastinum of the thoracic cavity. At that point, it enters the lowest part of the right atrium in an almost vertical line with the superior vena cava. The inferior vena cava returns blood from the region below the diaphragm.

Other Thoracic Structures

THYMUS

The **thymus** gland is located immediately behind the manubrium of the sternum in the superior mediastinum. In infancy and early childhood, the gland is a prominent mass of lymphoid tissue, which may extend downward into the anterior mediastinum. After puberty, it gradually decreases in size and is replaced by fatty tissue until it may be hardly recognizable in the adult. The thymus plays a major role in the development and maintenance of the immune system.

TRACHEA

Beginning as a continuation of the larynx in the neck, the **trachea** descends in front of the **esophagus** and enters the superior mediastinum a little to the right of the midline. At the level of the sternal angle, the trachea bifurcates into the **right** and the **left primary bronchi.** Each bronchus is posterior to its corresponding pulmonary artery. The region of bifurcation is known as the **carina.** A series of C-shaped cartilages keeps the trachea open for the passage of air. The posterior soft tissue of the trachea allows for expansion of the esophagus during swallowing.

ESOPHAGUS

The **esophagus** extends from the pharynx at the level of the cricoid cartilage (C-6) in the neck to the **stomach** in the upper left quadrant of the abdomen. As it descends through the neck and the superior mediastinum, the esophagus is in a near midline position between the trachea anteriorly and the vertebral bodies posteriorly. The **left bronchus** passes in front of, or anterior to, the esophagus. In the posterior mediastinum, the esophagus descends anterior to and to the right of the descending aorta. In this region, its anterior relationships are to the pericardium and the **left atrium.** In the lower regions of the posterior mediastinum, the esophagus curves to the left to penetrate the diaphragm at the **tenth thoracic vertebral level.** As it curves, the esophagus passes anterior to the aorta.

There are four clinically significant constrictions in the esophagus: at its beginning; at the level of the aortic arch; at the level of the carina, where the left bronchus crosses it; and where it passes through the diaphragm. Objects tend to lodge in these constricted areas.

THORACIC DUCT

The **thoracic duct** is the primary duct of the lymphatic system. It collects lymph from the entire body, except the upper right quadrant. The thoracic duct begins in the abdomen as the **cisterna chyli.** From there, it ascends on the right side of the aorta, just anterior to the vertebral column. At level T-12, it passes through the diaphragm in the same opening as does the descending aorta and enters the posterior mediastinum of the thorax. The relative position of the thoracic duct in the posterior mediastinum of the lower thoracic region is illustrated in Figure 2–4. At the level of the fifth thoracic vertebra, it deviates to the left, posterior to the esophagus, and enters the superior mediastinum; then it ascends into the neck on the left side. The thoracic duct empties into the venous system at the junction of the left internal jugular and the left subclavian veins.

AZYGOS VEINS

The **azygos vein** begins in the abdomen and enters the posterior mediastinum of the thorax along with the thoracic duct. In the thorax, the azygos vein ascends just anterior to the vertebral column. It is posterior to the esophagus and to the right of the aorta and the thoracic

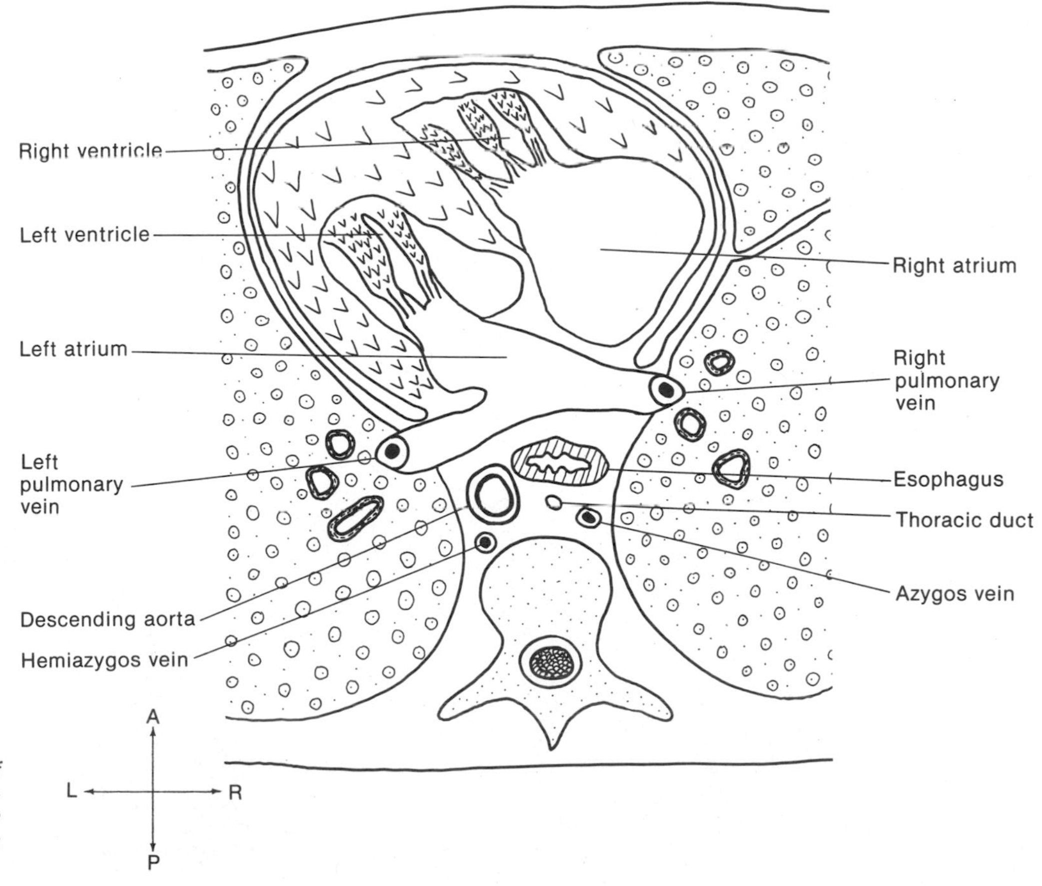

FIGURE 2–4. Relationships of structures in the posterior mediastinum. Note the descending aorta, the esophagus, the thoracic duct, and the azygos vein.

duct. The relative position of the azygos vein is illustrated in Figure 2–4. At the level of T-4, it arches over the root of the right lung to enter the **superior vena cava** in the superior mediastinum. A smaller **hemiazygos vein** ascends on the left side to the level of T-9 and then crosses the midline behind the aorta and esophagus and empties into the azygos. These veins drain the thoracic wall and the posterior abdominal wall.

Muscular and Skeletal Components Associated With the Thorax

Numerous muscles span the back and pectoral regions of the thorax but are functionally associated with the upper extremity. Many of these muscles anchor the arm to the trunk as well as being involved in movement. Sections of the thorax also show these muscles as well as the humerus and bones of the pectoral girdle. The **pectoral girdle** consists of the **clavicle,** or collar bone, anteriorly and the **scapula,** or shoulder blade, posteriorly. Since these skeletal and muscle components are clearly evident on thoracic sections, they are included in this chapter. The upper extremity and its associated articulations are more thoroughly discussed in Chapter 5.

MUSCLES OF THE PECTORAL REGION

The pectoral region is located on the anterior thoracic wall. Four muscles are associated with this region. These muscles help attach the upper limb to the thoracic skeleton. All are associated with movements of the arm, either by acting directly on the humerus or by acting on the bones of the pectoral girdle. Some of the muscles of this region are illustrated in Figure 2–5.

PECTORALIS MAJOR. The **pectoralis major** is a large, fan-shaped muscle that covers the anterior chest wall. From an extensive origin on the clavicle, the sternum, and the ribs, the muscle fibers converge to insert on the humerus. Near its insertion, this muscle forms the anterior wall of the axilla, which is the junction of the arm and thorax, or the region of the armpit (see Fig. 5–1).

PECTORALIS MINOR. The **pectoralis minor** is a smaller muscle, which lies deep to the pectoralis major. The origin of the pectoralis minor from ribs three, four, and five is much less extensive than that of the pectoralis major. The pectoralis minor inserts on the coracoid process of the scapula.

SUBCLAVIUS. A small **subclavius** muscle lies deep to the clavicle. It extends from its origin on the first rib to the posterior surface of the middle portion of the clavicle. It appears to stabilize the clavicle during shoulder movement. It also affords some protection for the subclavian vessels when the clavicle is fractured. It is mentioned because it is usually seen in sections of this region.

SERRATUS ANTERIOR. A fourth muscle, the **serratus anterior,** is associated with the pectoralis muscles in the axilla (see Fig. 5–1). The fan-shaped serratus anterior muscle originates from the first eight ribs and then passes posteriorly to insert on the medial border of the scapula. As it runs its course from the ribs to the scapula, the muscle is closely applied to the wall of the thorax. The serratus anterior primarily acts with other muscles to stabilize the scapula so that it can be used as a fixed point in producing movement of the humerus.

BACK AND SHOULDER REGION

The muscles of the back and shoulder region may be divided into three groups: the superficial back muscles,

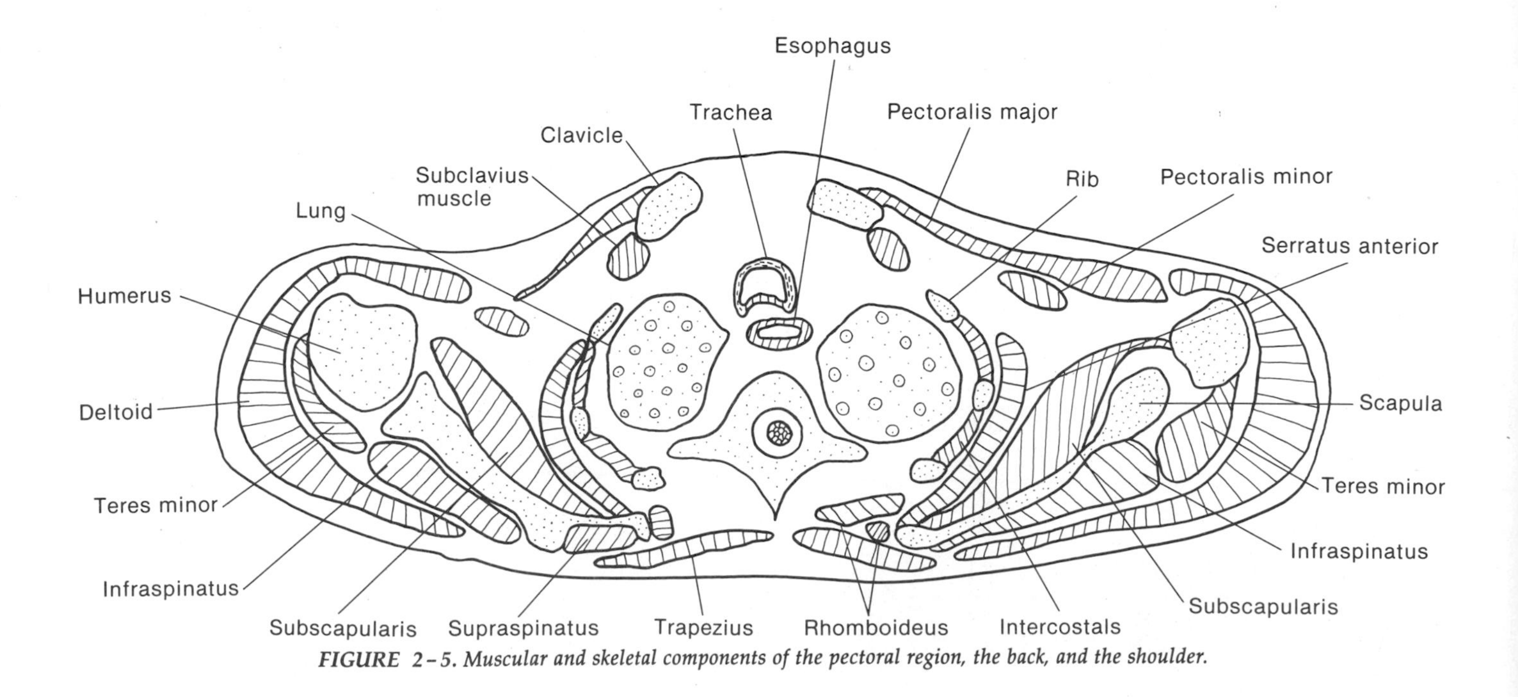

FIGURE 2–5. Muscular and skeletal components of the pectoral region, the back, and the shoulder.

the deep back muscles, and those muscles associated with the scapula. Several of these muscles are illustrated in Figure 2–5.

SUPERFICIAL BACK MUSCLES. The **trapezius** and the **latissimus dorsi** are the two superficial muscles of the back and shoulder region. The origin of the trapezius extends from the occipital bone to the spinous process of the twelfth thoracic vertebra. It inserts on both the clavicle and the scapula. Weakness of the trapezius muscle or damage to the accessory nerve that innervates the muscle results in a drooping shoulder.

The latissimus dorsi is an extensive muscle that originates from the spinous processes of the vertebrae from the seventh thoracic vertebra down through the sacrum. Some fibers also arise from the crest of the ilium. From this broad origin, the muscle fibers converge into a tendon that inserts in the intertubercular groove of the humerus.

DEEP BACK MUSCLES. Three relatively thin strap-like muscles lie deep to the trapezius. The **levator scapulae,** the **rhomboideus minor,** and the **rhomboideus major** all originate on the vertebral column and insert on the medial border of the scapula. These muscles act on the scapula to stabilize and control its position during active motion of the humerus. The fourth muscle, the **serratus anterior,** lies deep to the latissimus dorsi. This muscle was discussed previously with the pectoralis muscles because of its association with them in the axilla.

MUSCLES ASSOCIATED WITH THE SCAPULA. Six muscles are described in the scapular region. All of these muscles pass from the scapula to the humerus and act on the shoulder joint.

The **deltoid** is a superficial muscle that covers the shoulder. This muscle forms the lateral mass and the rounded contour of the shoulder. In addition to its origins on the spine and the acromion of the scapula, the deltoid also partially originates on the clavicle. The insertion is on the humerus.

The **supraspinatus muscle** fills the supraspinous fossa of the scapula, superior to the spine, whereas the **infraspinatus muscle** lies in the infraspinous fossa, inferior to the spine of the scapula. The **subscapularis muscle** occupies the subscapular fossa on the costal surface of the scapula. Located along the lateral margin of the infraspinatus muscle, the **teres minor** is frequently inseparable from the infraspinatus muscle. The supraspinatus, infraspinatus, subscapularis, and teres minor muscles all reinforce the fibrous capsule of the shoulder joint. These muscles, together with the fibrous capsule, are collectively referred to as the **rotator cuff** of the shoulder joint. The rotator cuff holds the head of the humerus in the glenoid cavity of the scapula; thus it protects and stabilizes the joint.

The **teres major** is an oval muscle that runs from the inferior angle of the scapula to the intertubercular groove of the humerus, where it inserts with the tendon of the latissimus dorsi. It, along with the latissimus dorsi, forms a portion of the posterior wall of the axilla. For further detail on the muscles of the back, refer to Chapter 5.

Brachial Plexus

The **brachial plexus** is a network of nerves formed by the ventral rami (branches) of spinal nerves C-5 to C-8 and T-1. It extends from the posterior triangle of the neck into the axilla and supplies innervation to the arm. Portions of the brachial plexus are clearly evident in transverse sections, between the anterior scalene and the middle scalene muscles (see Fig. 2–7). Five nerves, supplying innervation to the arm, emerge from the brachial plexus. These are the **musculocutaneous,** the **median,** the **ulnar,** the **axillary,** and the **radial nerves.**

Injuries to the brachial plexus may occur by disease, trauma, or stretching. These injuries result in paralysis and/or anesthesia. The extent of the signs and symptoms depends on the part of the plexus that is injured.

Breast

Both males and females have breasts. Mammary glands for the production of milk are located within the breast. Normally, these become well developed and functional only in the female.

GENERAL BREAST ANATOMY

The breast is located in the superficial fascia of the pectoral region overlying the pectoralis major muscle. It normally extends from the lateral margin of the sternum to the anterior border of the axilla, between the second and sixth ribs. Breast tissue is separated from the deep fascia of the underlying muscle by a retromammary layer of loose connective tissue. Some breast carcinomas may invade the loose connective tissue and deep fascia and may become fixed to the underlying muscle, restricting movement of the breast.

At birth and throughout early childhood, the male and female breasts are similar. Externally, near the center of the breast, a circular area of pigmented skin, known as the **areola,** surrounds an elevated nipple. Within the areola, numerous sebaceous glands, which appear as small nodules under the skin, secrete an oily substance to keep the tissues soft. Internally, there are a few rudimentary ducts, radiating from the nipple. During puberty, the mammary gland in the female undergoes developmental changes, but the male gland usually remains rudimentary.

In the adult male, the breast consists of some supporting adipose and fibrous tissue and of a few small ducts without any milk-secreting cells. These ducts, in the male, may become fibrous cords.

FEMALE BREAST STRUCTURE

The breast in the female includes two basic components: the **parenchyma,** which is glandular tissue, and the **stroma,** which is supportive connective tissue. The glandular parenchyma consists of 15 to 20 lobes of **alveoli,** or milk-producing cells, arranged radially around the centrally located nipple. An excretory **lactiferous duct** extends from each lobe to the nipple, where it terminates in a tiny opening at the surface. The glandular lobes are embedded in the connective tissue stroma, which contains variable amounts of fat. Condensations of connective tissue, known as **suspensory** or **Cooper's ligaments,** extend from the underlying deep fascia, through the breast, to the skin. These ligaments provide support for the breasts. The components of the female breast are illustrated in Figure 2–6.

HORMONAL CONTROL OF THE FEMALE BREAST

The development of the mammary gland in the female begins at puberty, when **estrogens** stimulate extensive growth of the lactiferous duct system along with the deposition of fat between the lobes. During pregnancy, the high level of **progesterone** stimulates the development of the milk-producing glands. However, no milk is actually produced until after parturition, when the sudden decrease in progesterone and estrogen signals the anterior pituitary gland to secrete **prolactin.** This hormone stimulates the production of the milk proteins and brings about the glandular secretion of milk. Stimulation by the suckling infant and by the hormone **oxytocin** results in the ejection of the milk from the breast.

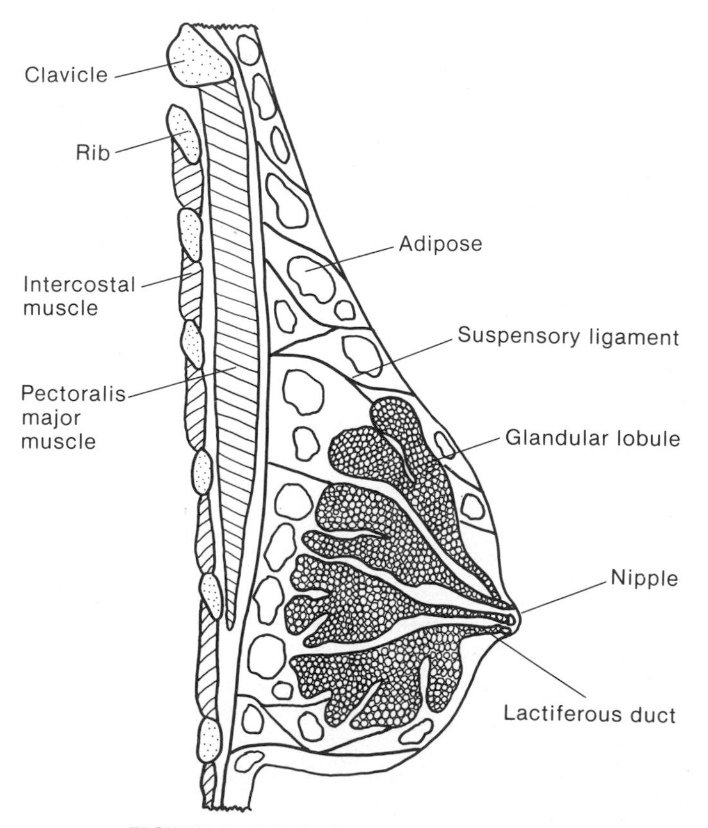

FIGURE 2–6. Sagittal section of a female breast.

LYMPHATIC DRAINAGE FROM THE BREAST

Although there are several routes of lymphatic drainage from the breast, about 75 percent of the drainage is by way of the **axillary lymph nodes.** In addition to the axillary nodes, other avenues of lymphatic drainage include the **parasternal,** the **supraclavicular,** and the **abdominal lymph nodes.** The drainage of one breast may also join the drainage of the opposite breast. The lymphatic drainage of the breast is of clinical significance in the diagnosis and treatment of breast cancer. As the cancer cells break loose, they travel through the lymphatic channels until they become trapped. Since 75 percent of the lymphatic drainage is through the axillary nodes, this is the most common site of metastases from breast carcinoma.

SECTIONAL ANATOMY OF THE THORAX

Transverse Sections

SECTION THROUGH THE THORACIC INLET

Sections through the superior thoracic aperture, or thoracic inlet, show the structures passing through the root of the neck into the thorax. Sections at this level also show the shoulder region with the pectoral girdle and the upper extremity (Fig. 2–7). Since the superior thoracic aperture slopes inferiorly from posterior to anterior, a transverse section through the first thoracic vertebra typically does not intersect the manubrium of the sternum.

Anteriorly, in sections through the thoracic inlet, the **sternocleidomastoid muscles** are seen as thin straps, anterior to the lobes of the **thyroid gland.** The horseshoe-shaped **tracheal cartilage** is between the thyroid lobes and the **esophagus.** The **internal jugular veins** are lateral to the **common carotid arteries.** Also of interest at this level are the clavicle, the scalene muscles, the brachial plexus, and the first rib.

The components of the humeroscapular joint are discussed in more detail in Chapter 5, but because of their relationship to the wall of the thorax, they are also mentioned here.

At the level of the thoracic inlet, the most superficial muscle of the back is the **trapezius,** extending from the vertebral column to the scapula. The **rhomboideus major** and **rhomboideus minor muscles** are directly beneath the trapezius. The fibers of the two muscles may appear as a single muscle. Throughout the length of the vertebral column the **erector spinae muscles** can be seen in the groove between the spinous and transverse processes. The large **deltoid muscle** forms a semicircle around the humerus, from the clavicle to the spine of the scapula. Depending on the angle of the slice, you can see either two or three muscles directly related to the scapula. At some angles, when only the spine of the scapula is seen,

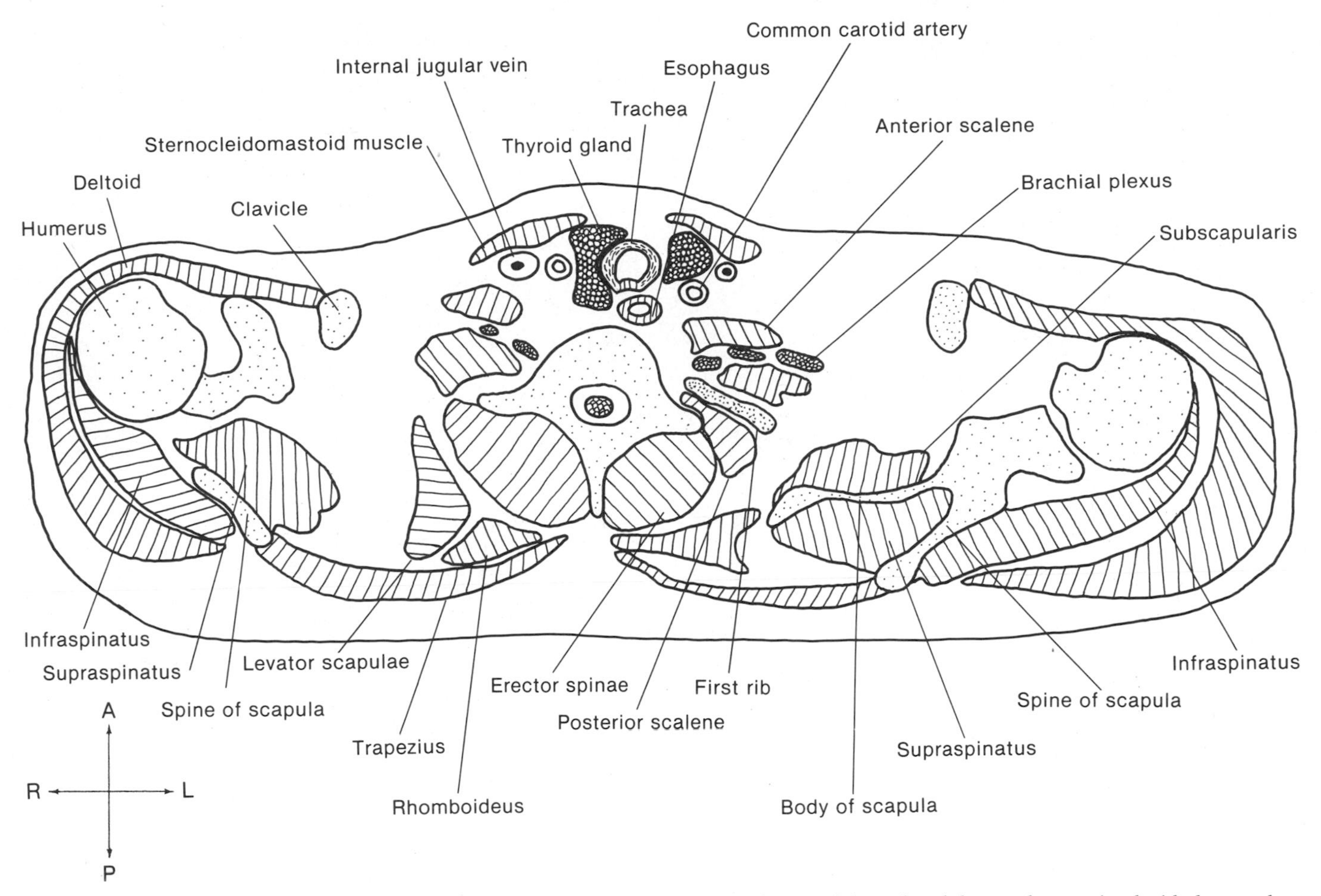

FIGURE 2-7. Transverse section through the thoracic inlet. Note the structures in the root of the neck and the muscles associated with the scapula.

there is a more superficial **infraspinatus muscle** and a deeper **supraspinatus muscle.** When the cut is below the scapular spine, through the blade or the body of the scapula, the more superficial muscle is still the infraspinatus but the muscle deep to the body is the **subscapularis muscle.** Occasionally, the angle of the slice is such that both the spine and the body of the scapula are seen. In this case, all three muscles are present. The infraspinatus appears superficial to the spine, the supraspinatus appears to be between the spine and the body, and the subscapularis is deep to the body of the scapula. These muscles are illustrated in Figure 2-7.

SECTION THROUGH THE APEX OF THE LUNG

Particular attention should be given to vascular relationships in this and the following few sections. As the aorta goes from the ascending to the descending portions, it arches in an anteroposterior direction as well as in a right-to-left direction. This means that the vessels closely associated with the aortic arch will assume corresponding arch-shaped positions. A cross-section at the level of the apex of the lung (Fig. 2-8) shows the **right common carotid artery** and the **right subclavian artery** as two separate vessels close together on the right side of the trachea. The close proximity of the two vessels indicates that this level is near the bifurcation of the brachiocephalic artery; thus in sections inferior to this (Fig. 2-9), the single **brachiocephalic (innominate) artery** is seen. The brachiocephalic artery is the first and most anterior branch from the arch of the aorta. On the left, the anterior vessel is the **left common carotid artery.** The **left subclavian artery** is also on the left, but it is more posterior. This is the third and most posterior branch from the aortic arch. The left **subclavian vein** and the **left internal jugular vein** are visible as they join to form the **left brachiocephalic (innominate) vein.** On the right, the brachiocephalic vein has already formed.

The anterior thoracic wall at the level of the lung apex is formed by the **pectoralis major** and **pectoralis minor muscles** and the two **clavicles** joined by an **interclavicular ligament.** A **subclavius muscle** is associated with each clavicle.

Two muscles that are associated with the rib cage, in addition to the intercostal muscles, are evident in Figures 2-8 and 2-9. The **serratus anterior muscles** that form the medial walls of the axillae appear to resemble parentheses that enclose the ribs and intercostal muscles. The **levator costarum muscle** connects a rib to its appropriate transverse vertebral process.

Muscles associated with the scapula and upper extremity include the **teres major** and the **teres minor.** The teres

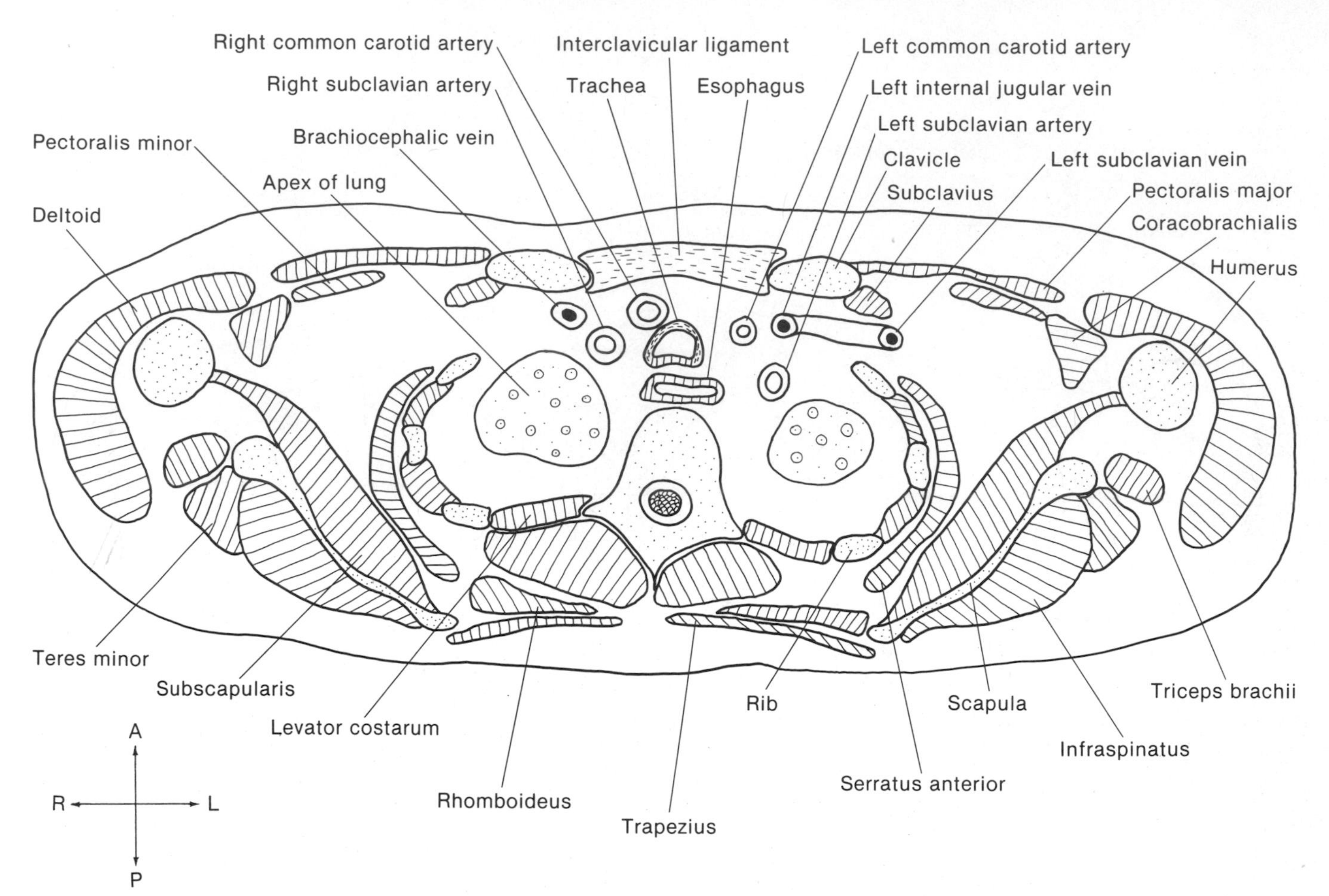

FIGURE 2-8. Transverse section through the apex of the lung. Note the separate right subclavian and right common carotid arteries.

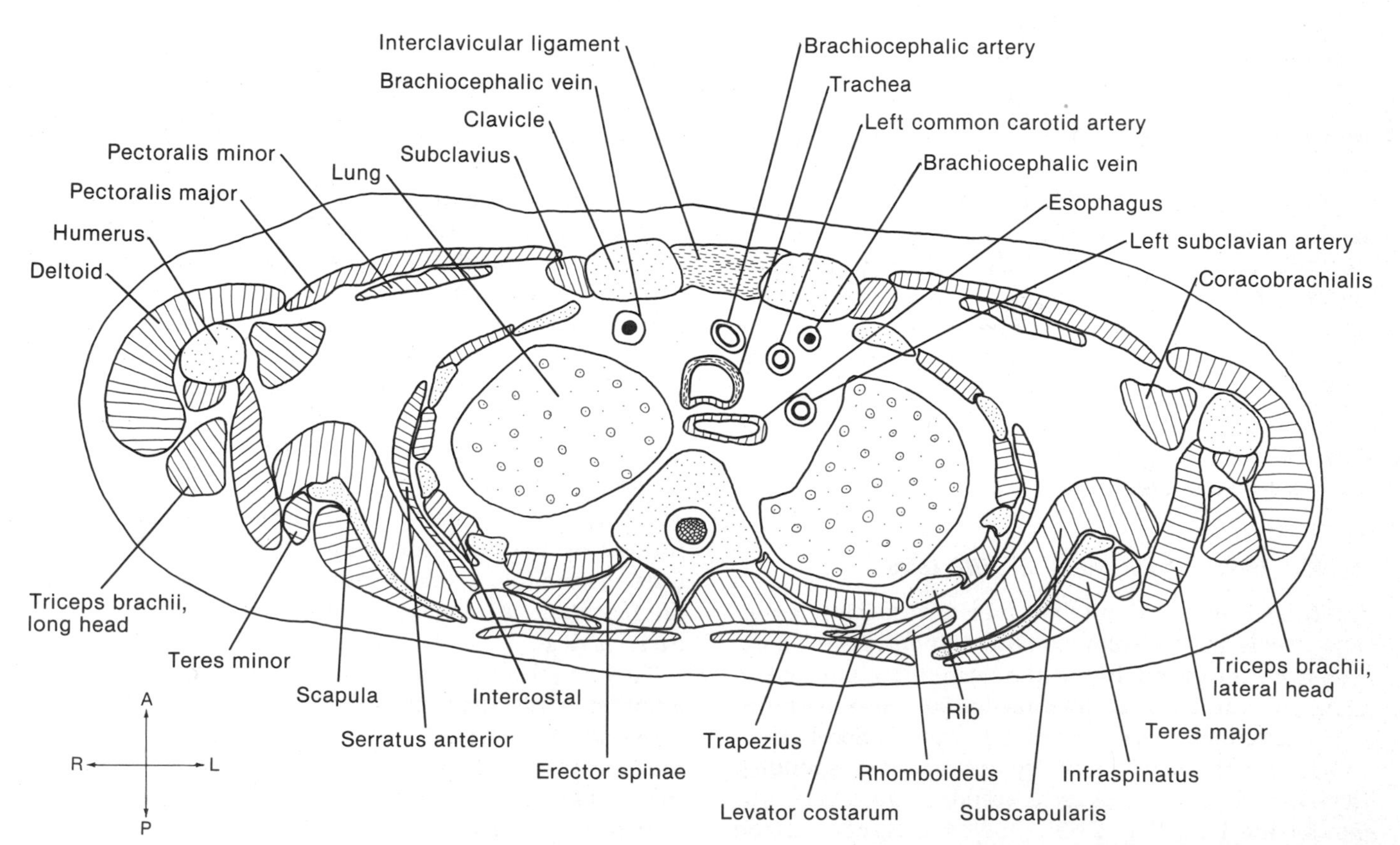

FIGURE 2-9. Transverse section through the apex of the lung but inferior to bifurcation of brachiocephalic artery. Note the single vessel, the brachiocephalic artery.

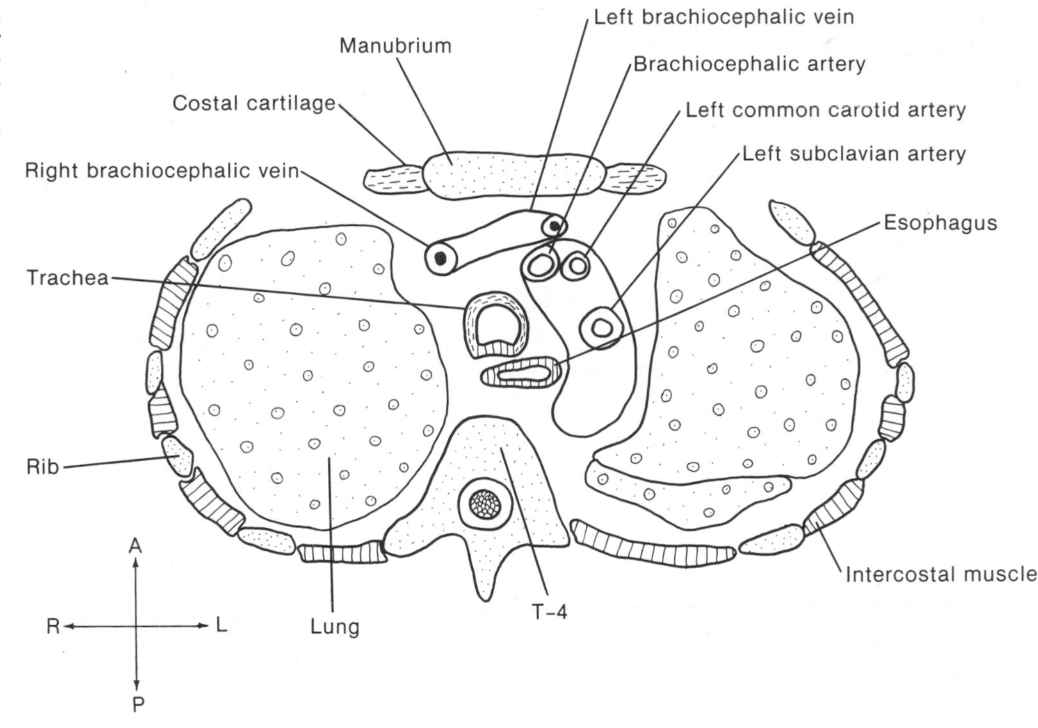

FIGURE 2-10. Transverse section through the aortic arch, inferior view. Note the three branches of the arch: anteriorly, the brachiocephalic artery; posteriorly, the left subclavian artery; and in the middle, the left common carotid artery.

minor is closely associated with the **infraspinatus muscle,** and it is sometimes difficult to distinguish between them, except by location. The teres minor is at the lateral end of the infraspinatus muscle. The teres major is lateral to the teres minor (see Fig. 2-9).

SECTION THROUGH THE AORTIC ARCH

A section through the arch of the aorta is likely to intersect the manubrium of the sternum. In an inferior view through the arch of the aorta (Fig. 2-10), there should be three holes in the arch representing the three branches: the anterior **brachiocephalic artery,** the posterior **left subclavian artery,** and the middle **left common carotid artery.** When looking down at a superior view of this level, as illustrated in Figure 2-11, there should be two large openings, the anterior **ascending aorta** and the posterior **descending aorta.** The **left brachiocephalic vein** crosses anterior to the aortic arch to join the **right brachiocephalic vein.** The right and left brachiocephalic veins join to form the **superior vena cava.**

SECTION THROUGH THE STERNAL ANGLE

The **sternal angle** (Louis angle), located between the T-4 and T-5 vertebral levels, is an easily palpable landmark.

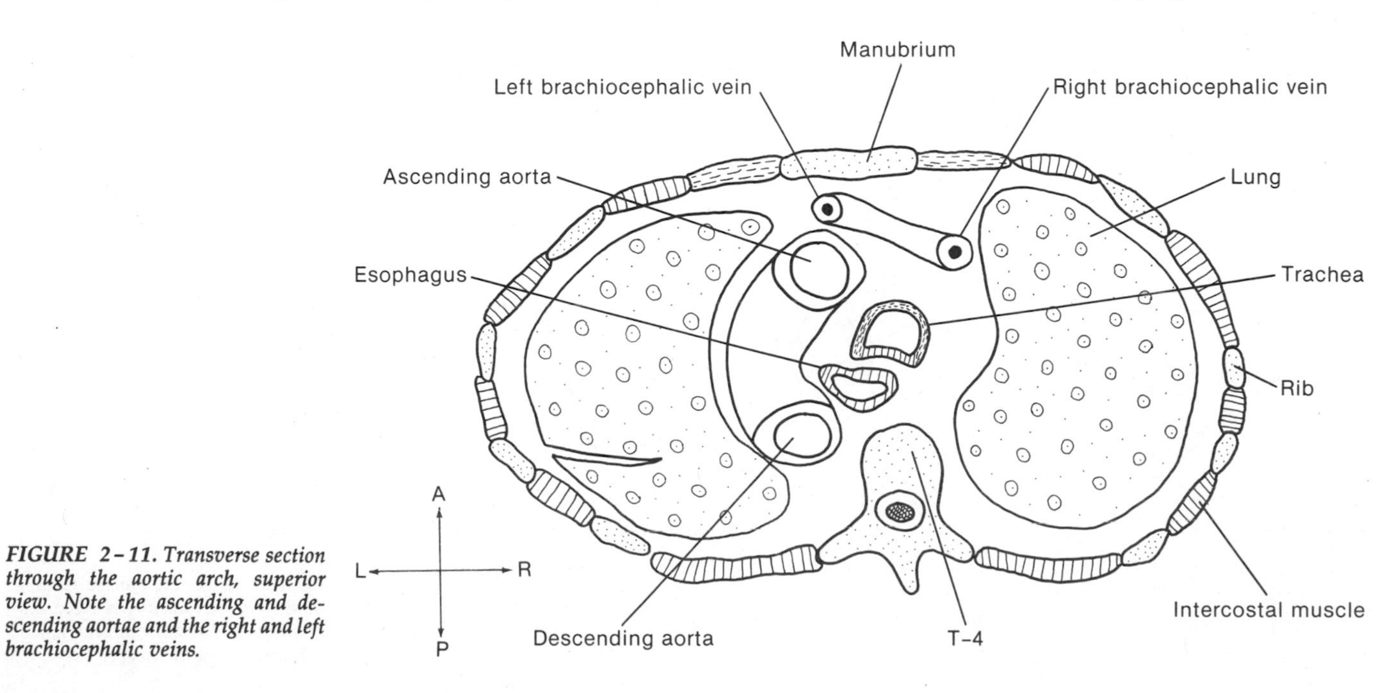

FIGURE 2-11. Transverse section through the aortic arch, superior view. Note the ascending and descending aortae and the right and left brachiocephalic veins.

Transverse sections at this level show an anteriorly located **ascending aorta** and a posteriorly situated **descending aorta** on the left side of the mediastinum (Fig. 2–12). The right-sided **superior vena cava** is opposite the ascending aorta. Posterior to the ascending aorta and the superior vena cava, the **trachea** begins to bifurcate into the **main stem (primary) bronchi.** The **esophagus** is between the tracheal bifurcation and the descending aorta. First seen at this level, the **azygos vein** makes a loop over the root of the right lung to enter the superior vena cava.

SECTION THROUGH THE PULMONARY TRUNK

The **pulmonary trunk** appears on the left side within the pericardium at approximately the T-5 vertebral level. In sequence from right to left, the great vessels are the **superior vena cava,** the **ascending aorta,** and the **pulmonary trunk,** as shown in Figure 2–13. At this level, the pulmonary trunk is posterior to the ascending aorta. At more inferior levels, it will become anterior. Even though the pulmonary trunk is the outflow from the right ventricle, it becomes the most left-sided of the three great vessels. The **right pulmonary artery,** as it branches from the pulmonary trunk, passes horizontally, posterior to the ascending aorta and to the superior vena cava, and then enters the right lung (see Fig. 2–13). The pulmonary trunk is on the left; thus, the right pulmonary artery is longer and more horizontal than the left. The right and left pulmonary arteries are anterior to the right and left main stem bronchi, respectively. In the posterior mediastinum, the large vessel on the left is the **descending aorta.** The **esophagus** is anterior to and slightly to the right of the descending aorta, and the **azygos vein** is to the right of the esophagus. The **thoracic duct** is usually between the azygos vein and the descending aorta, but it may be difficult to locate.

SECTION THROUGH THE BASE OF THE HEART

Sections through the base of the heart show the three great vessels at or near their attachment to the heart. The three great vessels maintain the same right-to-left order (superior vena cava, ascending aorta, pulmonary trunk) throughout the middle mediastinum, but the anterior-posterior relationship between the ascending aorta and the pulmonary trunk appears to change. At higher levels, the ascending aorta is anterior to the pulmonary trunk (see Fig. 2–13). At lower levels, the ascending aorta is posterior to the pulmonary trunk (Fig. 2–14). The pulmonary trunk courses in a posterior direction as it ascends from the right ventricle. The outflow orifice is guarded by the **pulmonary semilunar valve.** The superior vena cava remains on the right side as it enters the right atrium, the most right-sided chamber of the heart. The **left atrium** is the most posterior chamber of the heart, and the vessels entering the left atrium are the **pulmonary veins,** which carry freshly oxygenated blood from the lungs. The **esophagus,** the **azygos vein,** and the **descending aorta** are in the posterior mediastinum.

SECTION THROUGH THE CHAMBERS OF THE HEART

Features of the heart are illustrated in Figure 2–15. Transverse sections through the chambers of the heart show that the **left atrium** is the most posterior chamber. Posteriorly, the left atrium is related to the **esophagus.** Anteriorly, the left atrium is related to the centrally located

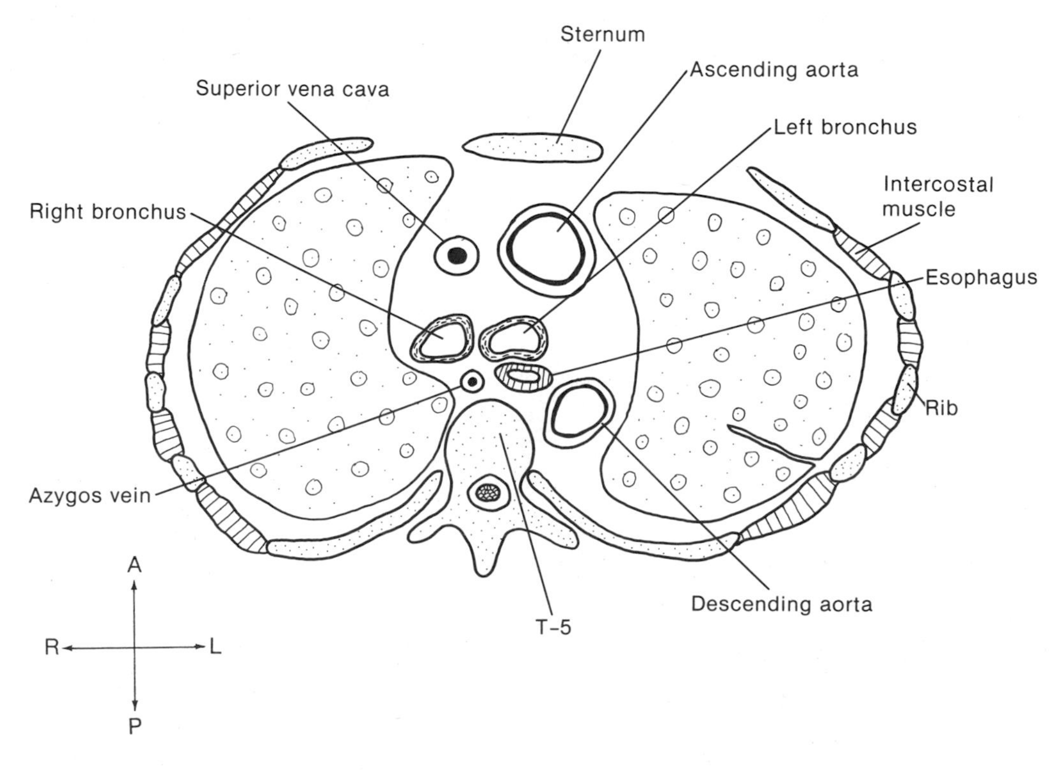

FIGURE 2–12. Transverse section through the sternal angle. Note the superior vena cava, the bifurcation of the trachea, and the azygos vein.

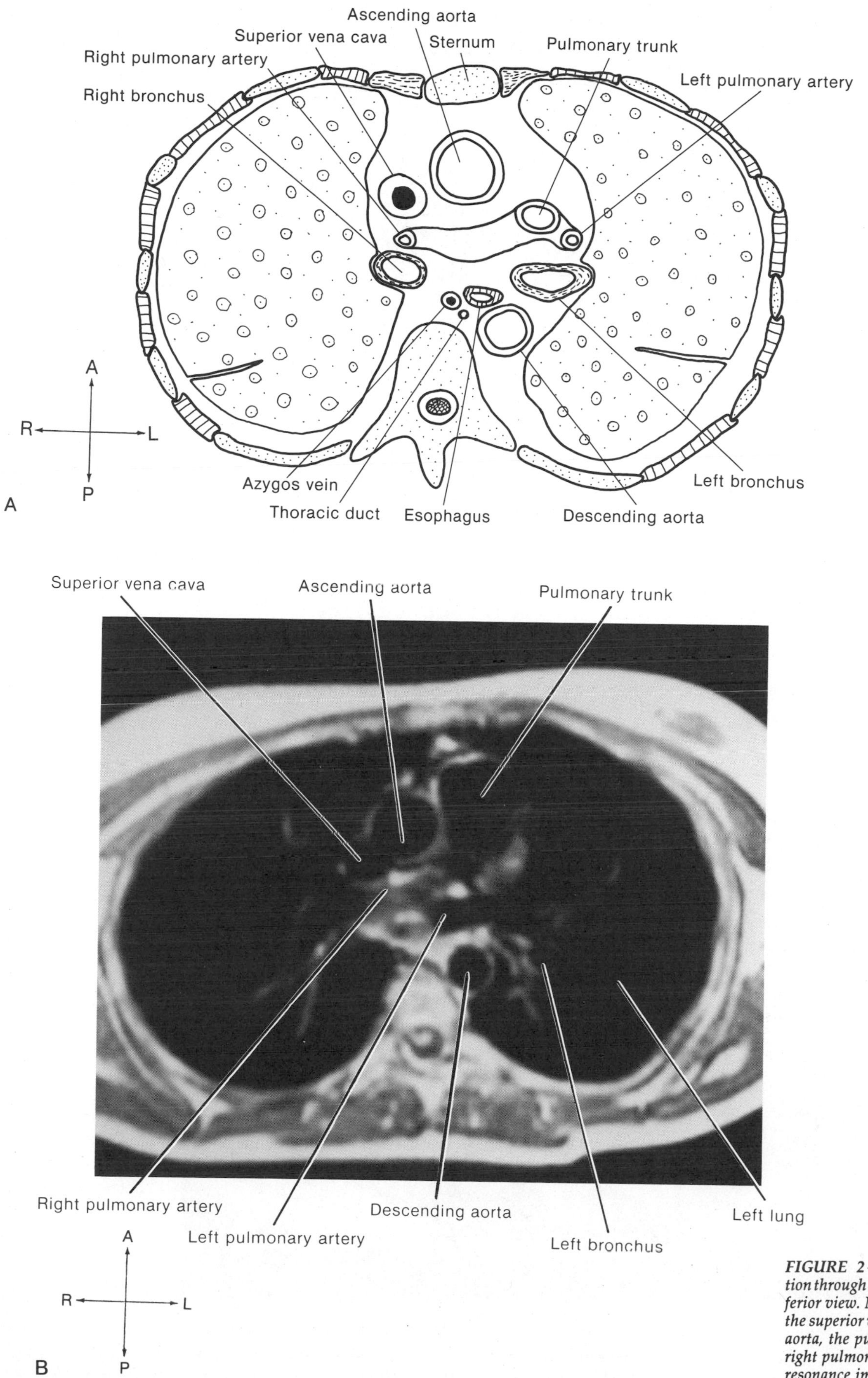

FIGURE 2-13. A, Transverse section through the pulmonary trunk, inferior view. Note the relationships of the superior vena cava, the ascending aorta, the pulmonary trunk, and the right pulmonary artery. B, Magnetic resonance image at similar location.

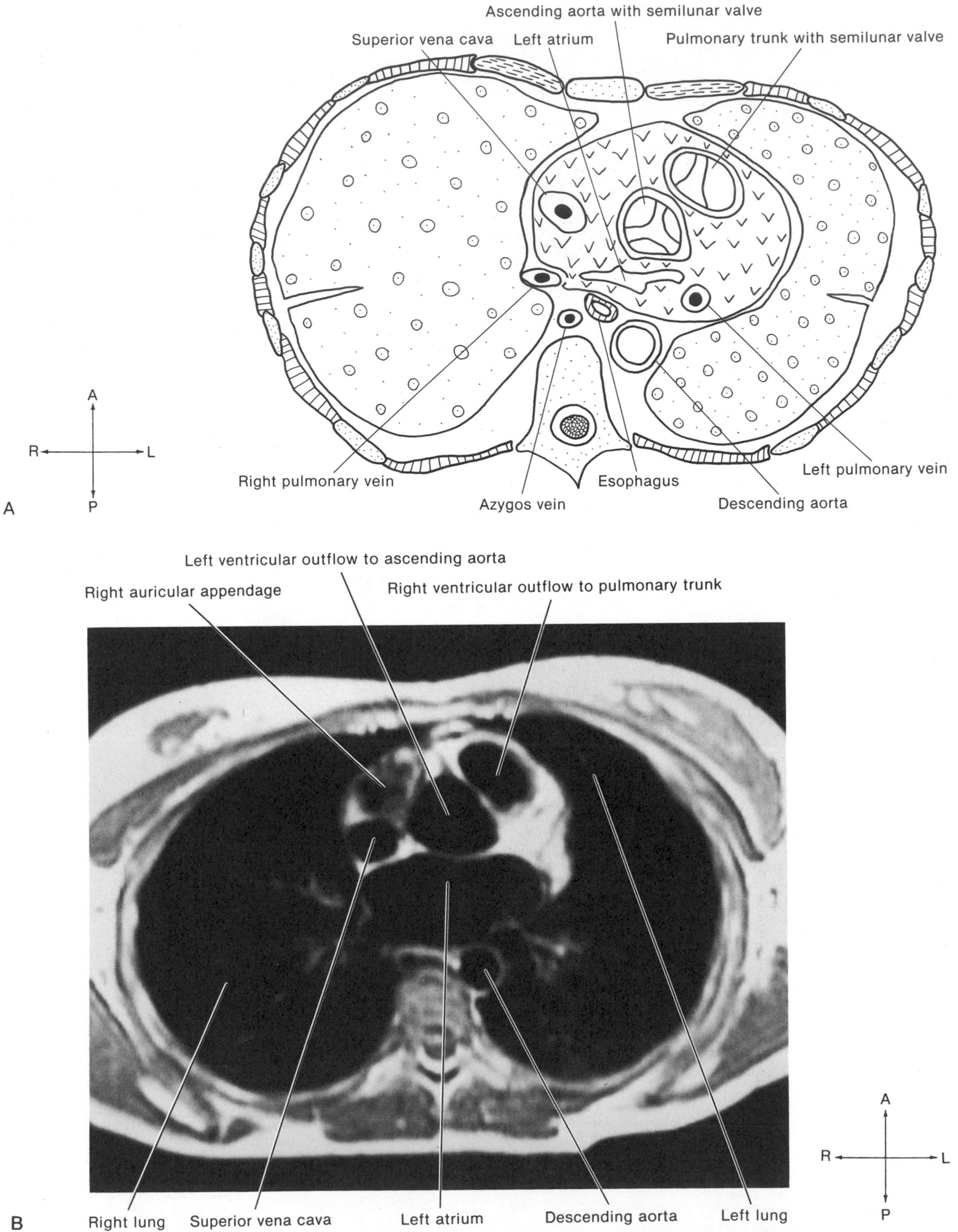

FIGURE 2-14. *A, Transverse section through the base of the heart, inferior view. Note the pulmonary semilunar valve, which is the most superior valve in the heart; and the left atrium, which is the most posterior chamber. B, Magnetic resonance image at similar level. Note the right auricular appendage.*

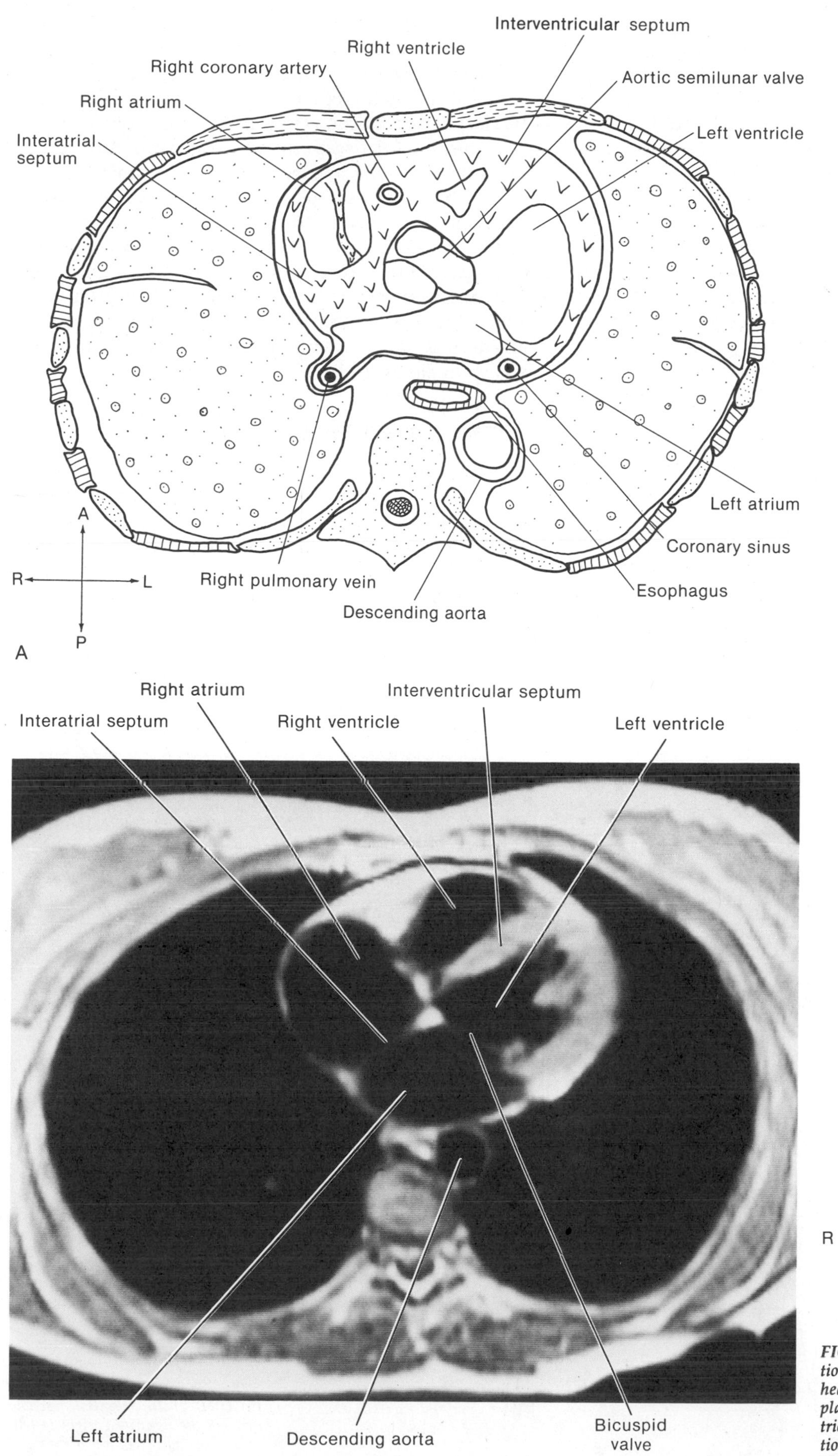

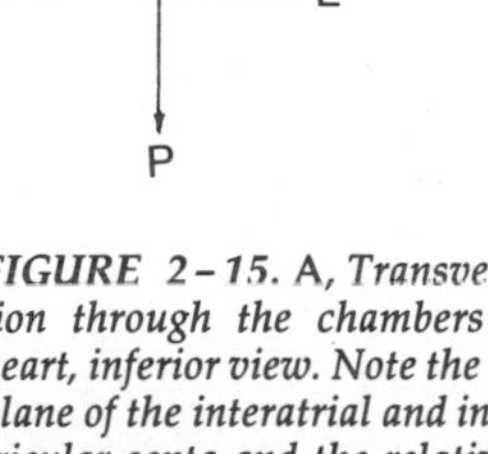

FIGURE 2-15. A, Transverse section through the chambers of the heart, inferior view. Note the coronal plane of the interatrial and interventricular septa and the relative positions of the chambers of the heart. B, Magnetic resonance image at similar level.

aortic outflow region of the left ventricle. The **right atrium** is the most right-sided chamber. The cavity of the right atrium is anterior and perpendicular to that of the left atrium; thus the **interatrial septum** is in a coronal plane. The most anterior chamber is the **right ventricle,** which is located immediately to the left of the right atrium. Blood flow from the right atrium, through the **tricuspid valve,** and into the **right ventricle** is primarily passive and is directed to the left and slightly anteriorly. The **bicuspid valve** between the left atrium and the **left ventricle** is directed inferiorly, anteriorly, and to the left. The thick **interventricular septum** is predominantly in a coronal plane. The **right coronary artery** is in the fat-filled sulcus between the right atrium and the right ventricle. The **coronary sinus,** a venous dilation that receives blood from the coronary circulation, empties into the right atrium. It is located in the left atrioventricular sulcus along the posterior surface of the heart.

Sagittal Sections

SECTION THROUGH THE RIGHT LUNG

Sagittal sections to the right of the midline that intersect the **right lung** show the upper lobe separated from the middle lobe by a **horizontal (minor) fissure.** The middle lobe is separated from the lower lobe by an **oblique (major) fissure.** Sections in this region usually show the relationship of the **scapula** to the **infraspinatus, supraspinatus,** and **subscapularis muscles.** The infraspinatus muscle fills the fossa below the scapular spine, and the supraspinatus muscle fills the space above the scapular spine. The subscapularis muscle is deep to the plate, or the body, of the scapula. Anteriorly, the **pectoralis major** and **minor muscles** form the thoracic wall. **Axillary vessels,** which are continuations of the subclavian vessels, are deep to the pectoralis muscles. These features are shown in Figure 2–16.

SECTION THROUGH THE RIGHT ATRIUM

The right atrium is the most right-sided chamber of the heart; thus it is the first chamber seen when making serial sagittal sections from the right to the left. Figure 2–17 illustrates a sagittal section approximately 2 cm to the right of the midline. It shows the **inferior vena cava** and the **superior vena cava** draining into the cavity of the **right atrium.** A continuation of the right atrium into the pectinate region of the **right auricle** is evident. The **left atrium** is seen as a posterior chamber with a **pulmonary**

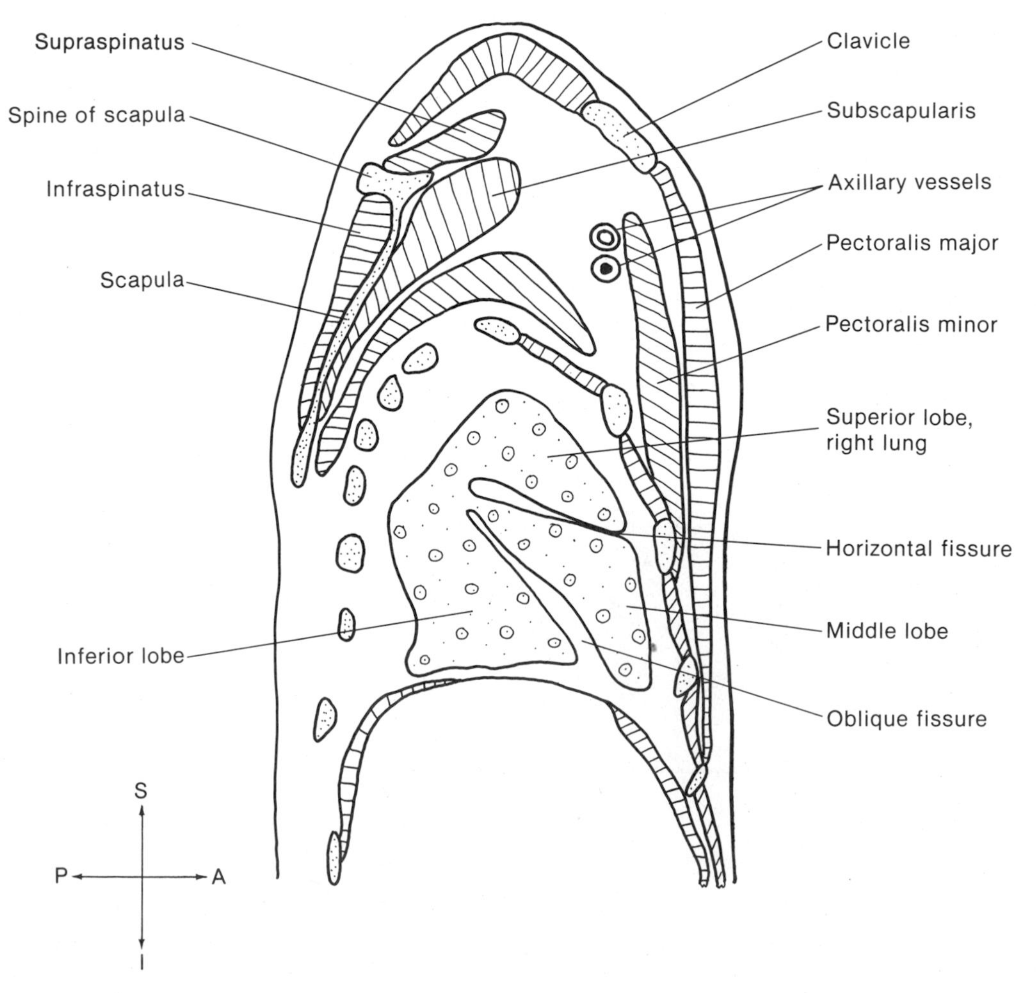

FIGURE 2–16. Sagittal section through the right lung, right-to-left view. Note the three lobes in the right lung and the muscles associated with the scapula.

FIGURE 2-17. Sagittal section through the right atrium, right-to-left view. Note the inferior and the superior venae cavae draining into the right atrium, the right pulmonary artery anterior to the right bronchus, and the right coronary artery in the right atrioventricular sulcus.

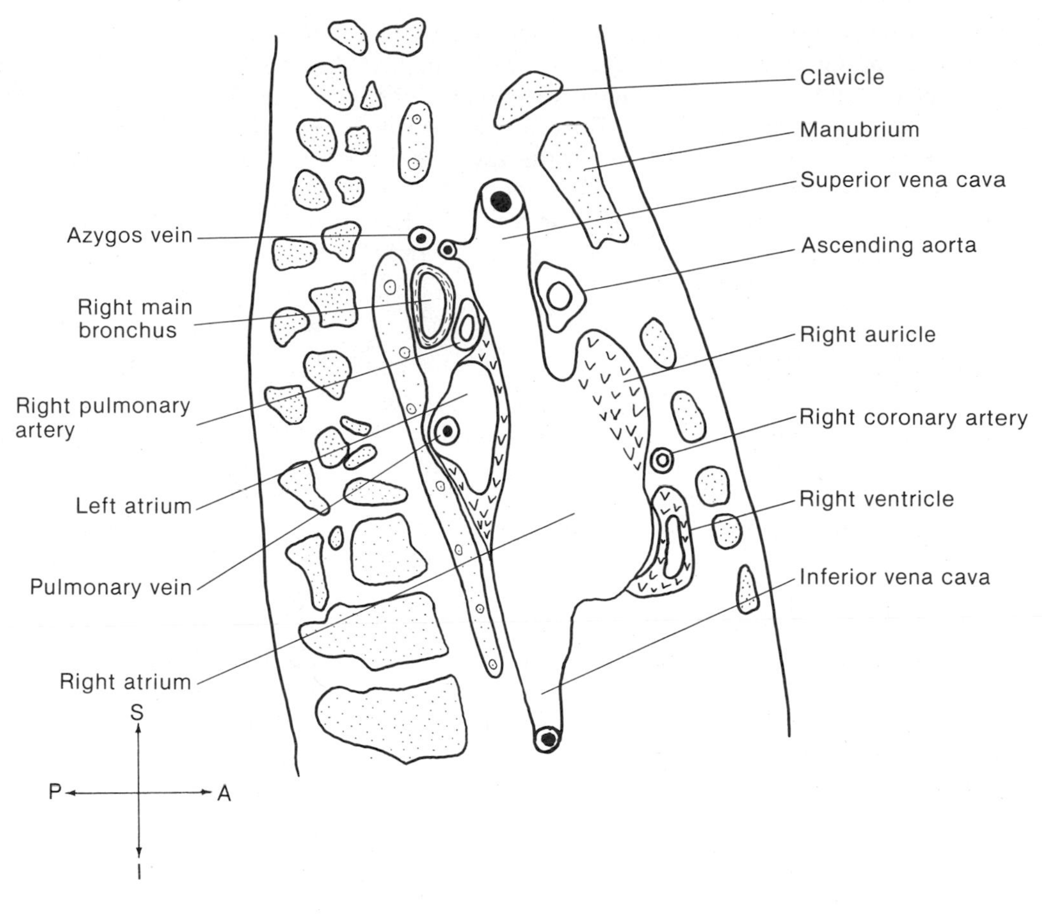

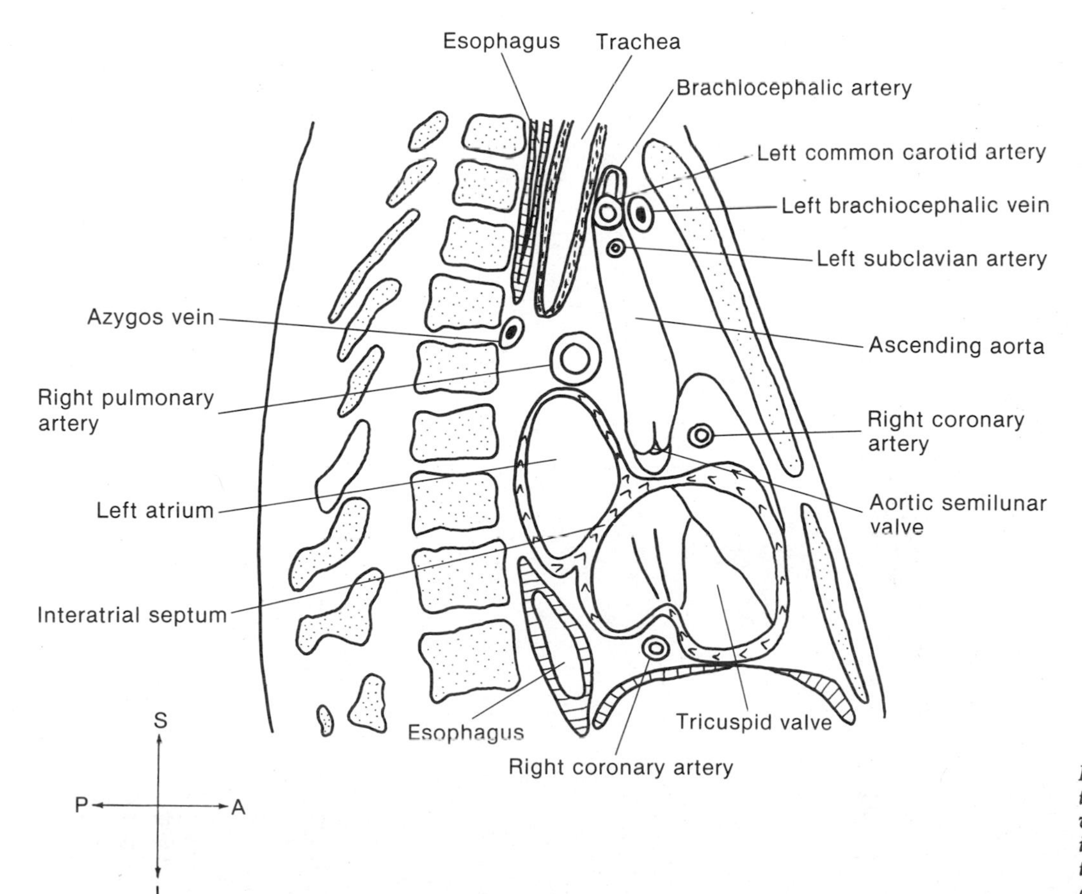

FIGURE 2-18. Midsagittal section through the thorax, right-to-left view. Note the ascending aorta with its semilunar valve and a cusp of the tricuspid valve between the right atrium and the right ventricle.

vein that drains into it. Just superior to the left atrium, the **right pulmonary artery** and the **right main bronchus** are sectioned at the point where they course to the right, posterior to the superior vena cava. The artery is anterior to the bronchus. This section also shows the **azygos vein** as it enters the superior vena cava. The myocardium in the anterior region of the heart is the **right ventricle.** The **right coronary artery** is in the sulcus between the right atrium and the right ventricle.

MIDSAGITTAL SECTION

A midsagittal section through the thorax (Fig. 2–18) usually cuts through the **aorta** as it ascends from the left ventricle. Remember that the aorta is to the right of the pulmonary trunk as they ascend from the heart. A portion of the **tricuspid valve,** between the right atrium and the right ventricle, should be seen, since it is either near the midline or slightly to the right. It is the most inferior of the four valves of the heart and is generally at the level of the intercostal space between the fourth and fifth ribs. The sound of the tricuspid valve is heard over the right half of the inferior portion of the body of the sternum. Near the bottom of the ascending aorta, at the level of the third intercostal space, the **aortic semilunar valve** guards the orifice between the left ventricle and the aorta. The aortic valve sound is heard on the right edge of the sternum in the second intercostal space. Openings for the **left subclavian,** the **left common carotid,** and the **brachiocephalic arteries** are located superiorly, where the aorta forms the arch. The **left brachiocephalic vein** is cut as it courses horizontally, anterior to the aorta, to meet its counterpart on the right side. The **right pulmonary artery** is located posterior to the aorta. Recall that as it goes toward the right lung, the right pulmonary artery passes posteriorly to the aorta and the superior vena cava.

SECTION THROUGH THE PULMONARY TRUNK

Just to the left of midline, the **pulmonary trunk** exits the **right ventricle** and curves posteriorly, over the **left atrium** (Fig. 2–19). Remember that the pulmonary trunk is to the left of the aorta as the vessels emerge from the heart. Since the aortic arch curves to the left, you may see a portion of the arch that is continuous with the **descending aorta.** The **left brachiocephalic vein** passes horizontally, anterior to the pulmonary trunk and the aortic arch; thus in sagittal sections the vein is seen in cross section as it is cut perpendicular to its axis. The **left bronchus** is also seen in cross-section as it courses on its path to the left lung. It is anterior to the descending aorta and posterior to the pulmonary trunk. After the short **left coronary artery** branches from the aorta, it follows a horizontal path posterior to the pulmonary trunk. The **coronary sinus** is found in the fat-filled sulcus, inferior to the posteriorly located left atrium (see Fig. 2–19).

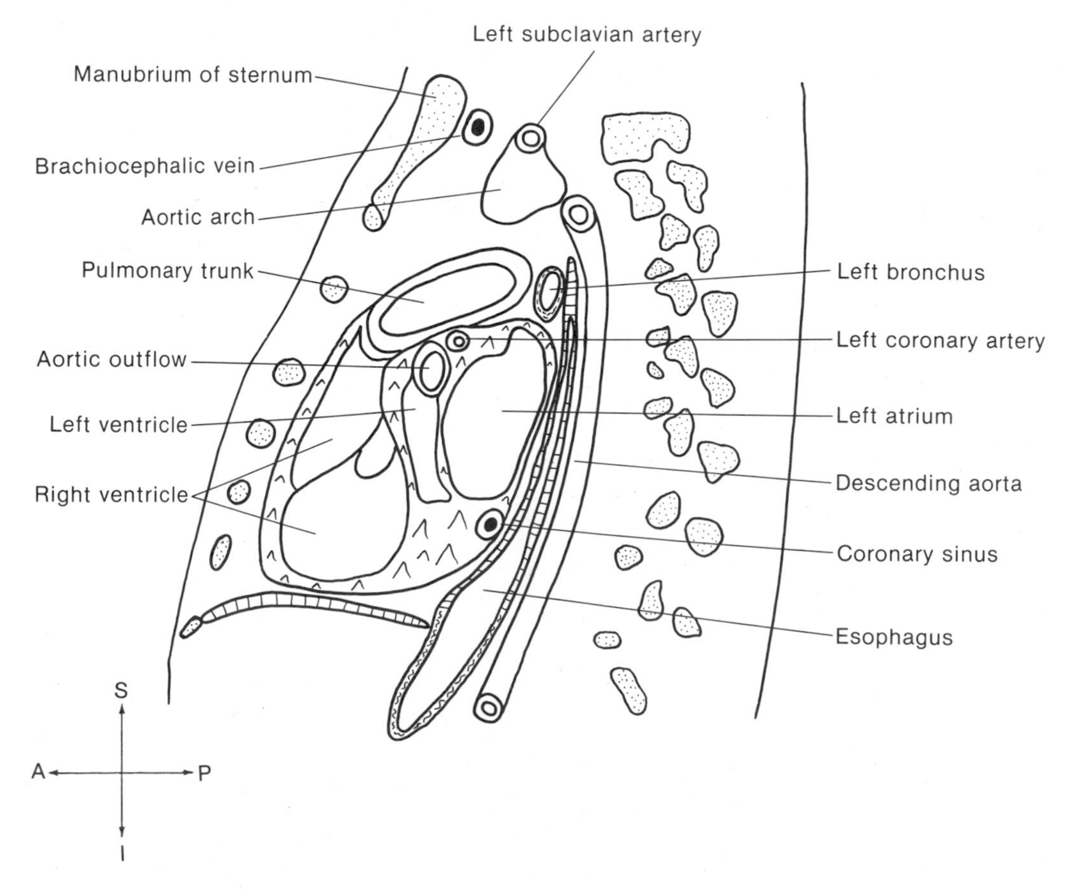

FIGURE 2–19. Sagittal section through the pulmonary trunk, left-to-right view. Note the anteroposterior orientation of the pulmonary trunk.

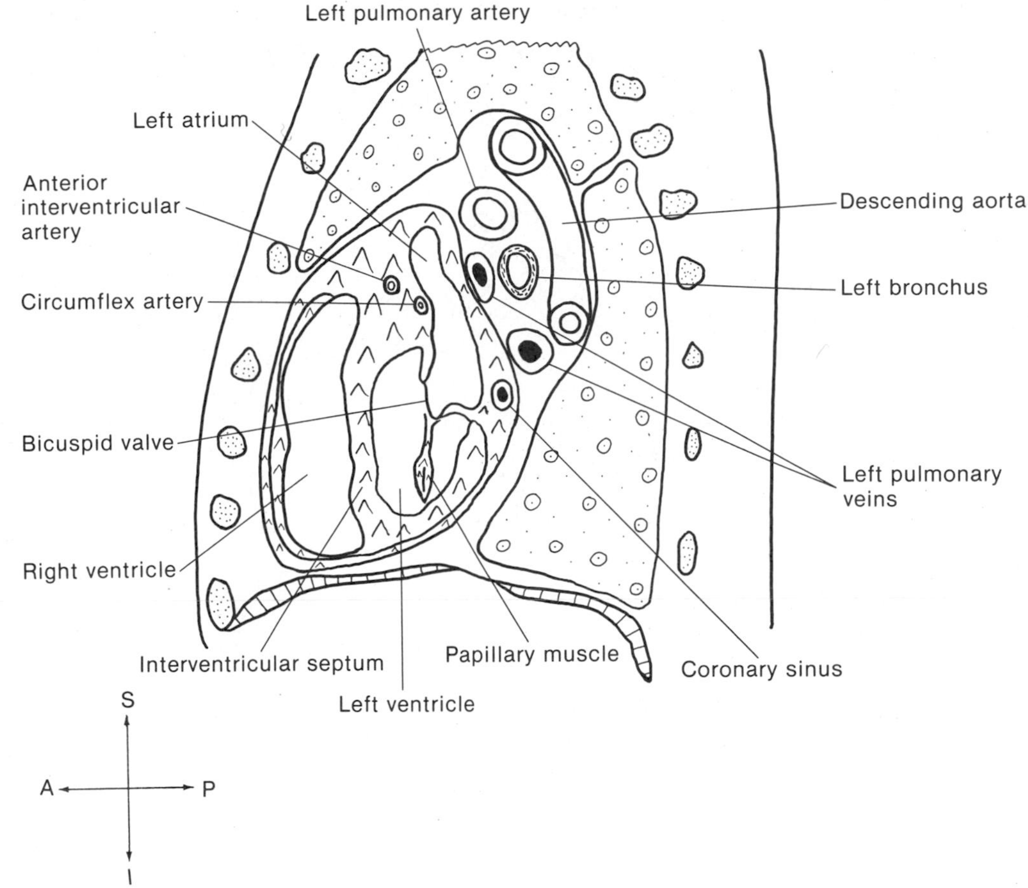

FIGURE 2-20. Sagittal section through the bicuspid valve between the left atrium and the left ventricle, left-to-right view. Note the left pulmonary artery and the left bronchus, two branches of the left coronary artery.

SECTION THROUGH THE BICUSPID VALVE

Sagittal sections showing the bicuspid valve are 4 to 5 cm left of the midline (Fig. 2-20). These sections show the **right ventricle** in an anterior position and its separation from the **left ventricle** by an **interventricular septum.** Two cusps (anterior and posterior) of the **bicuspid (mitral) valve** are between the left atrium and the left ventricle. Sections in this region usually show the two branches of the left coronary artery: the **anterior interventricular** (anterior descending) **branch,** located in the fat of the interventricular sulcus; and the **circumflex branch,** found in the anterior part of the atrioventricular sulcus. The **coronary sinus** is still evident on the posterior aspect of the heart. Since this section is to the left of the pulmonary trunk, the **left pulmonary artery** is seen anterior to the descending aorta and superior to the **left bronchus.** The elongated but narrow left lung is divided into two lobes by the oblique fissure.

Coronal Sections

The anteroposterior relationships of the mediastinal structures evidenced in the transverse and sagittal sections should be kept in mind when studying a coronal series. Structures that have a definite anteroposterior curvature (such as the **aortic arch** and the **azygos vein,** which curves over the root of the lung) will be cut perpendicular to their axes (Fig. 2-21). The posterior chamber of the heart, the left atrium, will be seen in the posterior coronal sections, but the right ventricle will not be seen because it is an anterior chamber. The muscles that may be present are discussed in a later chapter.

SECTION THROUGH THE LEFT ATRIUM

The **left atrium** is the most posterior chamber of the heart. It receives newly oxygenated blood from the lungs via the pulmonary veins. If the section is just right, these veins may be seen entering the left atrium (see Fig. 2-21). Superior to the left atrium, the **trachea** descends slightly to the right of midline. In some sections through the left atrium, the bifurcation of the trachea into the **right** and the **left main stem bronchi** may be evident. In more posterior sections, the esophagus is seen instead of the trachea. The posterior part of the **aortic arch** is evident to the left of the trachea. The most posterior of the three aortic arch tributaries, the **left subclavian artery,** may be seen as it branches from the arch. This vessel is situated close to the left lung. The **left vertebral artery** branches from the left subclavian artery and ascends in the neck.

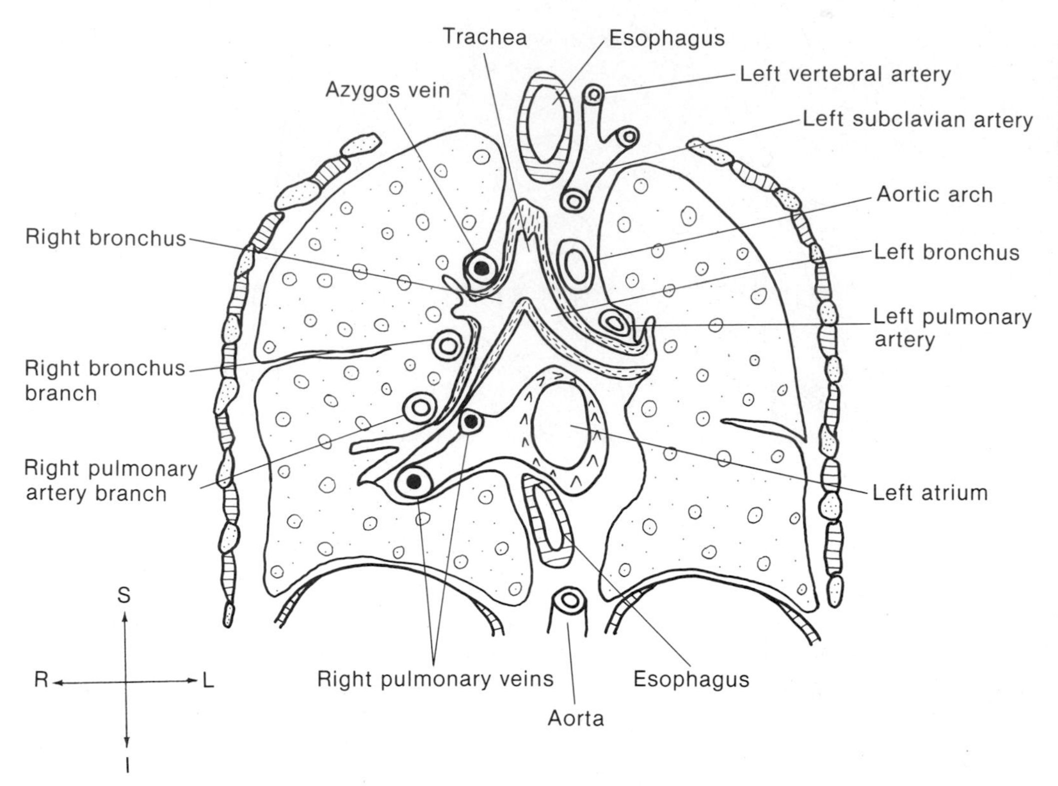

FIGURE 2-21. Coronal section through the left atrium, anteroposterior view. Note the pulmonary veins entering the left atrium, the bifurcation of the trachea, and the left vertebral artery that branches from the left subclavian artery.

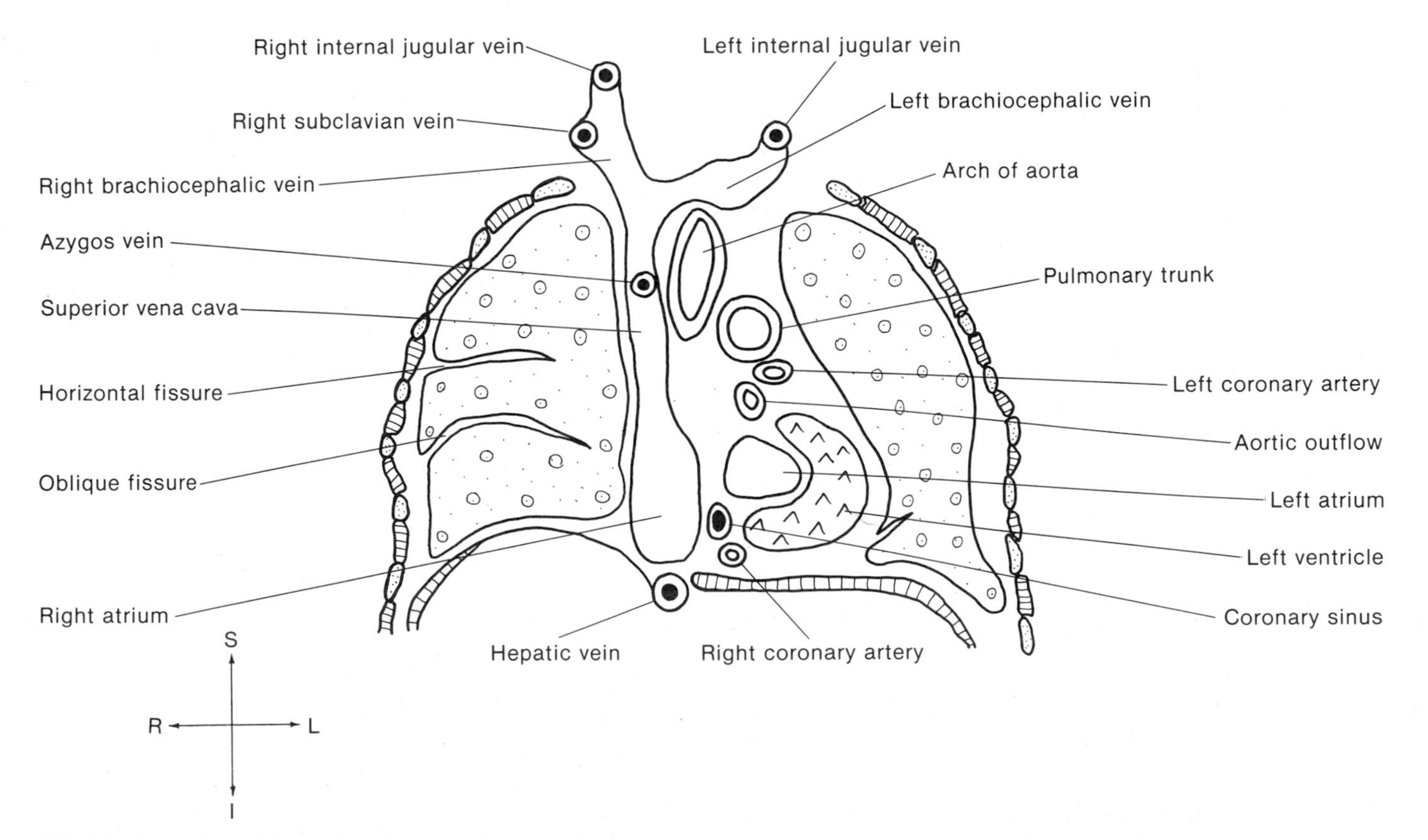

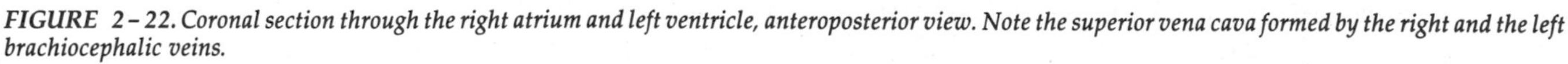
FIGURE 2-22. Coronal section through the right atrium and left ventricle, anteroposterior view. Note the superior vena cava formed by the right and the left brachiocephalic veins.

The **left pulmonary artery** is inferior to the aortic arch and to the left of the trachea. In some views, the right pulmonary artery may also be seen. The descending aorta is evident as it enters the abdomen through the diaphragm. The azygos vein parallels the aorta in this region. Superiorly, the azygos vein arches over the root of the right lung to empty into the superior vena cava.

SECTION THROUGH THE RIGHT ATRIUM AND THE LEFT VENTRICLE

Figure 2–22 illustrates a coronal plane through the right atrium and the left ventricle. It shows the **superior vena cava** in a straight line with the **right atrium** on the right side of the heart. The superior vena cava is formed when the **right** and **left brachiocephalic veins** join. The azygos vein empties into the superior vena cava. The apex of the heart, formed by the **left ventricle,** displaces the left lung; thus it is narrower than the right lung. The right-to-left arrangement of the superior vena cava, the ascending aorta, and the pulmonary trunk is evident.

REVIEW QUESTIONS

1. The thoracic cage is formed by the thoracic vertebrae, ______________, and ______________.
2. The sternal angle is located at the level of the second rib and ______________________ vertebrae.
3. The anterior boundary of the inferior thoracic aperture is the
   a. manubrium
   b. xiphisternal junction
   c. twelfth thoracic vertebra
   d. costal cartilages
4. The muscles between the ribs are called the ______________ muscles.
5. The pleura is a ______________ membrane consisting of two layers, the ______________ layer and the ______________ layer. The ______________ layer lines the thoracic wall.
6. **True** or **False:** The left lung is longer and narrower than the right lung and is divided into three lobes.
7. Which of the following represents an **INCORRECT** pairing?
   a. superior mediastinum—aortic arch
   b. middle mediastinum—heart
   c. posterior mediastinum—esophagus
   d. middle mediastinum—pulmonary veins
   e. posterior mediastinum—pulmonary arteries
8. **True** or **False:** The parietal layer of serous pericardium lines the fibrous pericardium of the pericardial sac.
9. The layer of the heart wall that contains cardiac muscle is called the ______________.
10. **True** or **False:** The right ventricle makes up most of the right border of the heart.
11. Which of the following statements concerning relationships of the heart is **NOT** correct?
    a. The superior and inferior venae cavae drain into the right atrium.
    b. The most posterior chamber of the heart is the left atrium, which receives blood from the pulmonary arteries.
    c. The bicuspid or mitral valve is in the anterior portion of the left atrium.
    d. The pulmonary semilunar valve is on the left side of the sternum at the level of the third costal cartilage.
    e. The tricuspid valve is anchored to papillary muscles in the right ventricle by chordae tendineae.
12. The two major branches of the left coronary artery are the ______________ and the ______________.
13. **True** or **False:** The normal "pacemaker" of the heart is the sinoatrial node.
14. Which of the following statements concerning the great vessels of the heart is **NOT** true?
    a. The superior vena cava enters on the right side of the heart.
    b. The vessel that passes horizontally, posterior to the aorta and the superior vena cava, is the right pulmonary artery.
    c. The origin of the pulmonary trunk is anterior to the origin of the ascending aorta.
    d. The most posterior branch of the aortic arch is the brachiocephalic artery.
    e. The ascending aorta is to the right of the pulmonary trunk.
15. Given the following numeric code, choose the sequence that represents the correct pathway for the flow of blood through the heart: (1) Right atrium; (2) left atrium; (3) right ventricle; (4) left ventricle; (5) tricuspid valve; (6) bicuspid valve; (7) aortic semilunar valve; (8) pulmonary semilunar valve; (9) pulmonary trunk; (10) ascending aorta; (11) lungs; (12) pulmonary veins.
    a. 1, 6, 3, 8, 9, 11, 12, 2, 5, 4, 7, 10
    b. 1, 5, 3, 8, 9, 11, 12, 2, 6, 4, 7, 10
    c. 3, 5, 1, 8, 9, 11, 12, 4, 6, 2, 7, 10
    d. 3, 5, 1, 8, 9, 11, 12, 4, 6, 2, 7, 10
    e. 1, 5, 2, 8, 9, 11, 12, 3, 6, 4, 7, 10
16. Which of the following is **NOT** a correct statement?
    a. The thymus is posterior to the manubrium.
    b. The left bronchus is posterior to the esophagus.
    c. The carina is at the level of the disc between the fifth and sixth thoracic vertebrae.
    d. The esophagus is positioned posterior to the left atrium.
    e. The azygos vein drains into the superior vena cava.

17. The two superficial muscles of the back are the ______________ and the ______________.

18. The ______________ plexus supplies innervation to the arm.

19. **True** or **False:** Suspensory ligaments in the breast are also known as Cooper's ligaments.

20. Over one half of the lymphatic drainage from the breast is by way of the ______________ lymph nodes.

21. Identify the indicated structures in Figure A.

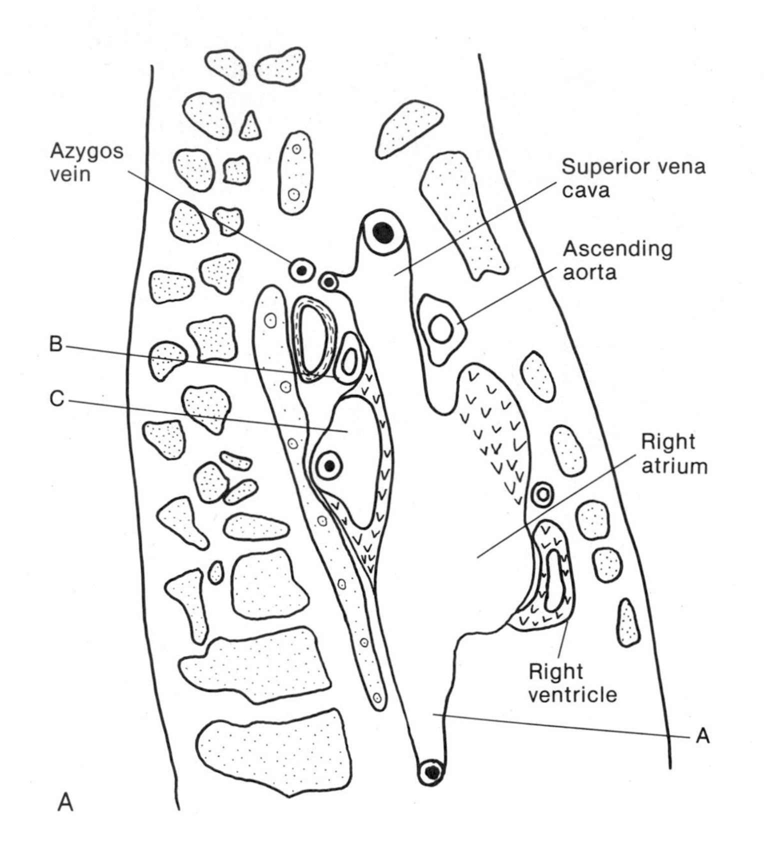

CHAPTER 3

ABDOMEN

OBJECTIVES

Upon completion of this chapter, the student should be able to do the following:

- State the boundaries of the abdomen.
- Define the transpyloric, subcostal, transumbilical, interiliac, median, and midclavicular planes and use these planes to divide the abdomen into four quadrants and nine regions.
- Describe the structure of the diaphragm, name and give the vertebral levels of the three major openings in the diaphragm, and identify the structures that pass through each opening.
- Name the four muscles that form the anterolateral abdominal wall and the three muscles that are associated with the posterior abdominal wall.
- Discuss the topography of the posterior abdominal wall and its effect on organ position and fluid accumulation.
- Discuss the peritoneum and its extensions, including the mesentery, omenta, ligaments, and cul-de-sacs.

- State the level of origin of the visceral branches of the abdominal aorta and identify the regions each one supplies.
- Identify and trace the pathway of the tributaries of the inferior vena cava.
- Trace the pathway of blood through the hepatic portal system of veins.
- Discuss the structure and the relationships of the liver, including its lobar subdivisions and its blood supply.
- Discuss the visceral relationships of the gallbladder.
- Describe the stomach's external features, its peritoneal extensions, its relationships, and its blood supply.
- Name the regions of the small intestine and discuss the relationships of each region.
- Identify the regions of the large intestine and discuss the relationships of each region.
- Describe the location and the relationships of the spleen.
- Discuss the location and the relationships of the head, neck, body, and tail of the pancreas.
- Describe the location and the relationships of the kidneys, ureters, and suprarenal glands.
- Identify the abdominal viscera, muscles, and blood vessels on transverse, sagittal, and coronal sections.

GENERAL ANATOMY OF THE ABDOMEN

Surface Markings

The ventral body cavity is divided into two distinct subdivisions that are separated by the dome-shaped diaphragm. The upper portion, superior to the diaphragm, is the thoracic cavity. The abdominopelvic cavity is the portion inferior to the diaphragm. For the sake of convenience, the large abdominopelvic cavity may be divided into an upper abdominal cavity and a lower pelvic cavity. This is an artificial division, since there is no partition between the two cavities and some structures may move from one region to the other.

BOUNDARIES

The **abdominal cavity,** the upper portion of the abdominopelvic cavity, is the largest cavity in the body. It extends from the **diaphragm** above to the **superior pelvic aperture** below. The dome of the diaphragm extends superiorly under the ribs to the level of the fifth intercostal space; thus the contents of the superior portion of the abdominal cavity are protected by the thoracic cage. Portions of the liver, stomach, and spleen are in this region. Inferiorly, the large wings, or **alae,** of the iliac bones offer some protection for the soft tissue. The superior peripheral boundaries are the **xiphoid process** of the sternum and the sloping **costal cartilages** of the false ribs. The inferior peripheral margins of the cavity are the right and left **iliac crests,** right and left **inguinal ligaments,** and the **symphysis pubis** in the center. The iliac crest is the highest portion, or the highest margin, of the ilium. It terminates anteriorly in the anterior superior iliac spine. The inguinal ligament is the folded inferior margin of the broad flat tendon, or **aponeurosis,** of the external oblique muscle. This ligament extends from the anterior superior iliac spine to the pubic tubercle, which is a small elevation about 2 cm lateral to the pubic symphysis. Just superior to the pubic tubercle, there is an opening in the aponeurosis that forms the inguinal canal.

UMBILICUS

The most obvious surface marking on the anterior abdominal wall is the **umbilicus,** or navel. The umbilicus is the scar that results from the closure of the umbilical cord shortly after birth. It represents the site of attachment of the umbilical cord in the fetus. The position of the umbilicus varies, depending on such factors as muscle tone, obesity, body build, and age. In general, however, it is located at the level of the intervertebral disc between the third and fourth lumbar vertebrae.

LINEA ALBA

When the skin of the abdomen is removed, the **linea alba,** a light-colored line extending from the xiphoid process to the symphysis pubis, is evident. The position of the linea alba is indicated on the surface of the abdomen by a shallow groove in the midline. The linea alba is formed by the fusion of the sheets of tendon that extend medially from the anterolateral muscles of the abdominal wall.

ABDOMINAL REGIONS AND PLANES

Superficial landmarks identify various abdominal planes that are used to indicate vertebral levels and to describe the location of deeper structures. Vertical and horizontal planes divide the abdomen into regions that are used to describe the location of organs or, in the clinical setting, the location of pain, tenderness, swelling, and abnormal growths. Five horizontal and three vertical planes will be described.

TRANSPYLORIC PLANE. The transpyloric plane is the most superior of the horizontal planes. It is located about halfway between the jugular notch and the symphysis pubis or, more simply, midway between the xiphoid and the umbilicus. This plane typically intersects the **pyloric region of the stomach,** which accounts for its name. Passing laterally to the right on this plane gives the location of the **first part of the duodenum** and the **top of the head of the pancreas.** Proceeding further to the right, this plane intersects the **ninth costal cartilage,** which gives the location of the **fundus of the gallbladder** and the upper portion of the hilum of the **right kidney.** Going to the left of the midline, the transpyloric plane gives the location of the **neck of the pancreas** and the middle portion of the **hilar region of the left kidney.** Usually this plane marks the level of the **first lumbar vertebra.**

SUBCOSTAL PLANE. A line through the most inferior point of the rib cage gives the position of the **subcostal plane,** which marks the level of the **third lumbar vertebra.** The subcostal plane intersects the **third part of the duodenum** and the **lower border of the pancreatic head.**

There is no good method of locating the L-2 vertebral level, except to note that it is halfway between the transpyloric and subcostal planes.

TRANSUMBILICAL PLANE. The **transumbilical plane** passes horizontally through the umbilicus. In individuals with relatively "normal" abdominal contour, this marks the level of the intervertebral disc between the third and fourth lumbar vertebrae.

INTERILIAC PLANE. The **fourth lumbar vertebral level** is located by the **interiliac plane** through the most superior point of the iliac crests. This is also called either the supracristal or the intercristal plane.

TRANSTUBERCULAR PLANE. The **transtubercular plane** passes through the tubercles of the iliac crests. The tubercles are small projections on the crests about 5 cm posterior to the anterior superior iliac spines. This plane marks the level of the **fifth lumbar vertebra.**

RIGHT MIDCLAVICULAR PLANE. The **right midclavicular plane** extends vertically from the midpoint of the right clavicle to the right midinguinal point. The midinguinal point is the midpoint of a line that joins the right anterior superior iliac spine and the symphysis pubis.

LEFT MIDCLAVICULAR PLANE. The **left midclavicular plane** is in the same position as the right midclavicular plane, except that it is on the left side. It extends from the midpoint of the left clavicle to the left midinguinal point.

MEDIAN PLANE. The **midsagittal plane,** or **median plane,** is a vertical plane through the umbilicus. It divides the body into right and left halves.

ABDOMINAL QUADRANTS. For descriptive purposes, the horizontal transumbilical plane and the vertical median plane are used to divide the abdomen into four quadrants. For example, the pain of acute appendicitis usually localizes in the lower right quadrant.

ABDOMINAL REGIONS. Nine abdominal regions are described using the horizontal subcostal and transtubercular planes and the vertical right and left midclavicular planes (Fig. 3–1). On the right and left sides, the three regions are, from superior to inferior, the **hypochondriac,** the **lumbar** or **lateral,** and the **iliac** or **inguinal** regions.

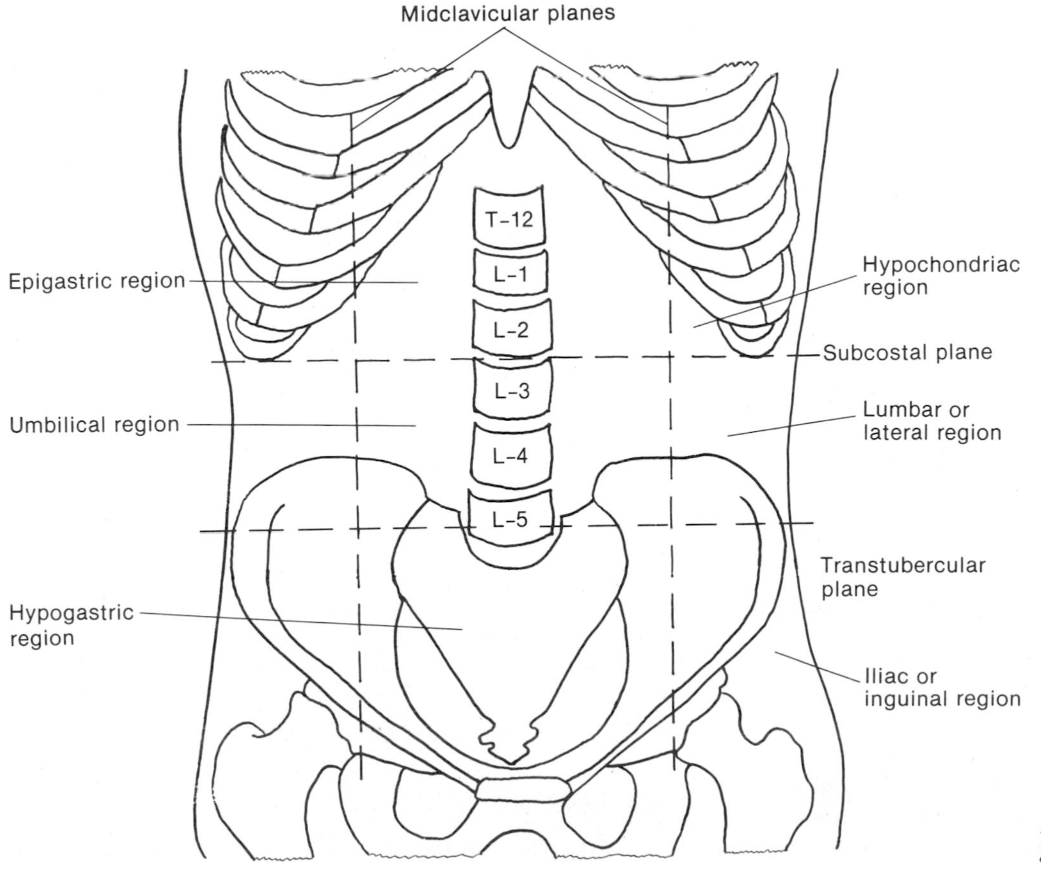

FIGURE 3–1. Nine abdominal regions.

The three regions in the midline are, from superior to inferior, the **epigastric**, the **umbilical**, and the **hypogastric** regions.

Diaphragm

The musculotendinous **diaphragm** forms a movable partition between the thoracic and the abdominal cavities. When the individual is supine, the diaphragm extends superiorly under the rib cage to the level of the fifth intercostal space. The large right lobe of the liver usually causes the diaphragm to rise to a slightly higher level on the right side than that on the left.

STRUCTURE OF THE DIAPHRAGM

The central portion of the diaphragm consists of tendinous fibers, which form a strong **central tendon.** All the muscle fibers of the diaphragm converge and insert on the central tendon. The muscular diaphragm is divided into three regions, according to the origin of its fibers. A short and narrow **sternal portion** arises from the back of the xiphoid process. The extensive **costal portion** originates from the inner surface of the lower six costal cartilages. These costal muscular fibers form the two domes, or **hemidiaphragms.** The **vertebral** or **lumbar portion** arises from the upper lumbar vertebrae as a pair of **muscular crura.** Each crus is a thick, fleshy, muscular bundle that tapers inferiorly and becomes tendinous. Fibers of each crus spread out and ascend, attaching to the central tendon. The right crus encircles the esophagus.

OPENINGS IN THE DIAPHRAGM

Structures passing from the thoracic cavity into the abdominal cavity must penetrate the diaphragm; thus the continuity of the diaphragm is interrupted by three large and several small apertures. Each opening is called a hiatus.

CAVAL HIATUS. At the level of the eighth thoracic vertebra, the wide **caval hiatus** for the inferior vena cava is located within the central tendon about 3 cm to the right of the median plane. The caval hiatus is not only the most superior of the three openings but also the most anterior. In addition to the inferior vena cava, this opening transmits the right phrenic nerve and some lymph vessels. Occasionally, the right hepatic vein passes through this opening before it enters the inferior vena cava.

ESOPHAGEAL HIATUS. The oval opening for the esophagus, the **esophageal hiatus,** is located in the muscular diaphragm posterior to the central tendon. It is 2 or 3 cm to the left of the midline, at the level of the tenth thoracic vertebra, and it is surrounded by the right crus of the diaphragm. In addition to the esophagus, the esophageal hiatus transmits the vagus nerve and the esophageal branches of the left gastric blood vessels.

AORTIC HIATUS. The long oblique **aortic hiatus** is located between the right and left crura and begins at the level of the twelfth thoracic vertebra. The aortic hiatus is the most posterior of the three large openings in the diaphragm. Technically, the aorta does not penetrate the diaphragm. Instead, it passes between the crura, slightly to the left of the midline. In addition to the aorta, this opening transmits the azygos vein and the thoracic duct.

Abdominal Wall

The abdominal wall is primarily composed of muscles covered with fascia and skin. Posteriorly the vertebrae form a midline ridge, which contributes to the internal configuration of the wall.

ANTEROLATERAL ABDOMINAL WALL

The anterolateral abdominal wall is formed by four muscles and their aponeuroses. These four muscles include the rectus abdominis and the three lateral muscles, the transversus abdominis and the external and internal oblique muscles (Fig. 3–2).

RECTUS ABDOMINIS MUSCLE. Anteriorly, the long, vertical **rectus abdominis muscles** extend the length of the abdominal wall, on either side of the linea alba, from the symphysis pubis to the xiphoid process. The rectus abdominis is enclosed in a rectus sheath formed by the aponeuroses of the three lateral muscles.

EXTERNAL OBLIQUE MUSCLE. The outermost layer of the lateral muscles is formed by the **external oblique muscle.** Fibers of this muscle originate from the ribs and extend downward and medially. Most of the fibers terminate in a broad aponeurosis, which inserts on the linea alba, the iliac crest, and the pubic tubercle. The inferior margin of this aponeurosis forms the inguinal ligament.

INTERNAL OBLIQUE MUSCLE. In contrast to the external oblique muscle, the fibers of the **internal oblique muscle** extend upward and medially, perpendicular to the external oblique muscle, from the iliac crest to the inferior borders of the ribs. Medially, the aponeurosis of the internal oblique splits into two layers and encloses the rectus abdominis.

TRANSVERSUS ABDOMINIS MUSCLE. The **transversus abdominis** is the innermost of the three flat muscles. These fibers pass in a transverse or horizontal direction. This arrangement of the three flat muscles provides maximum support for the abdominal viscera and diminishes the risk of tearing the muscles. Innervation of the skin and muscles of the anterolateral abdominal wall is derived from the **spinal nerves.**

POSTERIOR ABDOMINAL WALL

LUMBAR VERTEBRAE. Five large lumbar vertebrae with their intervertebral discs form the skeletal support

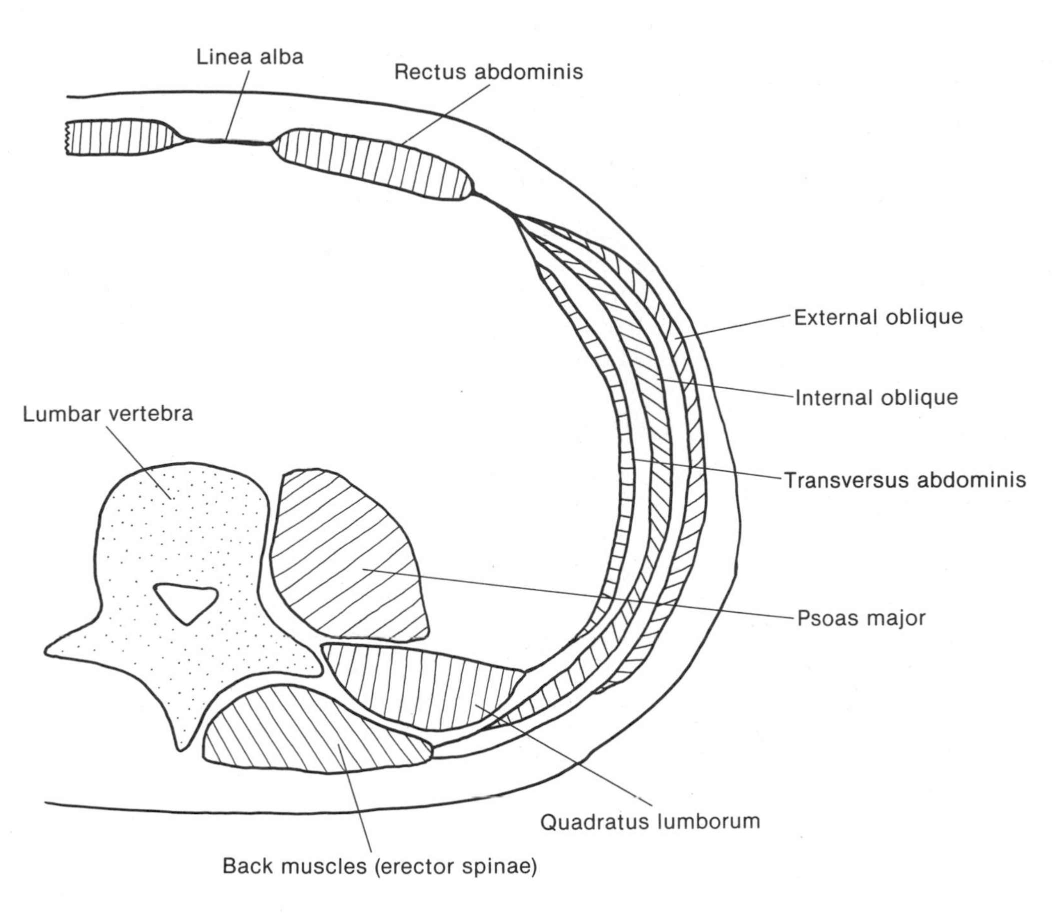

FIGURE 3-2. Musculature of the abdominal wall.

for the posterior abdominal wall. Lumbar vertebrae have large bodies with short, thick, blunt spinous processes. The transverse processes are also thicker than in other vertebrae. The vertebrae and discs form a normal lumbar curvature that is convex anteriorly. This curvature develops in a child during the second year as the child begins to walk and to put increased weight on the lumbar region. An exaggeration, or increase, in the convex curvature is called **lordosis.**

The spinal cord within the vertebral foramen ends at the level of the third lumbar vertebra at birth. As a result of the different growth rates of the cord and the vertebral column, the spinal cord ends at the level of the **second lumbar vertebra** in the adult. Even though the spinal cord ends at the L-2 level, the meninges, the subarachnoid space, and the cerebrospinal fluid continue to the second sacral vertebra. This is of clinical importance when performing a spinal tap to remove cerebrospinal fluid for laboratory examination. The needle used to remove fluid from the subarachnoid space is inserted between the third and fourth lumbar vertebrae, or sometimes between the fourth and fifth lumbar vertebrae. This minimizes the possibility of damage to the spinal cord. Remember that the location of L-4 is determined by the interiliac plane through the superior points of the iliac crests.

MUSCULATURE. The posterior abdominal wall is formed primarily by three pairs of muscles and their attachments to the vertebrae, the ribs, and the ilium.

Psoas Major Muscle. The long, thick **psoas major muscle** is lateral to the lumbar region of the vertebral column. Its fibers originate on the transverse processes, bodies, and intervertebral discs of the lumbar vertebrae, pass along the brim of the pelvis, and enter the thigh to insert on the lesser trochanter of the femur. In transverse sections, the psoas major will appear as large muscular masses adjacent to the lumbar vertebral body.

Iliacus Muscle. The **iliacus muscle** is a large triangular sheet of muscle in the iliac fossa on the medial side of the alae, or wings, of the ilium. It originates in the iliac fossa and inserts with the psoas major on the lesser trochanter. The iliacus and psoas major muscles are closely associated, and together they are often referred to as the **iliopsoas.** The iliopsoas is the most powerful flexor of the thigh.

Quadratus Lumborum Muscle. The third muscle associated with the posterior abdominal wall is the **quadratus lumborum.** This thick muscular sheet originates on the iliac crest and the transverse processes of the lower lumbar vertebrae. It ascends to insert on the transverse processes of the upper lumbar vertebrae and the twelfth rib. In transverse sections, this muscle appears somewhat lateral and posterior to the psoas major muscle. The arrangement of the posterior wall muscles is illustrated in Figure 3-2.

INNERVATION. Innervation of the skin and the muscles of the back is derived from branches of the lumbar nerves. The ventral rami of the lumbar nerves form the

lumbar plexus within the psoas major muscle. The largest and most important branches of the lumbar plexus are the **obturator** and **femoral nerves.** In the abdomen, the femoral nerve supplies both the psoas major and the iliacus muscles as it follows a path between the two muscles before entering the thigh. The obturator nerve passes through the obturator foramen of the pelvis to innervate muscles of the thigh.

TOPOGRAPHY. An examination of the posterior abdominal wall reveals one longitudinal and two oblique ridges. The **longitudinal ridge line,** sometimes called the longitudinal divide, is especially evident in transverse sections (see Fig. 3–2). It is formed by the **lumbar vertebrae,** the **normal lumbar lordosis,** the **psoas major muscles,** the **inferior vena cava,** and the **aorta.** On either side of the elevated longitudinal ridge, there is a paravertebral groove or gutter. In superior regions of the abdomen, this groove is occupied by the liver on the right and by the spleen on the left. The kidneys, ureters, and portions of the colon are also located in the paravertebral grooves. Whenever an organ crosses the midline, it is moved anteriorly as a result of the elevation of this longitudinal ridge. For example, the right lobe of the liver is rather posterior in position, whereas the left lobe is more anterior because it is moved forward by the longitudinal ridge. The pancreas offers another example of this anterior displacement. The tail of the pancreas is in a relatively posterior position near the spleen. As the gland courses to the right, it is pushed forward by the longitudinal ridge, which causes the head and neck to be more anteriorly situated.

Inferiorly, the longitudinal divide bifurcates to form two **oblique ridges** that mark the location of the pelvic inlet. These oblique ridges are formed by the **bony pelvic inlet,** the **psoas major muscles,** and the **iliac blood vessels.**

In addition to influencing organ position, the longitudinal and oblique ridges determine, to some extent, regions of fluid accumulation. Fluids tend to flow off the sides of the longitudinal ridges to accumulate in the paravertebral "valleys," or grooves, which are sometimes called paravertebral gutters. The paravertebral grooves slope posteriorly from the oblique ridges; thus the fluids tend to flow down the abdominal portions of the oblique ridges and to accumulate in the superior regions of the paravertebral grooves when the patient is supine.

VASCULATURE OF THE ABDOMINAL WALL

The principal arterial supply of the anterolateral abdominal wall comes from branches of the internal thoracic and internal iliac arteries. In addition to these vessels, branches of the intercostal and subcostal arteries contribute to the arterial supply at higher levels. All of these vessels branch and anastomose freely. Venous drainage is accomplished primarily through branches of the superficial epigastric and lateral thoracic veins. The posterior abdominal wall is supplied and drained by lumbar arteries and veins. These are direct tributaries of the aorta and inferior vena cava, respectively.

Peritoneum

The wall of the abdominal cavity is lined with a thin, translucent serous membrane called the **parietal peritoneum.** This forms a peritoneal sac and peritoneal cavity, both of which are enclosed in the abdominal cavity delineated by the muscular abdominal walls.

PERITONEAL SAC

During development, some organs of the abdominal cavity protrude from the wall into the peritoneal cavity and carry a covering of peritoneum with them. The layer around the organs is called the **visceral peritoneum** and is continuous with the parietal peritoneum that lines the walls. The space between the two layers of peritoneum is the **peritoneal cavity.** As the organs continue to develop, the peritoneal cavity is obliterated, leaving only a potential space between the visceral and parietal layers of peritoneum. The two layers of peritoneum are separated by only a capillary film of serous fluid for lubrication. This permits the organs to move against each other without friction. Normally, the peritoneal cavity has a very small volume and contains only a few drops of serous fluid. Abnormal accumulations of serous fluid, called **ascites,** may exaggerate the volume to form a real space of several liters in volume. In males, the peritoneal cavity is a closed cavity, but in females, the cavity communicates with the exterior through the uterine tubes, the uterus, and the vagina.

TERMINOLOGY RELATING TO THE PERITONEUM

The extent and character of the peritoneum are quite complex, and specific terms are used to describe different parts. The term **mesentery** is used to describe a double layer of peritoneum that encloses the intestine and attaches it to the abdominal wall. Blood and lymphatic vessels, nerves, lymph nodes, and fat cells are found between the two layers of a mesentery. An **omentum** is a mesentery, or double layer of peritoneum, that is attached to the stomach. The greater omentum hangs from the greater curvature of the stomach, over the intestines, like an apron. The lesser omentum joins the lesser curvature of the stomach and the proximal duodenum to the liver. The space behind the lesser omentum and the stomach is the **omental bursa,** or the **lesser sac.** The remainder of the peritoneal cavity is the **greater sac.** The omental bursa is closed at the end near the spleen but is open at the right edge. The opening into the omental bursa is the **epiploic foramen,** which permits communication between the lesser and the greater sacs of the peritoneal cavity. The **hepatic artery,** the **portal vein,** and the **bile duct** are enclosed within the layers of peritoneum at the free margin of the lesser omentum. Any peritoneal extension that is not a mesentery (or an omentum) is usually referred to as a **peritoneal ligament.** These are double layers of peritoneum that go from one organ to another, or to the anterior abdominal wall. An example of this is the **falciform ligament,** which extends from the liver to the anterior abdominal wall. Figure 3 – 3 illustrates the continuity of the peritoneum and its extensions, the mesenteries and the omenta.

There are numerous exceptions to these generalizations. This is particularly true with the peritoneal attachments associated with the stomach. Many of these, although called ligaments, are actually a part of the omenta.

CUL-DE-SACS

In certain places, peritoneal folds form blind pouches, or **cul-de-sacs.** The largest of these is the omental bursa, or lesser sac, which has already been described. In the pelvic region of the female peritoneal cavity there is a cul-de-sac between the rectum and the uterus. This is the **rectouterine pouch,** or pouch of Douglas. There is a **vesicouterine pouch** formed by the reflection of peritoneum from the uterus to the superior surface of the bladder. In males, there is a **rectovesical pouch** between the rectum and the posterior surface of the bladder.

RETROPERITONEAL STRUCTURES

Some organs of the abdominal cavity are not included within the peritoneal sac; instead, they are behind the peritoneum (see Fig. 3 – 3). Their location is described as being retroperitoneal. Only the anterior surfaces of these organs are covered with peritoneum. The kidneys, the pancreas, the duodenum, and the ascending and descending portions of the colon are all retroperitoneal.

Aorta and Arteries of the Abdomen

The abdominal aorta begins at the aortic hiatus in the diaphragm about 2.5 cm above the transpyloric line. At this level, the aorta is usually slightly left of midline, but as it descends, it assumes a more midline position. At the

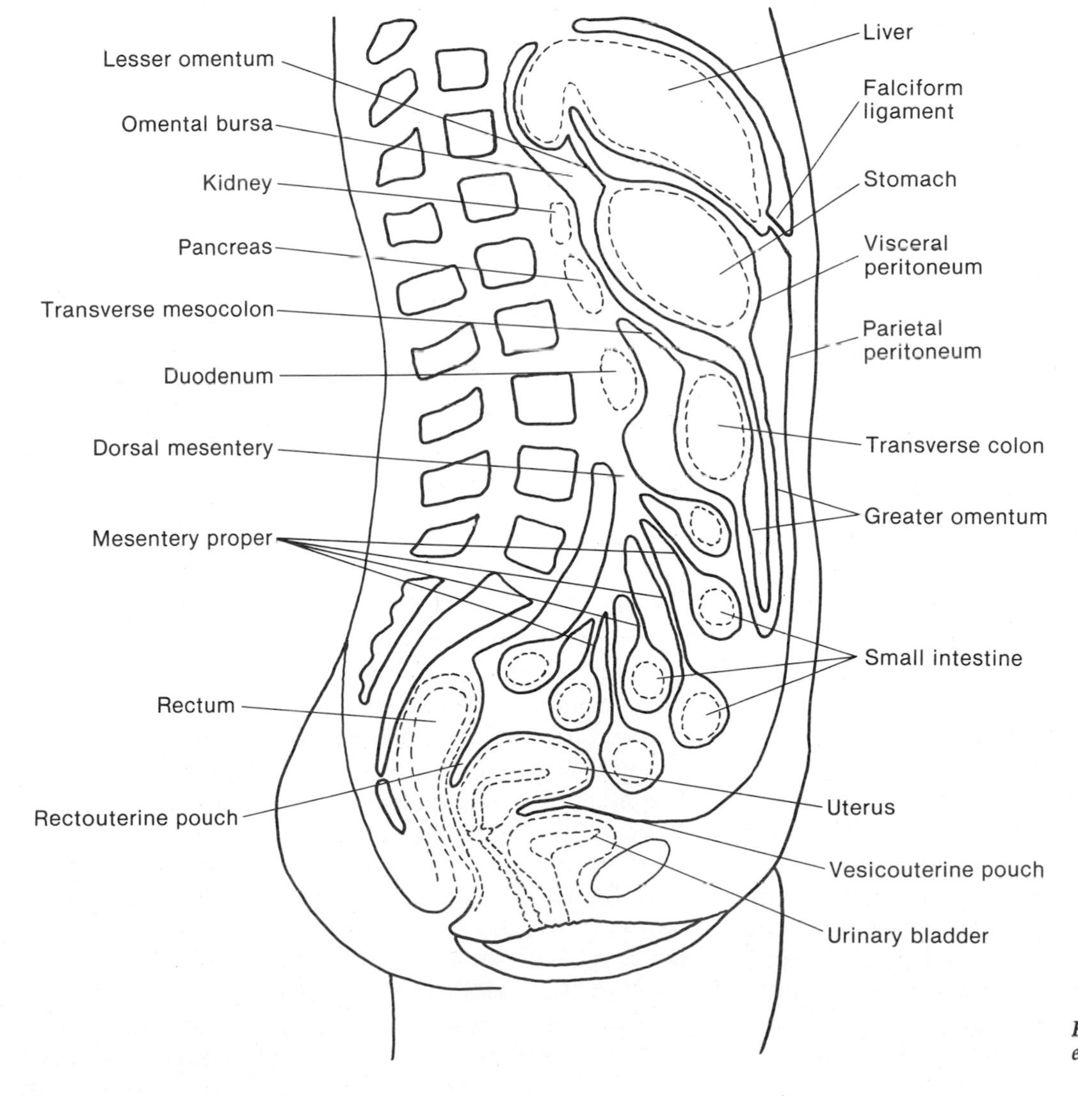

FIGURE 3 – 3. Peritoneum and its extensions.

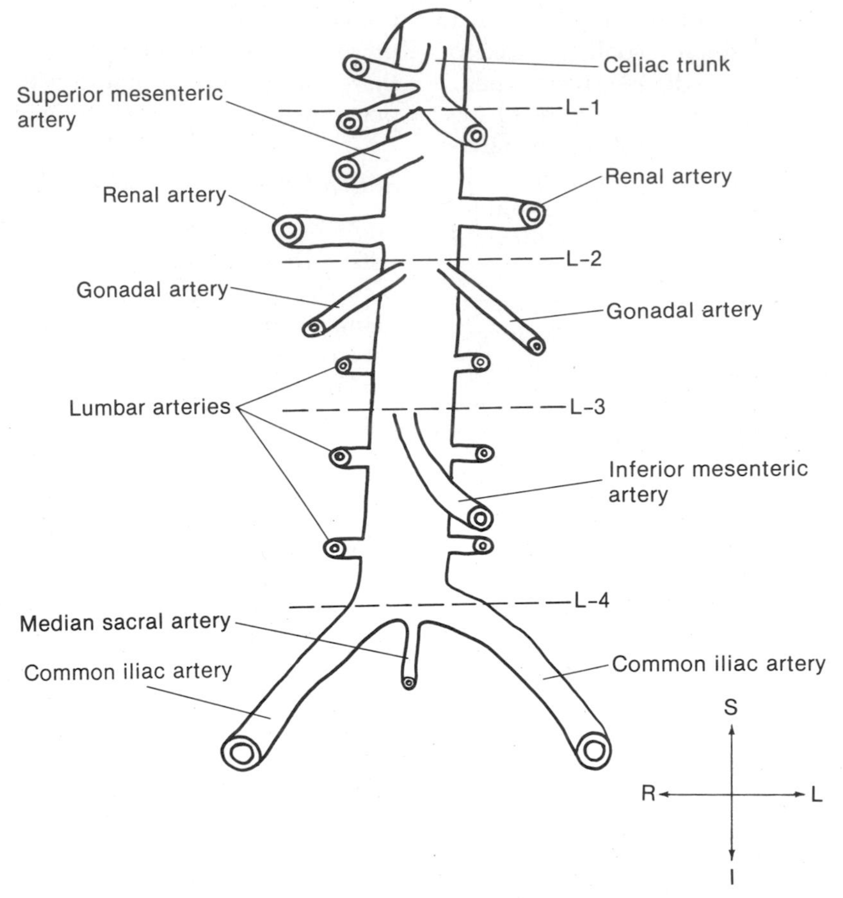

FIGURE 3–4. Branches of the abdominal aorta.

L-4 vertebral level, which is intersected by the interiliac line, the aorta bifurcates into the right and left common iliac arteries. The branches of the abdominal aorta may be divided into the following four groups: unpaired visceral, paired visceral, unpaired parietal, and paired parietal. The visceral branches supply the viscera or organs of the abdominal cavity, whereas the parietal branches supply the abdominal wall. The branches of the abdominal aorta are illustrated in Figure 3–4.

UNPAIRED VISCERAL BRANCHES

The unpaired visceral branches of the abdominal aorta are the celiac artery, the superior mesenteric artery, and the inferior mesenteric artery. Each of these will be considered separately.

CELIAC TRUNK. The celiac trunk (artery) is the first major branch of the abdominal aorta. It arises from the ventral surface of the aorta, just above the transpyloric line, near the upper margin of the first lumbar vertebra. The celiac trunk itself extends only 1 to 2 cm before it divides into the **left gastric, hepatic,** and **splenic arteries.** The **left gastric artery,** the smallest branch of the celiac trunk, passes to the left to supply the cardiac region of the stomach and then descends along the lesser curvature. Its branches anastomose with those of the right gastric artery, a branch of the hepatic artery. Intermediate in size, the **hepatic branch** of the celiac trunk travels to the right and enters the porta of the liver, where it divides into the right and left hepatic arteries. During its course, the hepatic artery gives off the **right gastric, gastroduodenal,** and **cystic arteries.** The third and largest branch of the celiac trunk is the **splenic artery.** This long, tortuous vessel passes horizontally to the left, behind the stomach, along the upper border of the pancreas, and enters the hilum of the spleen. As it passes along the upper border of the pancreas, the splenic artery gives off numerous **pancreatic branches.**

SUPERIOR MESENTERIC ARTERY. The second, unpaired, visceral branch of the aorta is the **superior mesenteric artery.** This vessel arises just below the transpyloric line at the level of the lower border of the first lumbar vertebra. The superior mesenteric artery branches and anastomoses freely to supply all of the small intestine, except for the duodenum. In addition, it supplies the cecum, the ascending colon, and most of the transverse colon. At its origin, the superior mesenteric artery is separated from the aorta by the left renal vein. The splenic vein and the body of the pancreas pass horizontally, anterior to the superior mesenteric artery.

INFERIOR MESENTERIC ARTERY. The third, unpaired, visceral vessel that arises from the aorta is the **inferior mesenteric artery.** This branch originates from the ventral surface of the abdominal aorta at the L-3 vertebral level, which is intersected by the subcostal plane. At first, the inferior mesenteric artery descends anterior to the aorta; it then passes to the left to supply the distal portion of the transverse colon and all of the descending colon, the sigmoid colon, and the rectum.

PAIRED VISCERAL BRANCHES

The paired visceral branches of the aorta are the suprarenal, the renal, and the gonadal arteries. Each of these is considered separately.

SUPRARENAL ARTERIES. A pair of small **suprarenal vessels** arises from the aorta, one vessel on each side, at the level of the superior mesenteric artery. These vessels course laterally and slightly superiorly to supply the suprarenal (adrenal) gland.

RENAL ARTERIES. Two large **renal arteries** arise from the sides of the aorta, at the upper L-2 vertebral level, just inferior to the superior mesenteric artery. Each vessel passes laterally, at right angles to the aorta, to enter the hilum of the kidney. Since the aorta is slightly to the left of the midline, the right renal artery is longer than the left. As it proceeds to the right kidney, the right renal artery passes posterior to the following: the inferior vena cava, the right renal vein, the head of the pancreas, and the second or descending part of the duodenum. As a result of the usually higher level of the left kidney, the left renal artery is generally slightly higher than the right. As it passes to the kidney, the left renal artery lies posterior to the body of the pancreas, the left renal vein, and the splenic vein. One or two accessory renal arteries may be present. These usually arise directly from the aorta.

GONADAL ARTERIES. The paired visceral **gonadal vessels**—the **testicular arteries** in the male and the **ovarian arteries** in the female—branch from the aorta just inferior to the renal vessels; thus they originate at the level of the lower margin of the second lumbar vertebra. Each testicular artery descends along the psoas muscle and passes over the ureters and lower part of the external iliac artery to reach the deep inguinal ring, where it enters the spermatic cord. Along with the other contents of the spermatic cord, the testicular artery enters the scrotum to supply the testes. In the female, the ovarian arteries descend along the psoas muscle to the pelvic brim. At the brim, they cross over the external iliac vessels to enter the pelvic cavity and continue in the suspensory ligament to supply the ovary.

UNPAIRED PARIETAL BRANCH

The unpaired parietal branch of the aorta is the **median sacral artery.** This vessel arises from the posterior surface of the aorta just proximal to its bifurcation. As it descends along the anterior surface of the L-4 and L-5 vertebrae, the median sacral artery gives off a pair of lumbar arteries, which supply a portion of the posterior abdominal wall.

PAIRED PARIETAL BRANCHES

Four pairs of **lumbar arteries** constitute the paired parietal branches of the aorta. These vessels arise from the posterolateral surface of the aorta along the upper four lumbar vertebrae. Lumbar arteries supply the posterolateral abdominal wall.

BIFURCATION OF THE AORTA

At the L-4 vertebral level, the aorta divides into the **right** and the **left common iliac arteries,** which diverge along the bodies of the fourth and fifth lumbar vertebrae. At the level of the disc between the fifth lumbar vertebra and the sacrum, each common iliac artery divides into internal and external branches. The **internal iliac artery** supplies the wall and viscera of the pelvis, the perineum, and the gluteal region. The **external iliac artery** supplies the lower limb.

Inferior Vena Cava and Veins of the Abdomen

The largest vein in the body is the **inferior vena cava.** The right and left common iliac veins join anterior to the fifth lumbar vertebra to form the inferior vena cava. The inferior vena cava receives tributaries as it ascends through the abdomen along the vertebral column. After passing along the posterior surface of the liver, the inferior vena cava goes through the caval hiatus of the diaphragm at the T-8 vertebral level. In the mediastinum, it penetrates the pericardium to drain into the lower part of the right atrium. There are no valves in the inferior vena cava, but there is a rudimentary semilunar valve at its atrial orifice.

In general, the inferior vena cava is slightly to the right of the aorta. In superior regions of the abdomen, it is anterior to the aorta. As the inferior vena cava descends, however, it becomes increasingly posterior, and by the L-4 and L-5 levels, it is posterior to the aorta. As the inferior vena cava ascends the abdomen, it is anterior to the right psoas muscle, the right renal artery, the right suprarenal gland, and the right crus of the diaphragm. The inferior vena cava is posterior to the peritoneum (retroperitoneal), the superior mesenteric vessels, the head of the pancreas, and the horizontal third part of the duodenum.

TRIBUTARIES OF THE INFERIOR VENA CAVA

Tributaries of the inferior vena cava include the **common iliac,** the **lumbar,** the **right gonadal,** the **renal,** the **right suprarenal,** the **inferior phrenic,** and the **hepatic veins.**

COMMON ILIAC VEINS. The external and internal iliac veins join anterior to the sacroiliac joint to form a common iliac vein. The right and left common iliac veins drain the same regions that are supplied by the arteries of the same name. The common iliac veins pass obliquely upward from the sacroiliac joint to the fifth lumbar vertebra, where they join to form the inferior vena cava.

LUMBAR VEINS. The lumbar veins consist of four or five pairs of vessels that collect blood from the muscles and skin of the abdominal wall. The arrangement of these veins varies. Some drain directly into the inferior vena cava, whereas others may enter the azygos system.

GONADAL VEINS. The testicular veins in the male begin on the dorsal side of the testes and ascend in the spermatic cord to enter the abdomen through the deep inguinal ring. The veins ascend retroperitoneally along the psoas muscle, anterior to the ureter. On the right side, the testicular vein opens directly into the inferior vena cava. On the left, it opens into the left renal vein. The ovarian veins in the female follow the same pattern as that of the testicular veins in the male, except the ovarian veins begin at the ovaries.

RENAL VEINS. The renal veins, which drain the kidneys, empty into the inferior vena cava at the L-2 vertebral level. They are usually anterior to the renal arteries because the inferior vena cava is anterior to the aorta at this level. The inferior vena cava is to the right side of the midline; thus the left renal vein is considerably longer than the right. From its origin in the kidney, the left renal vein passes to the right, posterior to the splenic vein and the body of the pancreas. As it crosses the midline to enter the inferior vena cava, the left renal vein passes in front of the aorta, just below the origin of the superior mesenteric artery. Sections at this level show the left renal vein, posterior to the superior mesenteric artery but anterior to the aorta. The left renal vein receives the left gonadal (either the testicular or ovarian) vein from below and the left suprarenal vein from above before it enters the inferior vena cava. The right renal vein is in a slightly more inferior position than is the left renal vein because the right kidney is at a lower level than the left kidney. The right renal vein passes posterior to the second or descending part of the duodenum.

SUPRARENAL VEINS. The right suprarenal vein is a short vessel, emerging from the right suprarenal gland and emptying directly into the posterior aspect of the inferior vena cava. The left suprarenal vein is usually longer than the right suprarenal vein and drains into the left renal vein. There is considerable variation in the arrangement of the suprarenal veins.

INFERIOR PHRENIC VEINS. These veins drain blood from the inferior or abdominal surface of the diaphragm. The left inferior phrenic vein usually joins the left suprarenal vein, but the right inferior phrenic vein generally drains directly into the inferior vena cava.

HEPATIC VEINS. The **hepatic veins** drain blood from the liver and then empty into the inferior vena cava. The **central veins** of the liver lobules collect blood from the **intralobular venous sinusoids.** The central veins merge, eventually forming the hepatic veins, which exit from the posterior surface of the liver and empty immediately into the inferior vena cava. Sometimes the right hepatic vein passes through the caval hiatus before entering the inferior vena cava.

HEPATIC PORTAL SYSTEM

A hepatic portal system of veins carries blood from the digestive system to the liver before the blood is returned to the inferior vena cava. Blood from the inferior mesenteric vein, the superior mesenteric vein, and the splenic vein enters the hepatic portal vein as illustrated in Figure 3–5. In the liver, the hepatic portal vein branches until it ends in small capillary-like spaces, called sinusoids, within the liver lobule. From the sinusoids, the blood enters the numerous central veins, which merge to form the hepatic veins.

INFERIOR MESENTERIC VEIN. The inferior mesenteric vein drains blood from the distal portions of the gastrointestinal tract, namely the descending colon, the sigmoid colon, and the rectum. The vessel begins at the rectum and ascends to the left of the inferior mesenteric artery. As it courses upward, it is retroperitoneal and anterior to the left psoas muscle. The inferior mesenteric vein usually empties into the splenic vein posterior to the body of the pancreas.

SUPERIOR MESENTERIC VEIN. The superior mesenteric vein collects blood from the small intestine, the cecum, the ascending colon, and the transverse colon. It begins in the right iliac fossa and ascends to the right of the superior mesenteric artery. Both the superior mesenteric artery and the superior mesenteric vein, along with their numerous branches, are enclosed within the layers of the mesentery. The superior mesenteric vein terminates behind the neck of the pancreas, where it joins the splenic vein, forming the hepatic portal vein.

SPLENIC VEIN. Four or five small vessels emerge from the hilum of the spleen and join, forming a single splenic vein. As it courses to the right, inferior to the splenic artery and posterior to the body of the pancreas, the splenic vein receives numerous tributaries from the pancreas. The splenic vein terminates behind the neck of the pancreas, where its junction with the superior mesenteric vein forms the hepatic portal vein.

There are some important vascular relationships to remember in this region. As the splenic vein courses to the right, it passes anteriorly to the superior mesenteric artery near this artery's origin from the aorta. It was mentioned earlier that at this level the superior mesenteric artery is separated from the aorta by the left renal vein. Sections at this level will then show, in order, from anterior to posterior, the **splenic vein,** the **superior mesenteric artery,** the **left renal vein,** and the **aorta.**

HEPATIC PORTAL VEIN. The hepatic portal vein is formed behind the neck of the pancreas at the L-2 vertebral level by the union of the **superior mesenteric** and the **splenic veins.** From this point, it ascends obliquely to the right, posterior to the duodenum and anterior to the inferior vena cava. It is usually 7 to 8 cm long and is

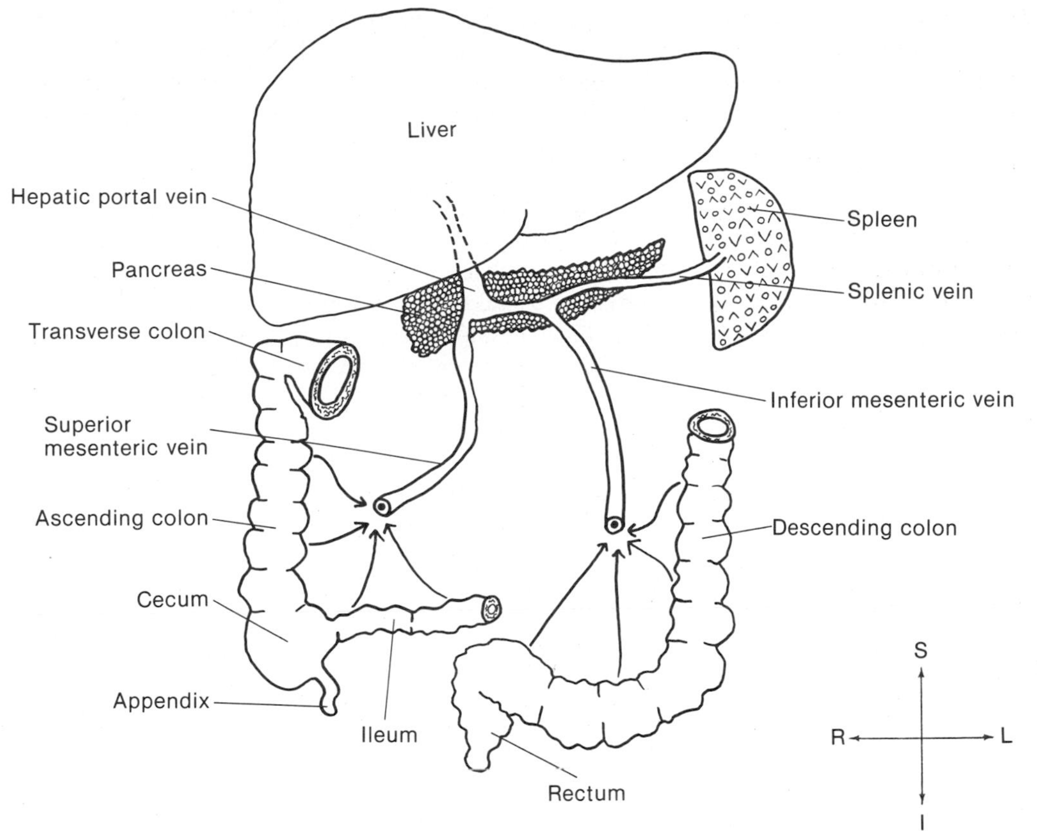

FIGURE 3-5. Hepatic portal circulation.

enclosed, along with the **hepatic artery** and the **bile duct,** in the free border of the lesser omentum. At the porta of the liver, the hepatic portal vein divides into right and left branches and enters the substance of the liver. Since the left lobe of the liver is more anterior than the right lobe, the left branch of the portal is more anteriorly located than the right one. In the lesser omentum, the portal vein is posterior to both the hepatic artery and the bile duct, and the bile duct is to the right of the hepatic artery. The portal vein is typically surrounded by connective tissue, which makes it quite echogenic and easy to identify.

Viscera of the Abdomen

LIVER

The **liver** is the largest organ in the abdominal cavity. The large right lobe occupies the right hypochondriac region and fills the right paravertebral groove. From the right hypochondriac region, the liver extends across the epigastric region into the left hypochondriac region. As the liver mass extends across the midline, it is moved forward by the longitudinal ridge line; thus, the smaller left lobe is more anteriorly positioned.

SURFACES AND ASPECTS. The anterior, superior, and posterior aspects of the liver are related to the diaphragm and follow its configuration. The convex superior aspect usually bulges upward more on the right side than it does on the left. The inferior or visceral surface is flatter than that of the superior aspect, but the inferior surface has depressions where it is in contact with the abdominal viscera. The inferior surface is not in a horizontal plane. Instead, it is situated at a 45 degree angle to both the longitudinal and the horizontal planes; thus, the more inferior portions are more anteriorly positioned and the more superior portions are more posteriorly located. The visceral surface of the right lobe is related to the right kidney, the right colic flexure, the gallbladder, and the duodenum. On the left, there is a large depression for the stomach and a smaller depression for the colon.

PERITONEAL RELATIONSHIPS. Most of the liver is enclosed in visceral peritoneum. An exception to this is a triangular space on the posterior surface, called the **bare area,** that is devoid of peritoneum and is in direct contact with the diaphragm. There is a deep groove in the bare area for the inferior vena cava. As the peritoneum around the bare area is reflected onto the diaphragm, it forms the **cardinal ligaments,** which represent the margins of the bare area.

There is a small space between the visceral peritoneum of the liver and the parietal peritoneum on the diaphragm, both anteriorly and posteriorly. Anteriorly, this is part of the **greater sac** and is called the **subphrenic recess.** Posteriorly, to the left of midline, the space is part of the **omental bursa,** or lesser sac. This is illustrated in the diagram of the peritoneum in Figure 3-3.

On the right side, the peritoneum is reflected from the liver over the surface of the right kidney, forming a **hepa-**

torenal recess. This recess is clinically significant because this represents the most posterior portion of the peritoneal cavity. Fluids and pus tend to accumulate in the hepatorenal recess when the patient is supine.

The lesser omentum, which extends from the lesser curvature of the stomach and the first part of the duodenum to the liver, is continuous with the visceral peritoneum of the liver. The left portion, between the stomach and the liver, is called the **gastrohepatic** (hepatogastric) **ligament.** The right portion, between the duodenum and the liver, is called the **duodenohepatic** (hepatoduodenal) **ligament.** To state this another way, the gastrohepatic and hepatoduodenal ligaments make up the lesser omentum. The portal vein, hepatic artery, and bile duct are enclosed with the right margin of the lesser omentum.

The **falciform ligament** is a thin, anteroposterior fold of peritoneum that is attached to the convex surface of the liver, to the diaphragm, and to the anterior abdominal wall down to the level of the umbilicus. The falciform ligament marks the superficial division between the right and the left lobes on the anterior surface.

CONFIGURATION OF THE VISCERAL SURFACE. An examination of the inferior, visceral surface of the liver reveals—with a little imagination—an H-shaped configuration, which separates the organ into four distinct regions. In addition to the right and left lobes, there are a caudate lobe and a quadrate lobe. The left arm of the H is formed inferiorly by the **ligamentum teres** and superiorly by the **ligamentum venosum.** The ligamentum teres represents the obliterated umbilical vein, which carries blood from the placenta to the liver in the fetal circulation. The ligamentum venosum is the remnant of the ductus venosus, which carries blood directly from the umbilical vein to the inferior vena cava and bypasses the liver in the fetal circulation. The **left lobe** is to the left of the line that is formed by the ligamentum teres and the ligamentum venosum. The right line of the H is formed inferiorly by the gallbladder and superiorly by the inferior vena cava; the **right lobe** is to the right of this line.

The **caudate** and **quadrate lobes** are located between the two vertical lines of the H and are separated by its crossbar, which is formed by the porta hepatis and includes the portal vein, the hepatic artery, and the hepatic duct. The caudate lobe is superior and posterior to the crossbar (porta hepatis), and the quadrate lobe is inferior and anterior to it (Fig. 3–6). The ligamentum venosum separates the left lobe from the caudate lobe, which is separated from the right lobe by the inferior vena cava. Inferiorly, the ligamentum teres separates the left lobe from the quadrate lobe, and the gallbladder separates the quadrate lobe from the right lobe. In the central region, the porta hepatis separates the caudate lobe from the quadrate lobe.

A superficial anatomic examination indicates that the caudate and quadrate lobes are a part of the right lobe. The falciform ligament, anteriorly, and the ligamenta teres and venosum, posteriorly, mark the surface division between the right and the left lobes. An investigation of the internal morphologic character and the vasculature of these lobes reveals that, functionally, the caudate and the quadrate lobes are more closely related to the left lobe than to the right.

BLOOD SUPPLY. Blood is brought to the liver by the **hepatic artery** and the **hepatic portal vein.** Approximately 70 percent of the blood supply to the liver is nutrient-rich venous blood, which is brought to the liver by the hepatic portal vein from the digestive system. The

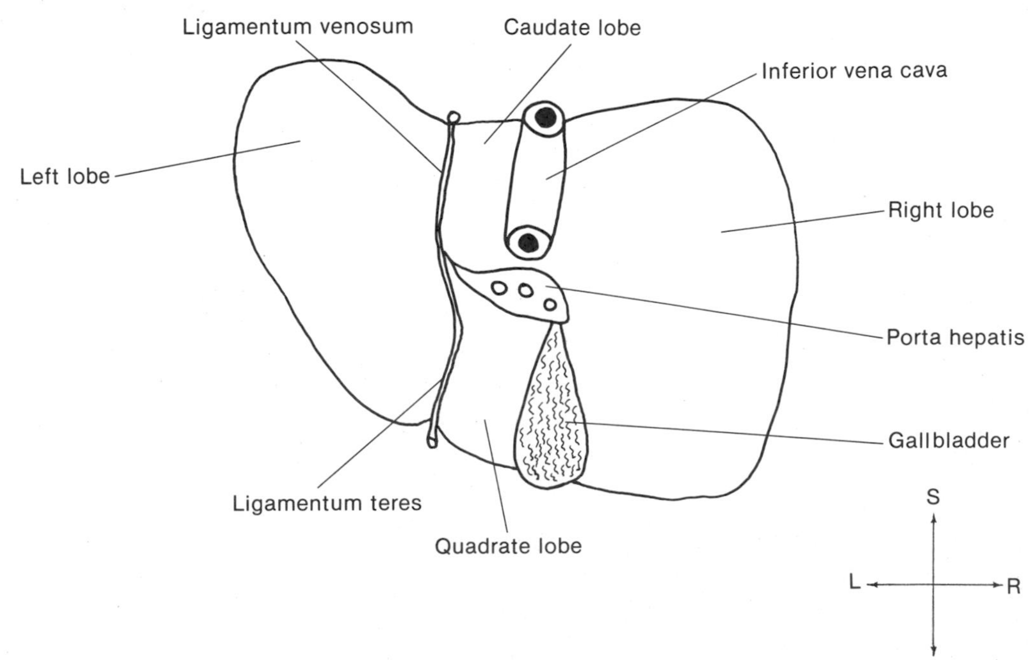

FIGURE 3–6. Configuration of the visceral surface of the liver.

remaining 30 percent is oxygenated blood that is supplied by the hepatic artery. The blood passes through the **sinusoids of the liver** to enter the **central vein** of a liver lobule. The central veins converge to form **hepatic veins,** which transport the blood to the inferior vena cava.

GALLBLADDER

The **gallbladder** is a saclike reservoir for bile. It lies along the right edge of the quadrate lobe of the liver. For descriptive purposes, it may be divided into the fundus, the body, and the neck. The fundus is the inferior extremity of the sac, which usually protrudes from the inferior margin of the liver. It is in contact with the anterior abdominal wall, the duodenum, and the transverse colon. The body extends upward from the fundus and is in direct contact with the visceral surface of the liver. It is also related to the duodenum and the transverse colon. The neck is a narrow and constricted portion directed toward the porta of the liver. It is continuous with the cystic duct, which joins the hepatic duct to form the common bile duct. A hormonal feedback system controls the flow of bile from the common bile duct into the duodenum.

ESOPHAGUS

Most of the **esophagus** lies in the thoracic cavity; only the terminal portion is located in the abdomen. While in the thoracic cavity, the esophagus lies anterior to the vertebral column and to the right of the descending aorta. At the T-7 vertebral level, it deviates to the left and passes anterior to the aorta on its way to the stomach. After penetrating the diaphragm at the T-10 vertebral level, the abdominal portion of the esophagus forms a groove in the left portion of the liver and enters the stomach at the **cardiac orifice.** The right margin of the esophagus is continuous with the lesser curvature of the stomach. A **cardiac notch** separates the left margin of the esophagus from the fundus of the stomach. At the cardiac orifice the **lower esophageal sphincter** slows the passage of food from the esophagus into the stomach and prevents reflux of gastric contents into the esophagus.

STOMACH

The **stomach** acts to pulverize food and to mix it with gastric juice. It is located in the upper left quadrant of the abdomen. Under normal conditions, the stomach cannot be palpated, because the walls are rather flat and flabby. Since the walls are distensible, the size and the shape of the stomach vary with circumstances.

CURVATURES. The stomach has two curvatures. The concave **lesser curvature** is directed to the right and superiorly and is continuous with the right margin of the esophagus. The convex **greater curvature** forms the left and inferior margins of the stomach.

REGIONS. For descriptive purposes, the stomach is divided into four regions. The limited **cardiac portion** lies adjacent to the cardiac orifice, where the esophagus enters. The **fundus** is the rounded portion above the gastroesophageal junction. The major portion is the **body** of the stomach, which is between the fundus and the pyloric regions. The distal portion, which empties into the small intestine, is the **pyloric region.** The junction of the body and pyloric region is marked by the **angular notch,** or the **incisura angularis;** this is a notch on the lesser curvature. The pyloric region is divided into a wider portion, the **pyloric antrum,** and a narrower **pyloric canal.** The sphincter region at the orifice between the stomach and duodenum is the **pylorus.**

PERITONEAL RELATIONSHIPS. The stomach is completely covered by peritoneum and is attached to other organs by peritoneal folds and ligaments. The **gastrohepatic portion** of the lesser omentum radiates from the lesser curvature to attach to the liver. From the inferior and left greater curvature, the **greater omentum** falls like an apron over the intestines. A portion of the greater omentum is attached to the transverse colon and is called the **gastrocolic ligament.** The anterior surface of the stomach is related to the diaphragm, the left lobe of the liver, and the anterior abdominal wall. The posterior surface is related to many abdominal viscera, including the spleen, the left kidney and suprarenal gland, the pancreas, and the transverse colon.

BLOOD SUPPLY. All three branches of the celiac trunk contribute to the vasculature of the stomach. The **left gastric artery** is the smallest branch of the celiac trunk. It courses to the left along the lesser curvature and gives off branches along the way to supply that region of the stomach. The **right gastric** and **right gastroepiploic arteries** are branches of the common hepatic artery. The splenic artery contributes the **left gastroepiploic artery** and numerous short **gastric arteries.** The veins of the stomach follow the same pattern as for the arteries, both in name and in distribution. The venous blood from the stomach, like blood from other parts of the digestive tract, is taken to the liver by way of the hepatic portal vein, but there is considerable variation in the way gastric veins enter the portal system. From the liver, the blood enters the hepatic veins, which carry it to the inferior vena cava.

SMALL INTESTINE

Most of the digestion of food and absorption of nutrients take place in the **small intestine.** This long, convoluted tube is usually 6 or 7 m in length and is divided into the **duodenum,** the **jejunum,** and the **ileum.** The jejunum and ileum are enclosed in peritoneum and are suspended from the posterior abdominal wall by a fan-shaped **dorsal mesentery.** The duodenum is retroperitoneal except for a small portion near the pylorus of the stomach.

DUODENUM. The first part of the small intestine, beginning at the pyloric valve, is the **duodenum.** About 25 cm long, this portion presents a C-shaped pattern as it curves around the head of the pancreas to become continuous with the jejunum. The duodenum is the most fixed part of the small intestine. For descriptive purposes, the duodenum may be divided into four parts (Fig. 3-7).

The **first** or **superior part** is horizontal, beginning at the pylorus and extending to the gallbladder. Anteriorly, the first part is related to the quadrate lobe of the liver and the gallbladder. The inferior vena cava, the portal vein, the bile duct, and the gastroduodenal artery course along the posterior surface. The inferior border of the first (superior horizontal) part of the duodenum courses along the upper margin of the pancreatic head.

Posterior to the gallbladder, the duodenum turns sharply downward to become the **second** or **descending part.** It is against the posterior abdominal wall, in the paravertebral groove, along the right side of the first three lumbar vertebrae. As it descends, the second part is to the right of and parallel to the inferior vena cava. As the second part descends, it passes anterior to the hilum of the right kidney and posterior to the transverse colon. The second part, then, is related anteriorly to the right lobe of the liver and the transverse colon, posteriorly to the right kidney, the renal vessels, the ureter, and the psoas muscle, and medially to the bile duct and the pancreas. The pancreatic and bile ducts enter the second part.

After descending to the lower margin of the L-3, or the upper margin of the L-4, the duodenum again makes a sharp turn, this time to the left, to become the **third** or **inferior horizontal part.** As this part extends across the midline, from right to left, it is moved forward by the longitudinal ridge. As it courses horizontally, the third part passes anterior to the inferior vena cava, the aorta, the right ureter, the right gonadal vessels, and the psoas muscles, but it is posterior to the superior mesenteric vessels. Superiorly, this part is related to the pancreas and inferiorly to the coils of jejunum.

The **ascending fourth part** ascends anterior and slightly to the left of the aorta. In addition to the aorta, the fourth part is related, on its posterior aspect, to the left psoas muscle and to the left renal and left gonadal vessels. The left kidney and the left ureter are to the left of the fourth part, and the pancreas is to its right. Superiorly, the fourth part is related to the body of the pancreas. The fourth or ascending part ascends only to the second lumbar vertebra and then ends abruptly in the duodenojejunal flexure. This flexure is directed anteriorly and is attached to the posterior abdominal wall by a fibromuscular band called the suspensory muscle of the duodenum (or the ligament of Treitz).

JEJUNUM. The second division of the small intestine is the **jejunum.** This is a highly coiled tube that begins at the duodenojejunal flexure and continues until it imperceptibly changes into the ileum. Most of the jejunum is located in the umbilical region of the abdomen.

ILEUM. Even though there are no obvious structural changes at the junction between the jejunum and the **ileum,** there are changes in the tissues that form the wall. There is a lack of circular folds in the mucosa of the ileum. Accumulations of lymphoid tissue called Peyer's patches are also evident in the wall of the ileum. Most of the ileum lies in the hypogastric region; only the distal parts are in the pelvis. The ileum terminates in the right iliac region by opening into the cecum through the ileocecal valve.

Both the ileum and the jejunum are enclosed in peritoneum and are suspended from the posterior abdominal wall by fanshaped folds of mesentery. The duodenojejunal flexure and the ileocecal valve are rather fixed points; however, between these two points, the jejunum and the ileum are highly mobile and fill any available space in the abdominopelvic cavity. Because of their mobility, the jejunum and ileum are the regions frequently involved in hernias.

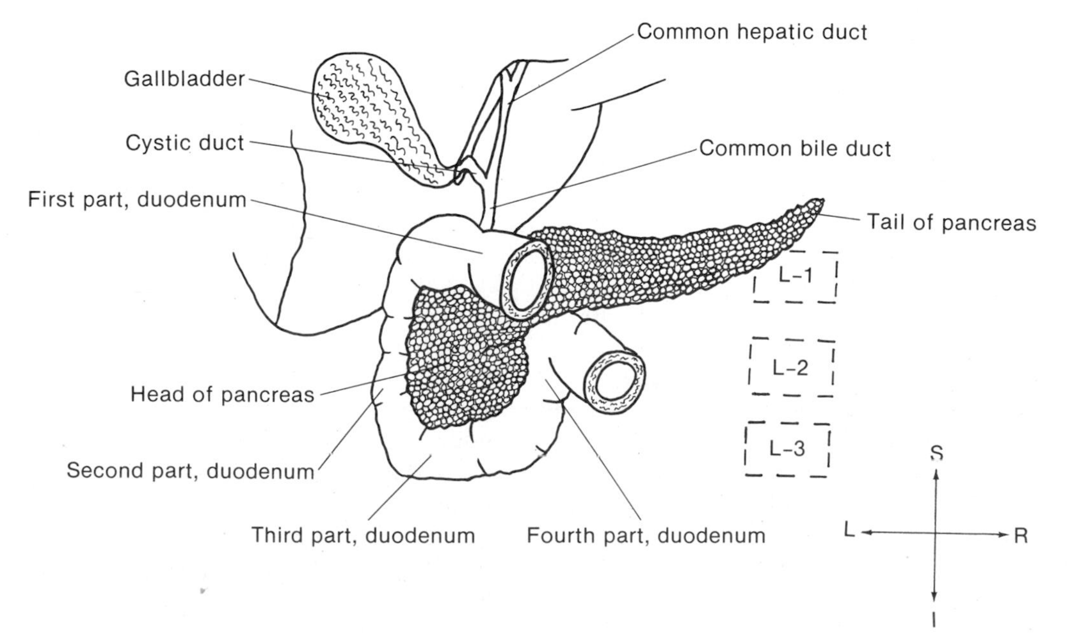

FIGURE 3-7. Divisions of the duodenum and its relationships with other organs and organ systems.

LARGE INTESTINE

The **large intestine** extends from the ileocecal valve to the anus, a length of approximately 1.5 m. It consists of the cecum, the colon, the rectum, and the anal canal. The large intestine forms an arch for the loops of the small intestine.

CECUM. Located in the right iliac region of the abdominopelvic cavity, the **cecum** is a blind saclike pouch that extends 5 to 7 cm below the ileocecal valve. The vermiform appendix is a blind tubular projection from the cecum. Although the position of the appendix varies considerably, it is commonly inferior and posterior to the cecum. The cecum is retroperitoneal and has no mesentery.

ASCENDING COLON. From the cecum, the large intestine passes superiorly on the right side of the abdominal cavity to the liver. When it reaches the visceral surface of the liver, the **ascending colon** bends sharply to the left in the **right colic (hepatic) flexure.** The ascending colon lies on the posterior abdominal wall, where it is separated from the muscles by the right kidney. It is retroperitoneal and has no mesentery.

TRANSVERSE COLON. The **transverse colon** extends across the abdomen from the right colic (hepatic) flexure to the **left colic (splenic) flexure.** It is the longest part of the large intestine. Suspended by a mesentery, called the **transverse mesocolon,** the transverse colon is the most movable part of the large intestine. Between the two flexures, the position of the transverse colon is variable; however, it typically forms a loop that is directed inferiorly. This loop may be on the transpyloric plane, or it may extend down to the pelvic brim. As the transverse colon extends across the abdominal cavity, it becomes more superior and more posterior; thus, the **left colic (splenic) flexure** is the most superior part of the large intestine and is also relatively posterior in position. At the left colic (splenic) flexure, the large intestine turns sharply downward, continuing as the descending colon.

DESCENDING COLON. The **descending colon** extends from the left colic (splenic) flexure to the iliac crest. As it descends, it passes along the lateral border of the left kidney. The descending colon is of a smaller diameter than either the ascending or the transverse colon. It is retroperitoneal and has no mesentery. At the iliac crest, the descending colon becomes continuous with the sigmoid colon.

SIGMOID COLON. The **sigmoid colon** begins at the pelvic brim, crosses the sacrum, and then curves to the midline at the third sacral segment. It is enclosed in peritoneum and has a long mesentery, the **sigmoid mesocolon.** Because of its mesentery, the sigmoid colon is quite movable. It is usually located in the pelvis, but it may extend upward into the abdomen.

RECTUM. The **rectum** extends from the third sacral segment to the pelvic diaphragm just below the tip of the coccyx, a length of about 12 cm. At this point, the rectum turns dorsally and becomes the anal canal. The rectum is partially covered with peritoneum but it has no mesentery; therefore, it is considered to be retroperitoneal.

ANAL CANAL. The **anal canal** is 2.5 to 4.0 cm in length and represents the final portion of the intestinal tract. As the canal penetrates the pelvic diaphragm to terminate at the anus in the perineum, it is supported by the **levator ani muscles.**

SPLEEN

The **spleen** is a highly vascular organ composed of lymphoid tissue in the left hypochondriac region of the abdomen. It is located posterior to the stomach and is protected by the ninth, tenth, and eleventh ribs. In addition to the stomach, the visceral surface of the spleen is related to the left kidney and the transverse colon. The hilum of the spleen is closely related to the tail of the pancreas. With the exception of the hilum, the spleen is enclosed in peritoneum, and it is held in place by two peritoneal ligaments. The gastrosplenic ligament attaches the hilum to the greater curvature of the stomach, and the splenorenal (lienorenal) ligament attaches the hilum to the left kidney. The spleen varies considerably in size and shape, depending on the distention of the stomach and the colon. Vascular needs of the spleen are supplied by the splenic artery and vein, which penetrate the organ at the hilum.

PANCREAS

The **pancreas** is an elongated, soft, pliable gland that has both exocrine and endocrine functions. Since it is covered with peritoneum only on its anterior surface, it is considered to be retroperitoneal. The pancreas extends across the posterior surface of the abdomen from the duodenum to the spleen. For descriptive purposes it is divided into the head, the neck, the body, and the tail.

The **head** is the broad, flattened, right extremity of the pancreas that lies within the curve of the duodenum. A small **uncinate process** projects inferiorly and medially, posterior to the duodenum; thus, it rests on the inferior vena cava and the left renal vein. The superior mesenteric vessels are anterior to the uncinate process. The **neck** is a constricted portion to the left of the head. The splenic and superior mesenteric veins join to form the hepatic portal vein posterior to the neck of the pancreas. The neck merges imperceptibly with the **body.** The body of the pancreas extends to the left and superiorly across the aorta, and its posterior surface is related to the following structures: the superior mesenteric artery, the splenic artery and vein, the left suprarenal (adrenal) gland, and the left kidney with its vessels. It is separated from the stomach by the omental bursa. The **tail** of the pancreas is the left extremity, which is close to the hilum of the spleen. The tail is the most superior and posterior portion of the pancreas.

The main **pancreatic duct,** or the **duct of Wirsung,** begins in the tail of the pancreas, runs through the substance of the gland, and then empties into the descending second part of the duodenum. Blood is supplied via the splenic artery, which is a branch of the celiac trunk. The splenic vein carries blood from the spleen to the hepatic portal circulation; then hepatic veins return it to the inferior vena cava.

KIDNEYS

LOCATION AND POSITION. The **kidneys** are paired, red-brown, bean-shaped organs just below the diaphragm in the superior part of the paravertebral grooves. They are behind the peritoneum (retroperitoneal) along the posterior body wall, against the psoas muscle, and adjacent to the vertebral column. The superior part of each kidney is protected by the ribs. Each adult kidney is approximately 12 cm long and 6 cm wide and extends from the level of the twelfth thoracic vertebra to the third lumbar vertebra. This level changes during respiratory movements and with variations in posture. Because it is pushed down by the liver, the right kidney is generally slightly lower than the left. A cushion of **perirenal fat** surrounds each kidney, and the ribs, muscles, and intestines serve as its protective barriers. Occasionally, a kidney slips from its normal position and is no longer held securely in place by adjacent organs or its covering of fat. This condition is known as a floating kidney, or ptosis of the kidney, and it may cause a kinking, or twisting, of the ureter with a subsequent obstruction of urine flow. The kidneys also become more susceptible to physical trauma when they drop below the rib cage.

HILUM. The region where the blood vessels enter and leave and where the ureter exits to descend to the bladder is called the **hilum of the kidney.** The hilum is generally at the level of the transpyloric plane, with the plane intersecting the midhilar region of the left kidney and upper hilar region of the right kidney.

RELATIONSHIPS. On its posterior or dorsal surface, the kidney is related superiorly to the diaphragm and inferiorly to the quadratus lumborum and psoas muscles. The anteriomedial surface of the superior pole of each kidney is covered by a **suprarenal (adrenal) gland.**

The anterior or ventral relationships are different for the right and left kidneys. On the right, the kidney forms a renal impression on the visceral surface of the liver. The second or descending part of the duodenum descends across the hilar region, and the right colic (hepatic) flexure of the colon covers the inferior pole of the kidney. The anterior or ventral surface of the left kidney is related to the left suprarenal gland, the stomach, the pancreas, the spleen, the left colic (splenic) flexure, and the coils of the small intestine.

URETERS. The **ureters** are muscular ducts that transport urine from the kidneys to the urinary bladder. Originating at the renal pelvis, the ureters descend retroperitoneally along the psoas muscle. The right ureter is closely related to the inferior vena cava as the ureter descends. The abdominal portion of the ureter crosses the pelvic brim and the external iliac artery just distal to the bifurcation of the common iliac artery; it then continues as the pelvic ureter and enters the posterior surface of the urinary bladder. The ureter is vulnerable during pelvic and abdominal surgery because it may resemble a blood vessel and may accidentally be ligated.

SUPRARENAL GLAND

A **suprarenal (adrenal) gland** lies on each side of the vertebral column in close relation to the superior pole of the corresponding kidney. The suprarenal gland is separated from the kidney by fatty connective tissue. The right gland is somewhat pyramidal, whereas the left is more semilunar. On the right, the suprarenal gland is limited by the liver laterally, the inferior vena cava anteriorly, and the diaphragm medially. The left suprarenal gland is related to the left crus of the diaphragm medially and to the left kidney posteriorly and laterally. Anteriorly, it is related to the stomach and pancreas. Each gland has an abundant blood supply from the suprarenal arteries, which branch directly from the aorta. Branches of the renal and inferior phrenic arteries also supply the suprarenal glands.

SECTIONAL ANATOMY OF THE ABDOMEN

Transverse Sections

SECTION THROUGH THE UPPER ABDOMEN, LEVEL T-9

Transverse sections through the superior regions of the abdomen also intersect portions of the **heart** in the pericardial sac and **lungs** in the pleural cavity (Fig. 3–8). The hiatus in the **diaphragm** for the **inferior vena cava** is at vertebral level T-8; thus, in this region of the abdominal cavity, the inferior vena cava may be embedded in the substance of the **liver.** The **right** and **left hepatic veins** are short vessels that drain into the inferior vena cava near the caval hiatus in the diaphragm. Figure 3–8 illustrates the right hepatic vein as it drains into the inferior vena cava. The left hepatic vein is slightly anterior to the right hepatic vein and inferior vena cava, because the left lobe of the liver is more anteriorly positioned.

SECTION THROUGH THE T-10 LEVEL

The three most noticeable structures in transverse sections at level T-10 are the **liver,** the **stomach,** and the **spleen.** The relationships of these organs are illustrated in Figure 3–9.

The **right lobe** of the liver fills the right paravertebral groove. The **left lobe** is moved anteriorly as it crosses the

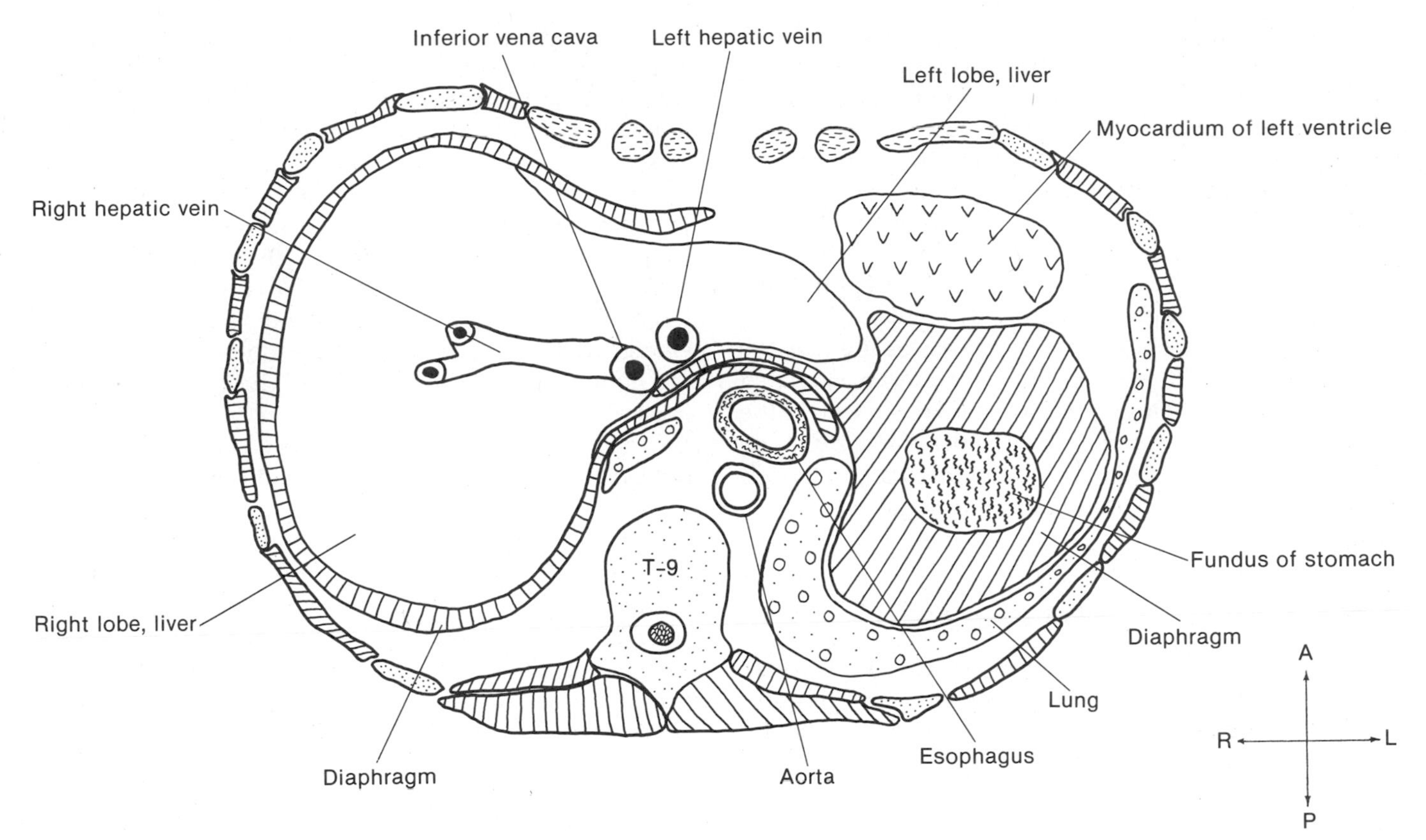

FIGURE 3-8. Transverse section through the upper abdomen, level T-9.

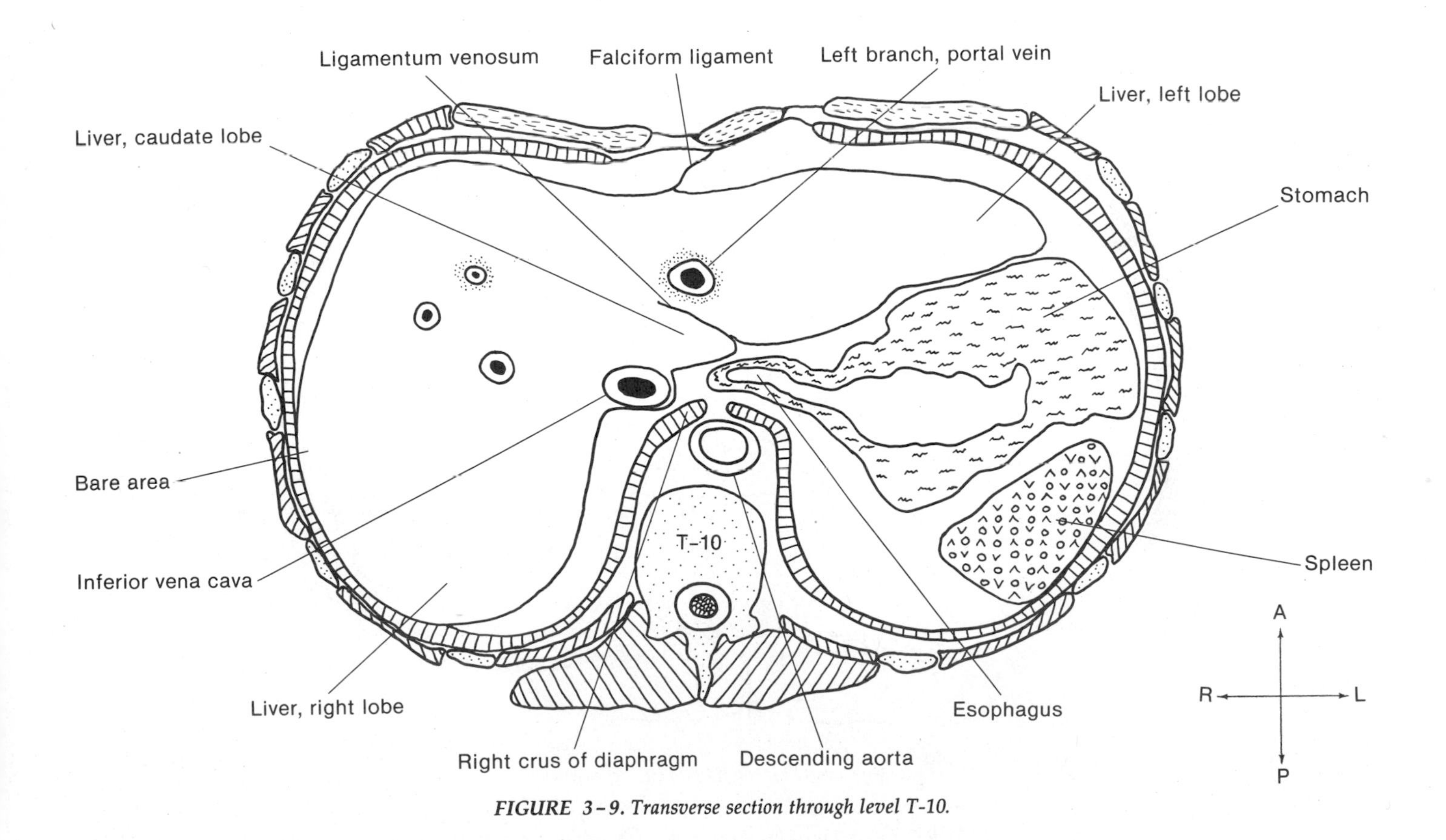

FIGURE 3-9. Transverse section through level T-10.

longitudinal ridge to the left side. The **falciform ligament** attaches the anterior surface of the liver to the anterior abdominal wall and separates the right and left lobes superficially. Posteriorly, the **bare area** of the liver is closely related to the diaphragm. The **inferior vena cava** is on the posterior surface, just right of the midline, near the vertebral body. The fissure for the **ligamentum venosum** separates the **caudate lobe** from the left lobe. Within the parenchyma of the liver, branches of the **portal vein** are surrounded by connective tissue.

Figure 3–9 also shows the **esophagus** as it enters the cardiac region of the **stomach.** The stomach is related anteriorly to the left lobe of the liver and posteriorly to the **spleen.** The lateral surface is next to the diaphragm. The spleen is against the posterior abdominal wall, in the left paravertebral groove, posterior to the stomach.

SECTION THROUGH THE PORTA HEPATIS

The relationships of the structures evident in transverse sections through the porta of the liver are of significance. The most obvious structure at the porta hepatis is the **hepatic portal vein.** The portal vein is formed behind the neck of the pancreas and then courses obliquely toward the liver. It is enclosed in the free margin of the lesser omentum. At the porta hepatis, the portal vein is posterior to the **hepatic arteries** and the **hepatic ducts.** The common hepatic artery, a branch of the celiac trunk, also ascends toward the liver in the right free margin of the lesser omentum. Near the porta, the common hepatic artery divides, branching to form the right and left hepatic arteries, which are anterior to the portal vein and are to the left of the hepatic ducts. Within the liver parenchyma, bile canaliculi merge, forming progressively larger hepatic ducts until a right and a left hepatic duct emerge from the porta. The hepatic ducts are anterior to the portal vein and to the right of the hepatic arteries. Posterior to the portal vein, a narrow section of the liver, the caudate process, connects the caudate lobe with the right lobe. These relationships are illustrated in Figure 3–10.

Transverse sections at the level of the porta hepatis may also show the **suprarenal (adrenal) gland.** The right suprarenal gland is somewhat pyramidal. It is limited anteriorly by the **inferior vena cava,** posteriorly by the kidney, laterally by the liver, and medially by the right crus of the diaphragm. Draped along the margin of the kidney, the gland on the left is thinner and more semilunar than that on the right. Anteriorly and laterally, it is related to the stomach and the pancreas. Medially, it is related to the left crus of the diaphragm.

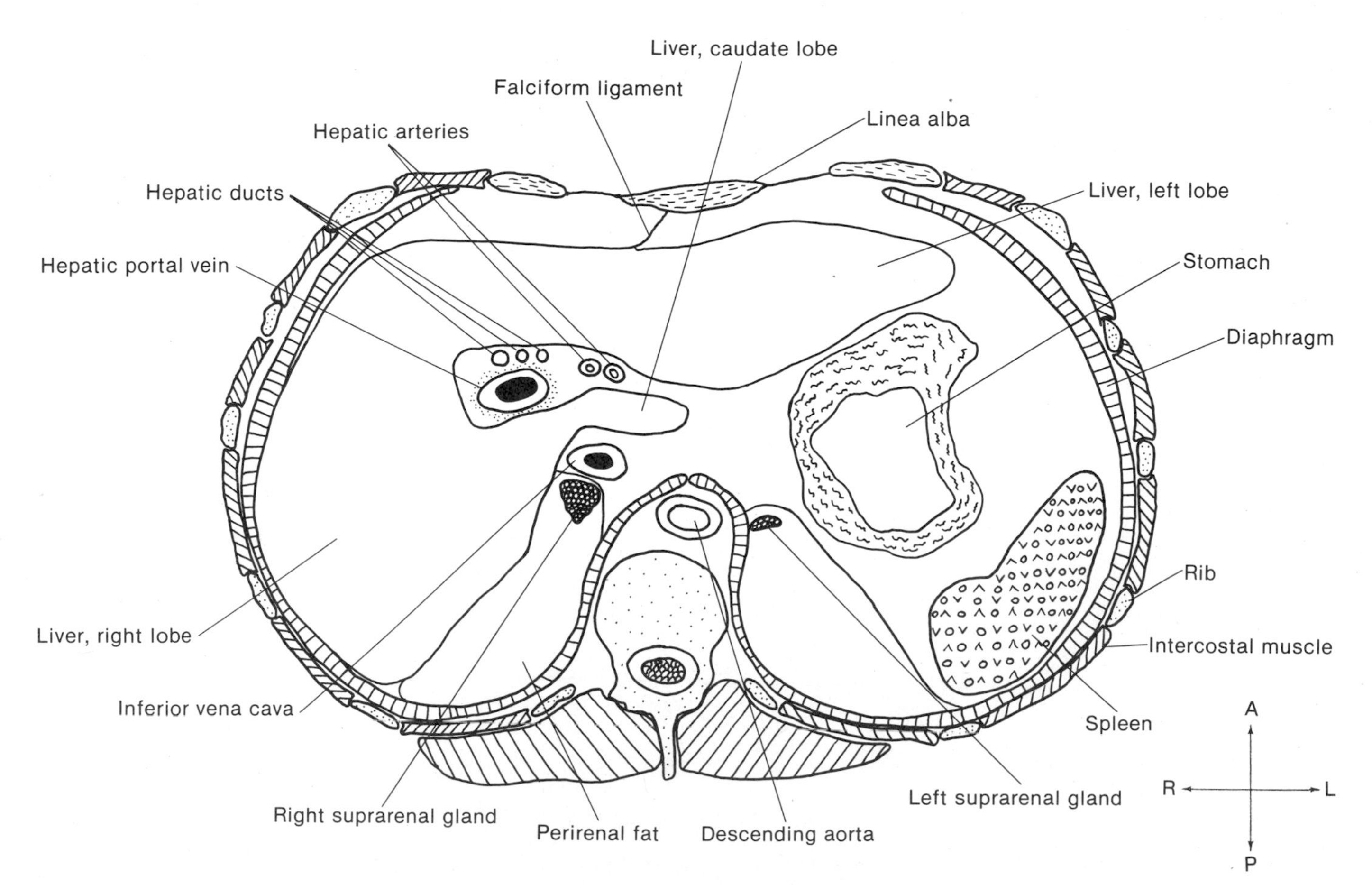

FIGURE 3–10. Transverse section through the porta hepatis.

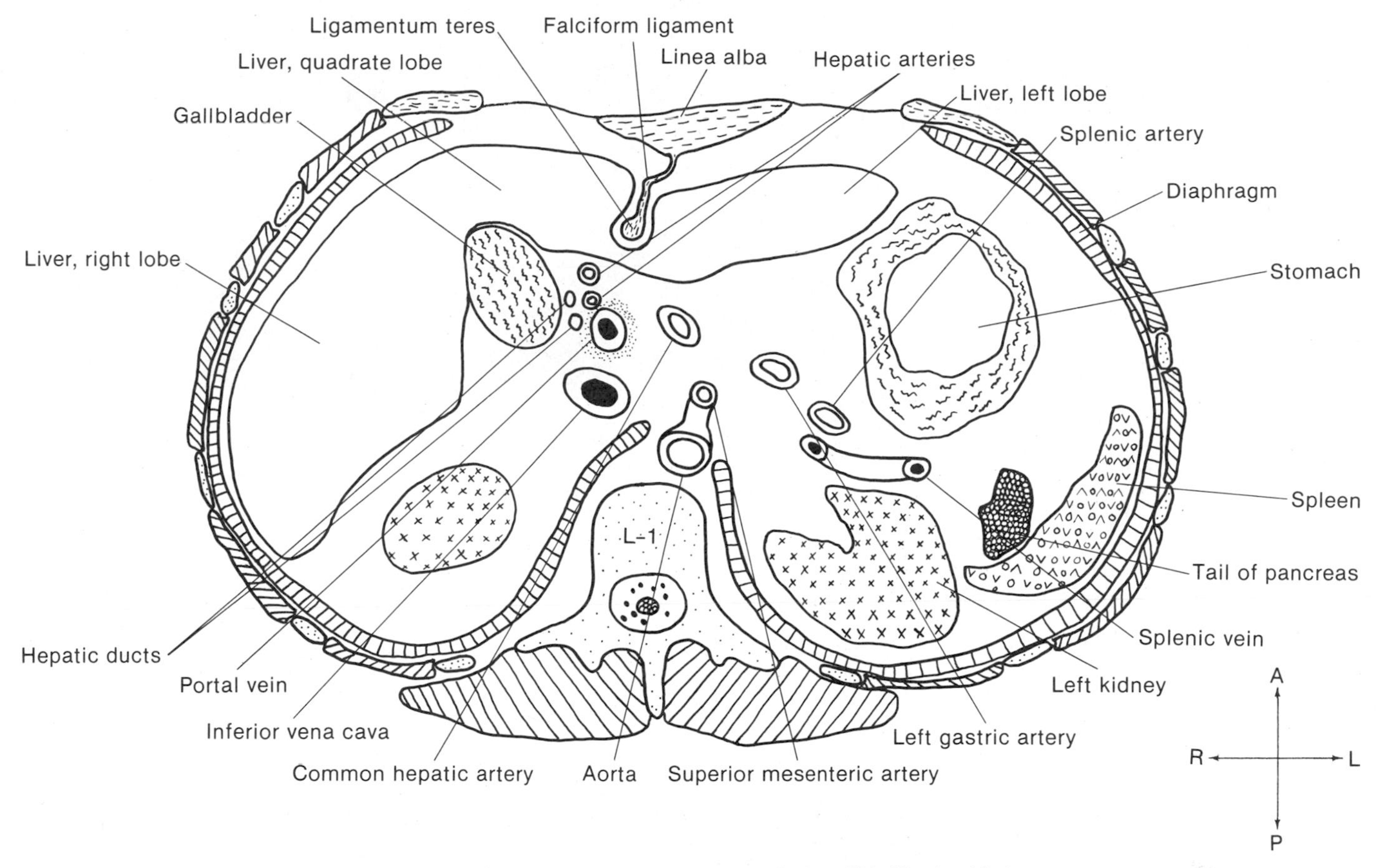

FIGURE 3-11. Transverse section through the gallbladder, level L-1.

SECTION THROUGH THE GALLBLADDER

Transverse sections inferior to the porta, through the **gallbladder,** show the anteriorly located **quadrate lobe** of the liver delineated by the gallbladder and the **ligamentum teres.** Relationships at this level are illustrated in Figure 3-11. Since the **celiac trunk** branches from the aorta just above the transpyloric line, which intersects the gallbladder, the trunk, or some of its three branches, is probably evident at this level. The **tail of the pancreas** frequently reaches this level, since the tail is the most superior portion of the pancreas.

SECTION THROUGH THE L-1/L-2 LEVEL

There are numerous interesting relationships to be noted at the level of the lower part of the first vertebra and at the upper part of the second lumbar vertebra. The **duodenum** is evident next to the **gallbladder.** The head of the **pancreas** is closely butted against the duodenum. The **common bile duct** forms a groove along the posterior surface of the pancreatic head, whereas the **gastroduodenal artery,** a branch of the common hepatic artery, courses along the anterior surface. The **portal vein** is formed posterior to the neck of the pancreas at the junction of the **splenic** and **superior mesenteric veins.** These relationships are illustrated in Figure 3-12.

A classic vascular arrangement evident at the L-1/L-2 level is also shown in Figure 3-12. The long **left renal vein** passes between the aorta and the superior mesenteric artery as the vein extends from the left kidney to the inferior vena cava. The left renal vein is anterior to the aorta and posterior to the superior mesenteric artery.

SECTION THROUGH THE HEAD OF THE PANCREAS

Figure 3-13 shows the head of the **pancreas** between the second and fourth parts of the **duodenum.** The third part appears in sections inferior to this level. The uncinate process of the pancreas projects from the pancreatic head and extends posterior to the superior mesenteric vessels. The **descending colon** appears in the left paravertebral groove. The larger **ascending colon,** in the right paravertebral groove, is more anterior than the descending colon. The kidneys are against the posterior abdominal wall and are related to the **psoas** and **quadratus lumborum** muscles. Notice also that the aorta and the inferior vena cava add height to the longitudinal ridge line.

Sagittal Sections

SECTION THROUGH THE ASCENDING COLON

The most lateral sections on the right side of the abdomen show the right lobe of the liver conforming to the shape of the diaphragm and protected by the rib cage. Sections taken approximately 2 cm more medially intersect the

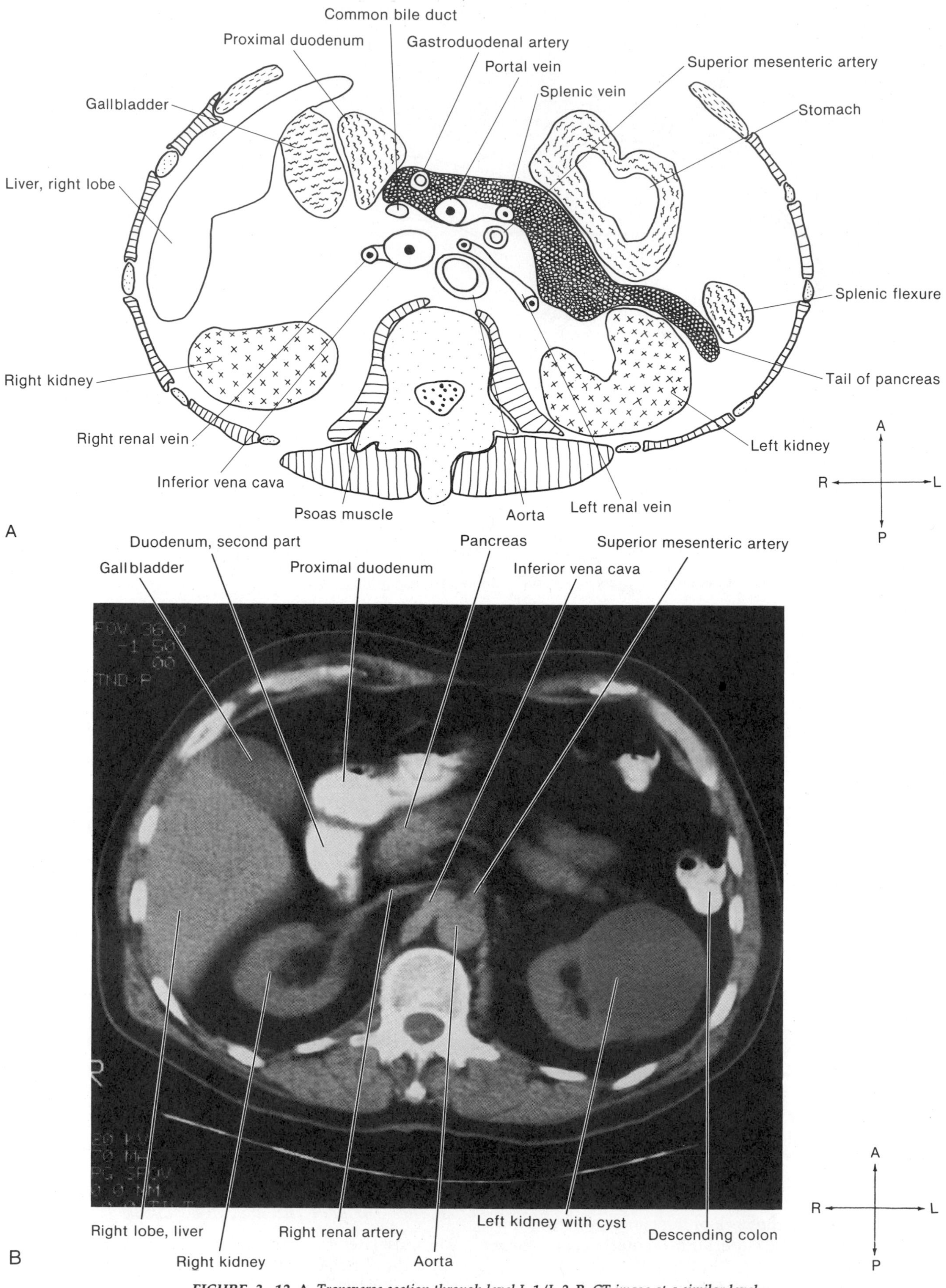

FIGURE 3-12. *A, Transverse section through level L-1/L-2. B, CT image at a similar level.*

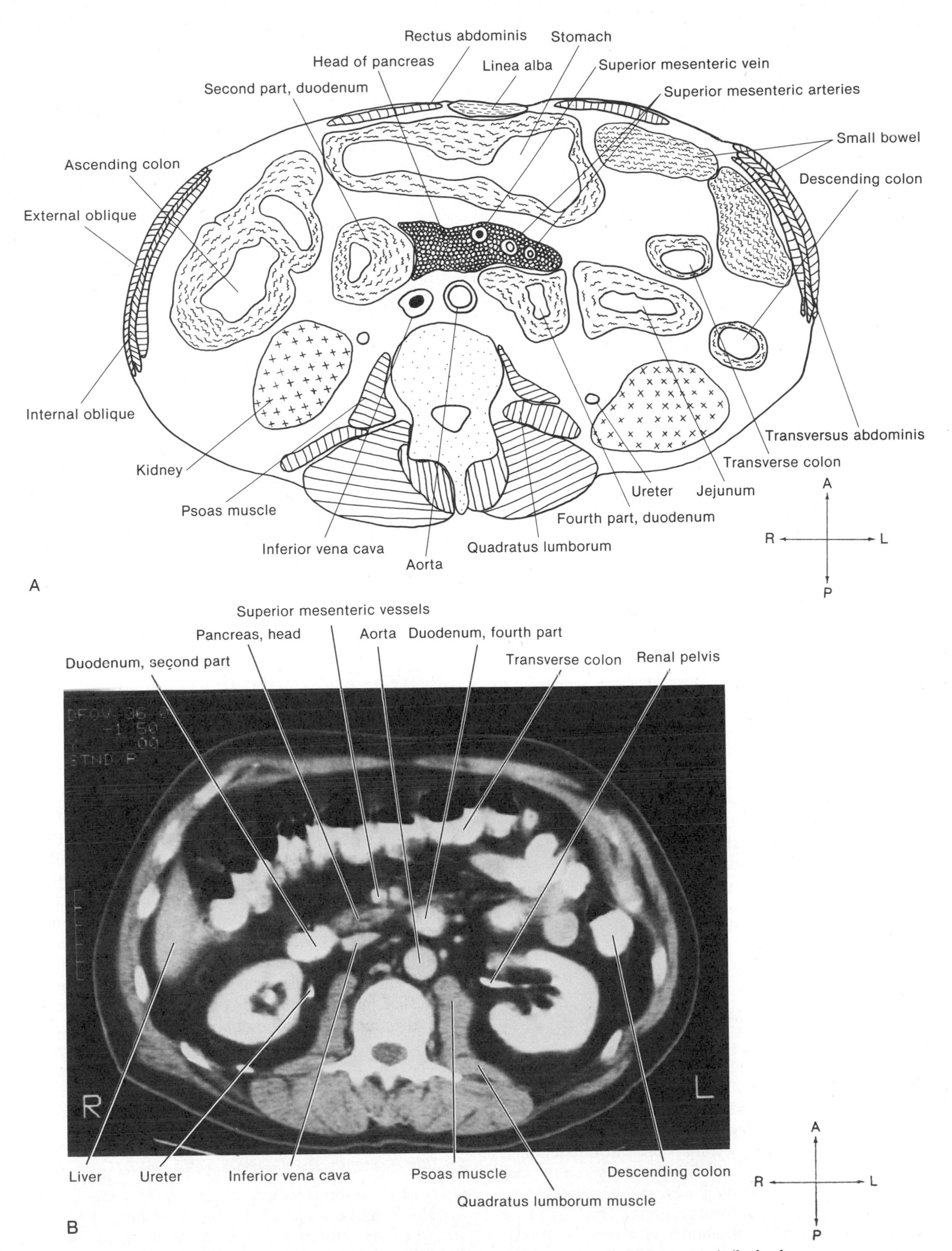

FIGURE 3-13. A, Transverse section through the head of the pancreas. B, CT image at a similar level.

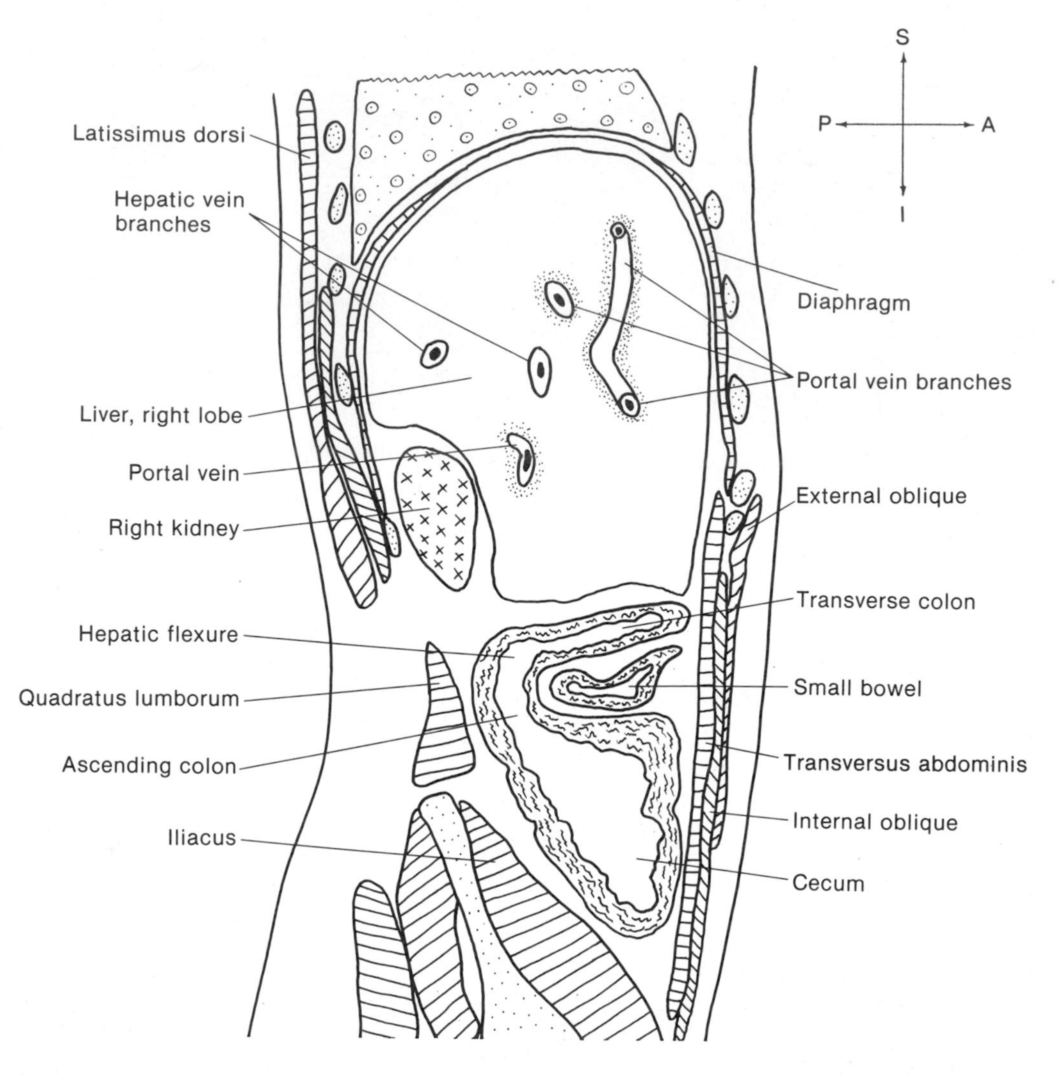

FIGURE 3–14. Sagittal section through the ascending colon.

kidney and the ascending colon. Figure 3–14 illustrates the organ relationships in this region. The predominant structure is the **right lobe** of the liver. The **right kidney** forms an indentation on the posterior visceral surface of the liver. The **ascending colon** extends from the **cecum** to the **hepatic flexure.** The **quadratus lumborum** and **iliacus** muscles form the posterior wall of the abdomen. Anteriorly the **external oblique, internal oblique,** and **transversus abdominis** muscles form the abdominal wall.

SECTION THROUGH THE GALLBLADDER

Sagittal sections approximately 6 or 7 cm to the right of the midline typically intersect the **gallbladder.** A large portion of the right lobe of the liver, with the right branch of the **portal vein,** is visible. Figure 3–15 illustrates the relationship of the **duodenum** to the **gallbladder** and the **kidney.** The descending second part of the duodenum is related to the gallbladder anteriorly and rests on the kidney posteriorly. This section illustrates the posterior position of the second part of the duodenum. The large **psoas muscle,** adjacent to the vertebral bodies, is evident in this region.

SECTION THROUGH THE HEAD OF THE PANCREAS

Since the head of the pancreas is surrounded by the four parts of the duodenum, the sagittal sections through this region are medial to the descending second portion of the duodenum but intersect the horizontal first and third parts. This is illustrated in Figure 3–16, which shows the first part of the **duodenum** superior to the **pancreatic head** and the third part inferior to it. Also in this region, the **right adrenal gland** is delineated by the **right kidney** and the **liver.** The **right renal artery** and **vein** course to the right of and posterior to the pancreatic head. Sections in this region show the anterior position of the **transverse colon** and the segmented nature of the **rectus abdominis muscle.** The porta of the liver separates the posterior **caudate lobe** from the anterior **quadrate lobe.**

SECTION THROUGH THE INFERIOR VENA CAVA

One or two centimeters to the right of the midline, the **inferior vena cava** ascends to penetrate the diaphragm and to empty into the right atrium of the heart. Figure 3–17 illustrates the relationships of the inferior vena

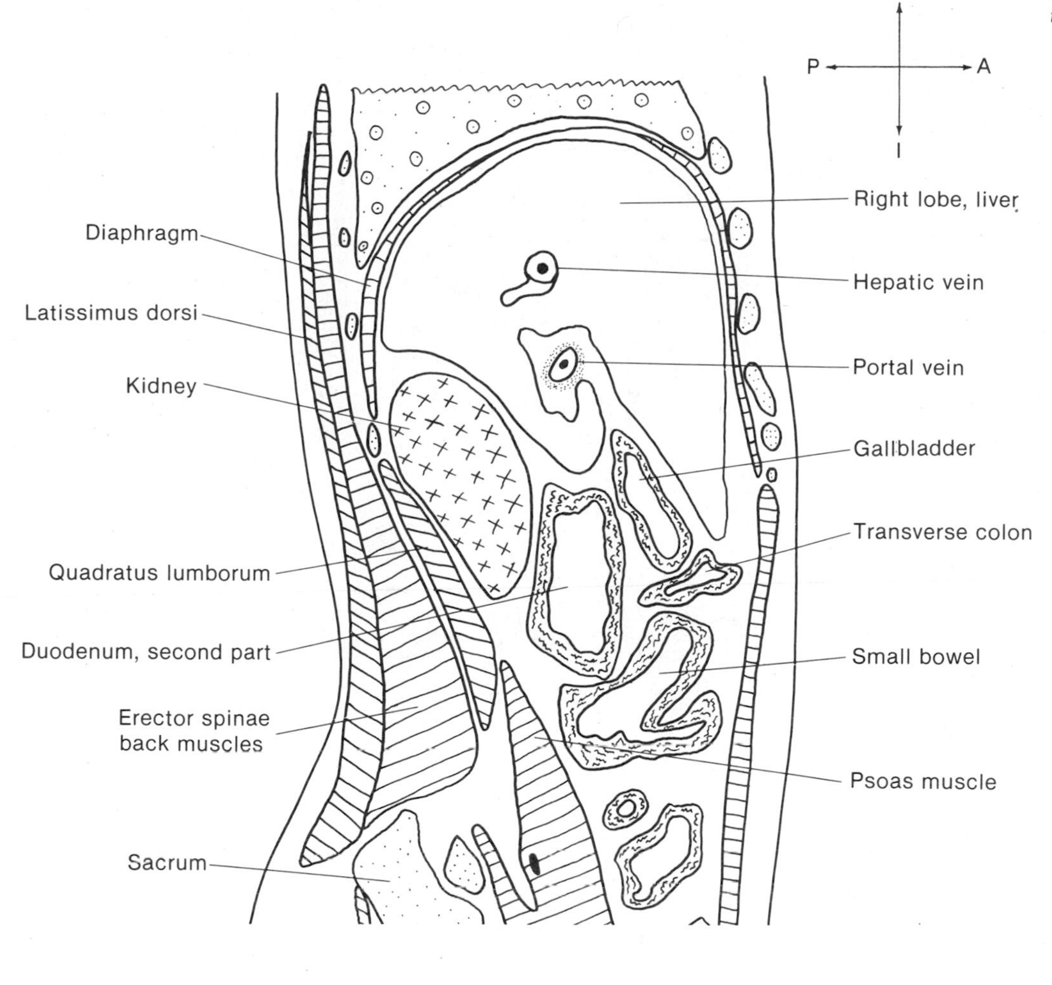

FIGURE 3-15. Sagittal section through the gallbladder.

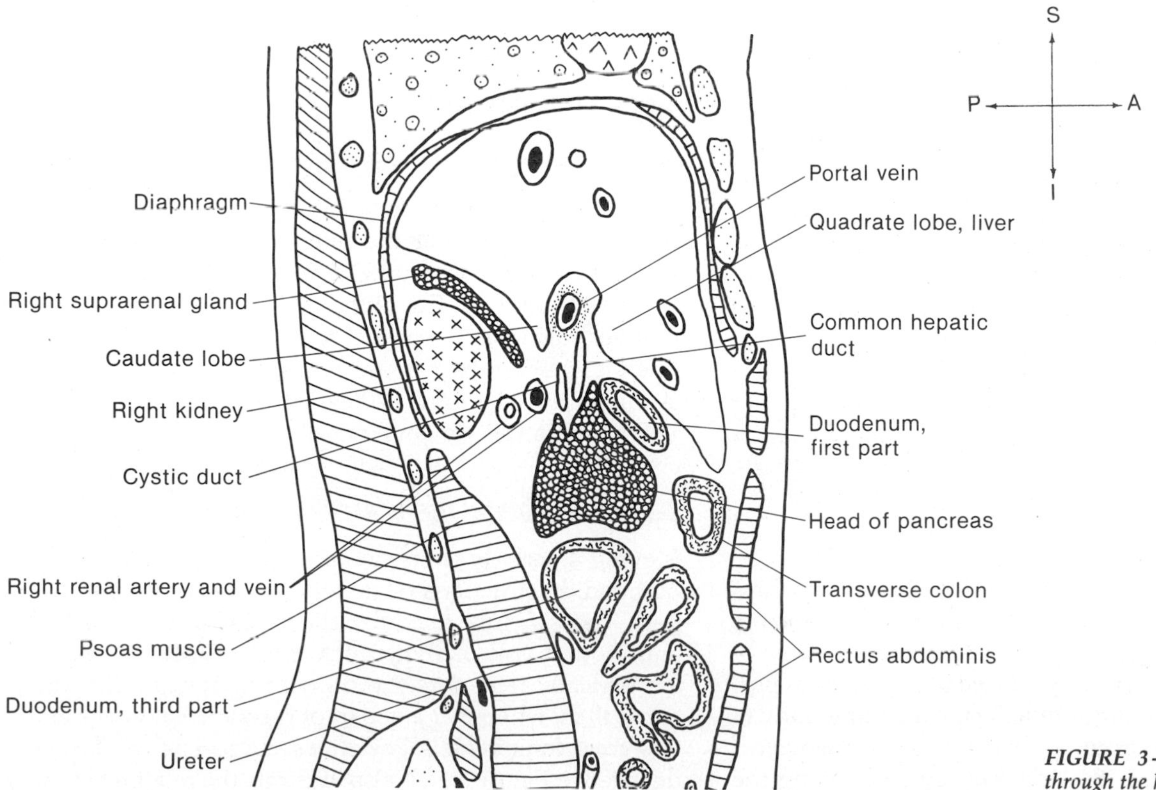

FIGURE 3-16. Sagittal section through the head of the pancreas.

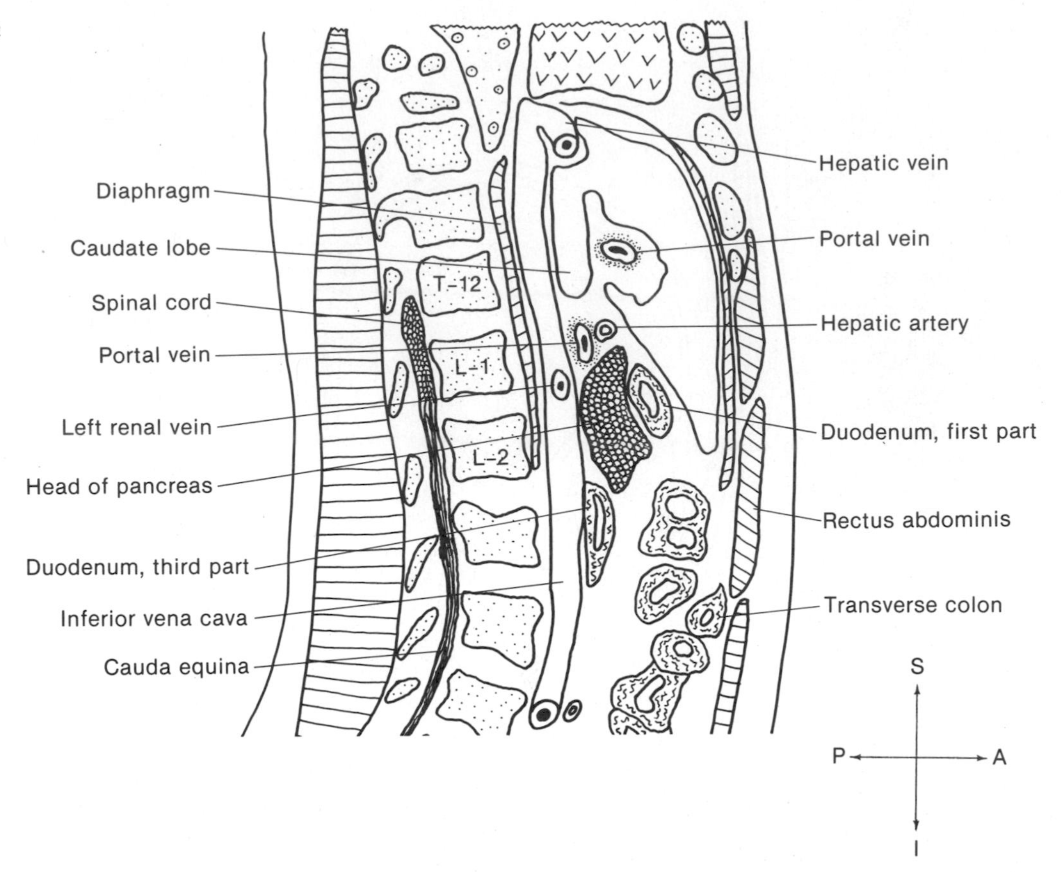

FIGURE 3-17. Sagittal section through the inferior vena cava.

cava. It is related anteriorly to the **caudate lobe** of the liver, the **pancreas,** and the horizontal third part of the **duodenum.**

MIDSAGITTAL SECTION

A little to the left of the inferior vena cava, in a position closer to the midline, the **aorta** descends along the vertebral column. The classic relationship of the aorta, the **superior mesenteric artery,** and the **left renal vein** is usually evident in this region. Watch for the superior mesenteric artery as it branches from the aorta at an angle of approximately 60 degrees. The left renal vein is neatly tucked in the angle between the two vessels, as illustrated in Figure 3-18. The horizontal third part of the **duodenum,** or possibly the ascending fourth portion, is just inferior to the left renal vein. Anteriorly, the pylorus of the **stomach** is related to the inferior portion of the **left lobe** of the liver.

LEFT PARASAGITTAL SECTION

Proceeding to the left from the midline, the relationships of the liver, the stomach, the spleen, the pancreas, and the kidney become evident. Figure 3-19 illustrates some of these relationships. Centrally located in the superior portion of the abdomen, the **stomach** is related anteriorly to the **left lobe** of the liver and posteriorly to the **spleen.** Just inferior to the spleen, the **left kidney** rests along the posterior abdominal wall. The body of the **pancreas** courses to the left, posterior to the stomach and anterior to the kidney. The **splenic artery** and **vein** are located along the posterior and superior margins of the pancreas.

Coronal Sections

SECTION THROUGH THE VERTEBRAL CANAL

When considering coronal sections of the abdomen, it is important to remember the anteroposterior relationships of the abdominal viscera. The **spleen,** the **kidneys,** and the **right lobe of the liver** are in the paravertebral grooves, against the posterior abdominal wall; consequently, they are seen in the most posterior sections. This is illustrated in Figure 3-20, which depicts a coronal section through the vertebral canal.

SECTION THROUGH THE VERTEBRAL BODIES

Moving anteriorly from the vertebral canal to the vertebral bodies, structures related to the spleen become evident. These include the **tail of the pancreas,** the **stomach,** and the **left colic (splenic) flexure.** The **splenic vessels** are typically related to the superior margin of the pancreas; thus, whenever any part of the pancreas is present, look for splenic vessels. Posterior sections, such as the one depicted in Figure 3-21, also illustrate the relationship of

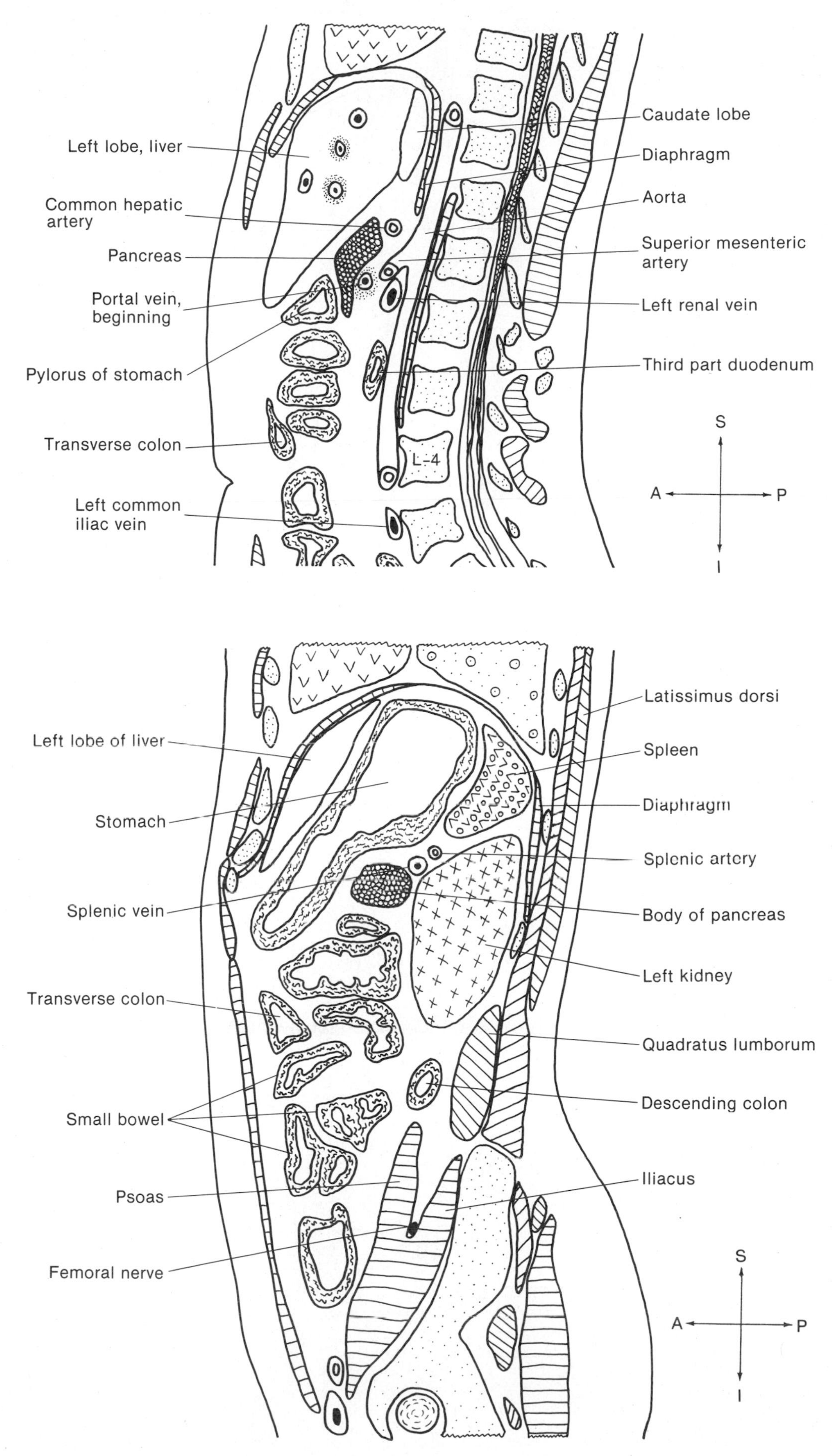

FIGURE 3-18. Midsagittal section through the abdomen, left-to-right view.

FIGURE 3-19. Left parasagittal section through the abdomen, left-to-right view.

FIGURE 3–20. Coronal section through the vertebral canal.

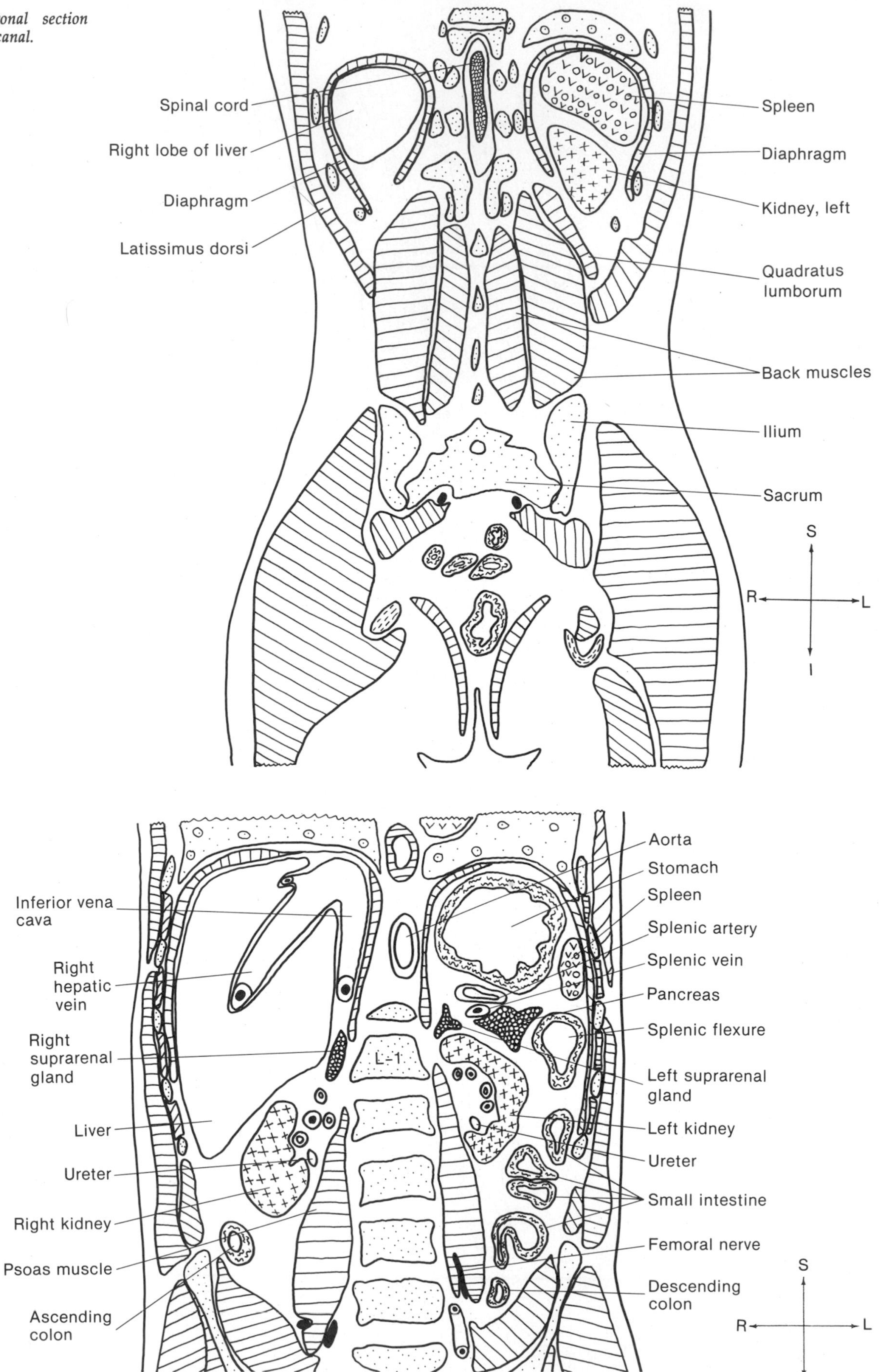

FIGURE 3–21. Coronal section through the lumbar vertebral bodies.

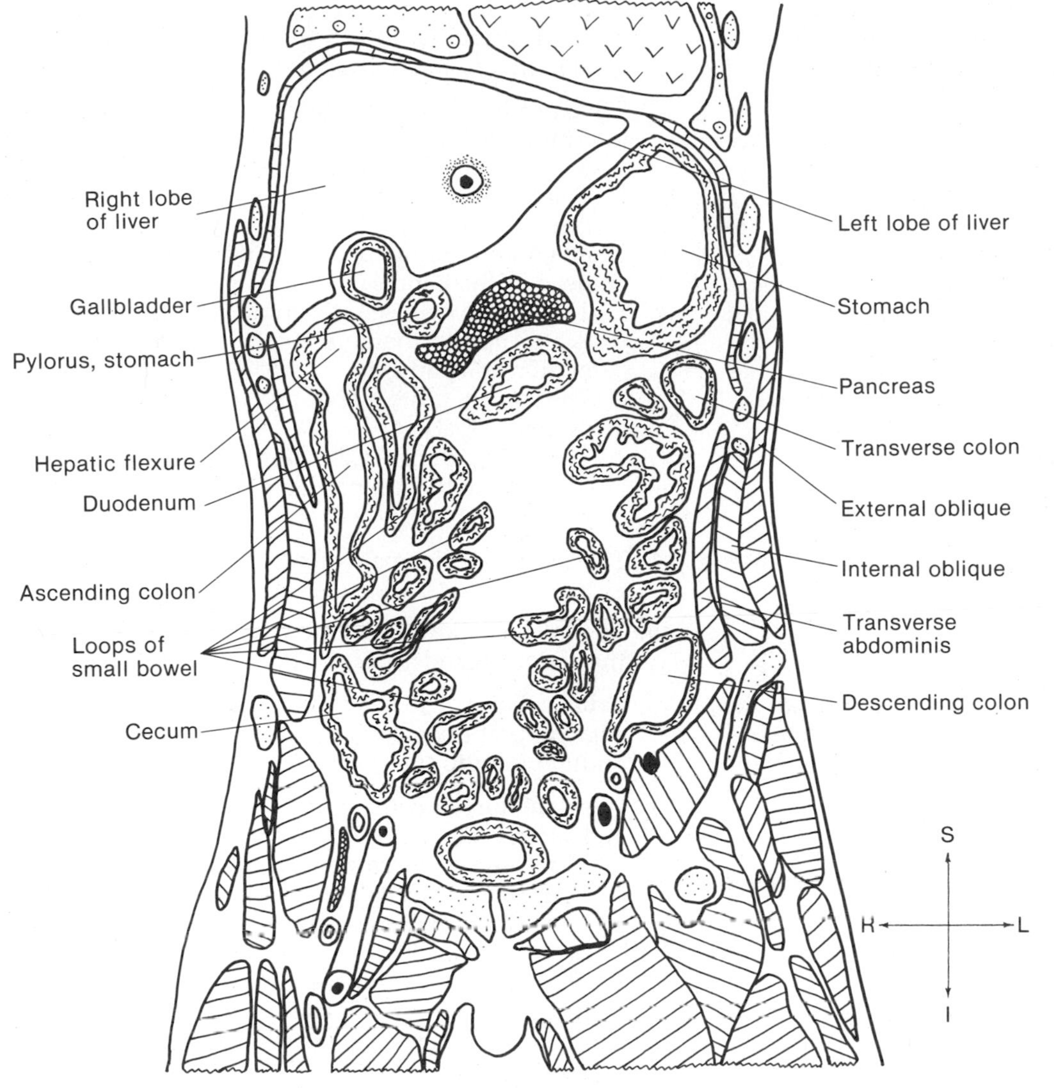

FIGURE 3-22. Coronal section through the anterior abdomen.

the **kidneys** to the **psoas muscle.** The normal lordosis of the vertebral column causes coronal sections through lumbar vertebral bodies to be more anterior than those through thoracic vertebrae.

SECTION THROUGH THE ANTERIOR ABDOMEN

Anterior sections of the abdomen, for example, through the gallbladder, are likely to show both the right and the left lobes of the liver and the structures related to them. The fundus of the gallbladder is the most anterior part. The body is slightly more posterior and is in contact with the visceral surface of the liver (Fig. 3-22). The gallbladder is related to the **right colic (hepatic) flexure** and to the right side of the **transverse colon.** Medially, the body of the gallbladder is related to the **pylorus** of the stomach and the **duodenum.** Since the **pancreatic head** is closely associated with the duodenum, it is usually seen in sections with the gallbladder. Coronal sections anterior to the gallbladder show only the liver.

REVIEW QUESTIONS

1. **True** or **False:** Portions of the liver, stomach, and spleen extend superiorly under the ribs and are protected by the thoracic cage.
2. The superior abdominal region in the midline is the ______________ region.
3. Which of the following statements about the diaphragm is **NOT** true?
   a. The costal portion of the diaphragm arises from the inferior six costal cartilages.
   b. The caval hiatus is the most anterior of the three openings in the diaphragm.
   c. The vagus nerve passes through the esophageal hiatus.
   d. The aortic hiatus is located at the level of the tenth thoracic vertebra.
   e. The azygos vein and the thoracic duct both pass through the aortic hiatus.
4. The innermost muscle of the anterolateral abdominal wall is the ______________ muscle.

5. **True** or **False:** The psoas muscles, the inferior vena cava, the lumbar vertebral bodies, and the aorta create a longitudinal ridge in the posterior abdominal wall.

6. Which of the following statements about the peritoneum is **NOT** true?
   a. The peritoneum is a serous membrane.
   b. The kidneys, the pancreas, and the stomach are retroperitoneal.
   c. The falciform ligament is a peritoneal ligament that attaches the liver to the anterior abdominal wall.
   d. The hepatic artery, the hepatic portal vein, and the bile duct are enclosed in the free margin of the lesser omentum.
   e. An omentum is associated with the stomach.

7. Which of the following statements about the abdominal aorta and its visceral branches is **NOT** true?
   a. The abdominal aorta extends from the aortic hiatus to the fourth lumbar vertebra.
   b. The celiac trunk arises from the aorta just above the transpyloric plane near the superior margin of the first lumbar vertebra.
   c. The renal arteries are paired and arise from the aorta at the level of the third lumbar vertebra.
   d. The superior mesenteric artery arises from the aorta at a level inferior to the celiac trunk but superior to the renal arteries.
   e. The celiac trunk branches into the common hepatic artery, the left gastric artery, and the splenic artery.

8. Which of the following veins flow directly into the inferior vena cava?
   a. hepatic, renal, and common iliac veins
   b. hepatic, external iliac, and right gonadal veins
   c. common iliac, renal, and left gonadal veins
   d. renal, left gonadal, and left suprarenal veins
   e. hepatic, renal, and left gonadal veins

9. Given the following numeric code, which of the following sequences represents a correct pathway for the flow of blood through the hepatic portal circulation: (1) portal vein; (2) hepatic vein; (3) liver sinusoids; (4) splenic vein; (5) central vein; (6) inferior mesenteric vein; (7) superior mesenteric vein.
   a. 6, 7, 2, 3, 5, 1
   b. 4, 7, 2, 3, 5, 1
   c. 7, 5, 1, 3, 2
   d. 4, 1, 5, 3, 2
   e. 6, 4, 1, 3, 5, 2

10. On the visceral surface, the right lobe is separated from the rest of the liver by the inferior vena cava and ______________.

11. In addition to the liver, the gallbladder is related to the ______________ and ______________.

12. **True** or **False:** The greater curvature of the stomach is directed inferiorly and to the right, and it has the greater omentum attached to it.

13. Which of the following statements about the small intestine is **NOT** correct?
    a. Most of the duodenum is retroperitoneal.
    b. The first part of the duodenum is related posteriorly to the inferior vena cava, anteriorly to the quadrate lobe of the liver, and inferiorly to the head of the pancreas.
    c. The flexure between the duodenum and jejunum is attached to the anterior abdominal wall by a peritoneal ligament.
    d. The ileum is quite mobile and tends to go into any available space.
    e. The third part of the duodenum is horizontal and is related to the inferior margin of the head of the pancreas.

14. The most superior part of the colon is usually the ______________.

15. The spleen is located in the ______________ region of the abdomen and is ______________ to the stomach.

16. **True** or **False:** The head of the pancreas is the most superior portion and is located on the right side of the midline, next to the spleen.

17. Which of the following statements about the kidneys and suprarenal glands is **NOT** correct?
    a. The hilum of the kidney is at the same level as the transpyloric plane.
    b. The right suprarenal gland is posterior to the inferior vena cava.
    c. The ureters are retroperitoneal and are usually related to the psoas muscles.
    d. The left suprarenal gland is related to the hepatic flexure of the colon and the left kidney.
    e. The right kidney forms a depression on the visceral surface of the liver, and its hilum is related to the second part of the duodenum.

18. Identify the indicated structures on Figure *A*.

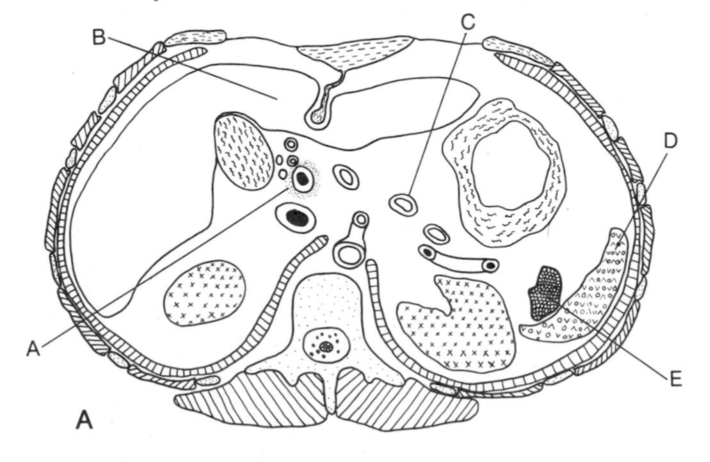

A

CHAPTER 4

PELVIS

OBJECTIVES

Upon completion of this chapter, the student should be able to do the following:

- Define the term "pelvis."
- Compare the structure of the os coxa in the child with that in the adult.
- Differentiate between the greater or false pelvis and the lesser or true pelvis.
- Identify the two principal muscles that line the wall of the true pelvis.
- Describe the structures that support the pelvic viscera and prevent it from falling through the pelvic outlet.
- Name two nerves that emerge from the sacral plexus, and state the importance of each.
- Describe the anterior relationships of the rectum in the male and in the female.
- Compare the relationships of the urinary organs in the male and in the female.
- Trace the pathway of sperm from their site of origin to their exit from the body.
- Name three accessory glands of the male reproductive system, describe their location, and state their significance.
- Describe the normal location and attachments of the ovaries and the factors that may alter the location.
- Identify the principal ligament that supports the uterus.
- Describe the normal position and relationships of the uterus.
- Identify and compare the three muscles of the urogenital region of the perineum in the male and in the female.
- Compare the dorsal and ventral columns of erectile tissue in both the body and the root of the penis.
- Describe and discuss the relationships of the female external genitalia.

- Compare vessel and muscle relationships in transverse sections through the sacroiliac joint and in transverse sections through the lower part of the sacrum.
- Discuss the relationships of the seminal vesicles and the prostate glands.
- Distinguish between the rectovesical pouch, the vesicouterine pouch, and the rectouterine pouch.
- Describe the relationships of the fornix of the vagina to the cervix.
- Identify the muscles, viscera, blood vessels, and skeletal components of the male pelvis in transverse, sagittal, and coronal sections.
- Identify the muscles, viscera, blood vessels, and skeletal components of the female pelvis in transverse, sagittal, and coronal sections.

GENERAL ANATOMY OF THE PELVIS

The term "pelvis" is ambiguous and confusing because in modern usage it means different things to different people. It can be used to describe a loosely defined region of the body where the trunk meets the lower limb. The term is also applied to the bony ring formed by the sacrum, the coccyx, and the two hip bones; this structure is sometimes referred to as the "bony pelvis." Finally, the term can describe the cavity enclosed by the bony pelvis, which is also called the "pelvic cavity." Generally, the context of the discussion clarifies the meaning.

The Pelvic Cavity

The bony pelvis encloses a funnel-shaped, or basin-shaped, cavity that is the inferior portion of the larger abdominopelvic cavity. The **pelvic cavity** is divided into a pelvis major (a false pelvis) and a pelvis minor (a true pelvis). The cavity of the pelvis major is the space between the iliac fossae, and its inferior boundary is defined by the pelvic brim. It is considered to be a part of the abdominal cavity and it contains abdominal viscera, such as portions of the small intestine and the sigmoid colon. The iliac crest is such an obvious dividing point on transverse sections that the pelvis major is included in the discussion of the pelvis rather than in that of the abdomen. The minor pelvis is the space below the pelvic brim and is enclosed by the sacrum, the ischium, the pubis, and the pelvic portions of the ilium. It contains the urinary bladder, the rectum, and the internal reproductive organs, as well as the portions of the mobile intestinal tract that may be able to reach it.

Bones of the Pelvis

The bony framework of the pelvis is called the **bony pelvis** and is formed by the sacrum, the coccyx, and the paired os coxae, or the hip bones.

SACRUM

The sacrum and coccyx make up the posterior midline portion of the bony pelvis. The **sacrum** serves to transmit the weight of the body to the hip bones and then to the lower extremities. Normally, five sacral vertebrae fuse into one triangular mass, called the sacrum, which articulates with the fifth lumbar vertebra superiorly, the coccyx inferiorly, and the hip bones laterally. The sacrum does not meet the fifth lumbar vertebra in a straight line; instead, the sacrum is tilted posteriorly to form a **lumbosacral angle.** In some individuals, either the first sacral vertebra remains separate from the other four, or the fifth lumbar vertebra may fuse with the sacrum. Both conditions put a strain on the nearest intervertebral articulation, and this may result in a degeneration of the joint and may cause low back pain.

When the five sacral vertebrae fuse together, they form a normal curvature wherein the anterior or pelvic surface is concave. Anteriorly, the upper margin of the first sacral vertebra forms the **sacral promontory,** which marks the posterior portion of the true pelvic inlet, or the pelvic brim. Both the anterior and the posterior surfaces of the sacrum have two rows of four holes, or openings, called **sacral foramina.** Branches of the sacral nerves pass through the sacral foramina. On the posterior surface, between the sacral foramina, poorly defined and fused spinous processes form the **median crest.** The fifth sacral vertebra has no spinous process and no lamina. This deficiency in the neural arch leaves a midline opening called the **sacral hiatus.** Local anesthetics may be injected through the sacral hiatus. This is called extradural, epidural, or caudal anesthesia.

On each side of the upper portion of the sacrum there is a rather large **articular surface** for articulation with the iliac bones. This forms the **sacroiliac joint.** The connection between the bones is further enhanced by strong **interosseous ligaments** that act as cords to bind the bones together.

COCCYX

The most inferior portion of the vertebral column is the **coccyx,** or tailbone. It usually consists of four rudimentary vertebrae, although there may be one more or one less. Vertebrae in the coccyx are just pieces of bone with no processes and no foramina. The coccyx offers no support for the vertebral column, but it does provide attachment for a portion of the gluteus maximus muscle and some of the muscles of the pelvic floor. In middle life, the bones of the coccyx usually fuse together to form a single structure. The coccyx is joined to the sacrum by cartilage. During childbirth or after a fall, the sacrum and coccyx may separate, which results in pain, especially when sitting.

OS COXAE

The **os coxae** (innominate) are commonly called the hip bones. Each os coxa consists of an **ilium,** a **pubis,** and an **ischium.** In the child, these are three separate bones that are joined together by hyaline cartilage. Each bone has its own ossification center within the cartilage. Throughout childhood, ossification continues and the cartilage is replaced by bone. By puberty, only a small Y-shaped region of cartilage remains in the acetabulum where the three bones meet. This is illustrated in Figure 4–1. By late teens, ossification is complete and the ilium, ischium, and pubis have fused together and formed a single unit called the os coxa.

ILIUM. The **ilium** is the largest of the three bones of the os coxa. The superior part of the ilium presents a large, flaring, winglike surface, called the **ala.** The inner aspect of the ala is the **iliac fossa,** which is the origin of the iliacus muscle. The **iliac crest,** the most superior portion of the hip bone, is the superior margin of the ala. The crest terminates both anteriorly and posteriorly in short projections known as the **superior** and the **inferior iliac spines.** Posteriorly, the ilium articulates with the sacrum at the **sacroiliac joint.**

PUBIS. The bodies of the two **pubic bones** meet in the anterior midline at the **symphysis pubis.** A small projection, just lateral to the body, forms the **pubic tubercle,** and from this point, the **superior pubic ramus** extends laterally to meet the ilium. The pelvic surface of the upper margin of the superior pubic ramus is sharp and forms the **pectineal line,** which continues with the arcuate line of the ilium and the sacral promontory to mark the pelvic brim. The **inferior pubic ramus** extends inferiorly from the body of the pubis to meet the ischium. The inferior rami of the two pubic bones meet at the symphysis pubis to form the **subpubic angle.** This angle is usually less than 70 degrees in the male and greater than 80 degrees in the female.

ISCHIUM. The inferior portion of the os coxa is formed by the **ischium.** Anteriorly, the ramus of the ischium meets the inferior pubic ramus at an indistinct point; consequently, the two are often referred to together as the **ischiopubic ramus.** The posteroinferior border of the ischium is formed by a bulky, rough area called the **ischial tuberosity.** A sharp, pointed **ischial spine** divides the space between the ischial tuberosity and the ilium into the **lesser** and the **greater sciatic notches.** The lesser sciatic notch is between the ischial tuberosity and the ischial spine, whereas the greater sciatic notch is between the ischial spine and the ilium and the sacrum. The notches are made into foramina by ligaments. These openings are closed by muscles.

The ilium, ischium, and pubis meet in the **acetabulum,**

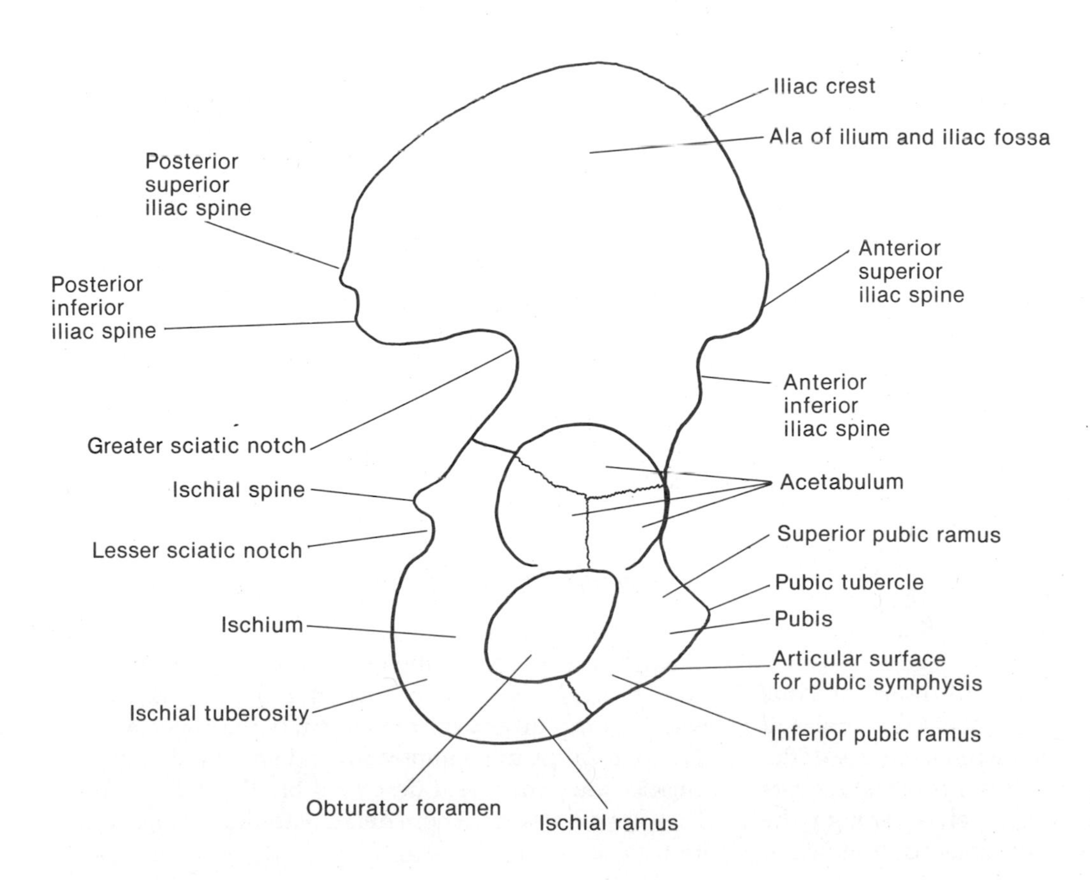

FIGURE 4–1. Features of the os coxa.

a deep fossa for articulation with the head of the femur. The ischium and pubis surround an opening called the **obturator foramen,** which is directed inferiorly. This foramen is closed by the obturator membrane and the obturator muscles. The **pelvic outlet,** or the inferior pelvic aperture, must also be covered with muscles and connective tissue membranes to give support and to maintain the pelvic viscera in position. Refer to Figure 4-1 to identify the features of the os coxae.

Musculature of the Pelvis

The muscles of the pelvic wall are functionally associated with movements of the thigh. Other muscles, such as the gluteal muscles and the anterior thigh muscles, are external to the pelvis but are seen in sections through the pelvis. These are mentioned here but are described in greater detail with the lower extremity.

MUSCLES IN THE WALL OF THE GREATER (FALSE) PELVIS

The muscles in the wall of the greater or false pelvis are actually abdominal muscles. The two principal muscles, the **psoas** and the **iliacus,** extend throughout the whole pelvic region and into the anterior thigh; thus, they are mentioned here. See Figure 4-9 for an illustration of these two muscles.

PSOAS. The **psoas,** a long, fleshy muscle, continues from the abdomen, into the pelvis, and then into the thigh, deep to the inguinal ligament, where it inserts on the lesser trochanter of the femur. It appears as a muscle mass lateral to the vertebral bodies. It acts with the iliacus muscle as a powerful flexor of the thigh. Lumbar nerves provide the innervation.

ILIACUS. The **iliacus** is a fan-shaped muscle that originates along the crest and the fossa of the ilium. In the pelvis, it lies along the lateral side of the psoas muscle. Fibers of the iliacus muscle insert on the psoas tendon and on the femur. The iliacus and psoas muscles appear to merge into one muscle, and consequently, they have a close functional relationship. These two muscles are often referred to as the single iliopsoas muscle. The iliacus is innervated by the femoral nerve.

MUSCLES IN THE WALL OF THE TRUE PELVIS

Most of the inner surface of the bony true pelvis is lined with muscle. The **obturator internus** and the **piriformis** are the principal muscles that form this lining and that make up the wall of the true pelvis (see Fig. 4-10).

OBTURATOR INTERNUS. The **obturator internus** is a fan-shaped muscle that covers most of the lateral wall of the true pelvis. It originates on the inner surface of the pelvic bones, crosses over the **obturator foramen,** covers this opening, and then leaves the pelvis through the **lesser sciatic notch.** As it leaves the lesser sciatic notch, it becomes tendinous and makes a sharp turn to insert on the medial surface of the greater trochanter. The obturator internus rotates the thigh.

PIRIFORMIS. The **piriformis** is located partially on the posterior wall of the true pelvis and partially external to the pelvis, posterior to the hip joint. It originates on the anterior surface of the sacrum, and it passes through the **greater sciatic notch** to insert on the greater trochanter of the femur. The piriformis is closely associated with the sacral nerve plexus, and it is innervated by the sacral nerves. It rotates and abducts the thigh.

MUSCLES OF THE PELVIC FLOOR

The pelvic outlet (inferior pelvic aperture) must be covered to give support and to maintain the pelvic viscera in position. The floor of the pelvis includes all structures contributing to this support, that is, the peritoneum, the pelvic diaphragm, and the urogenital diaphragm. The principal structure supporting the pelvic viscera is the **pelvic diaphragm,** which forms a muscular pelvic floor. The pelvic diaphragm is a hammocklike structure that primarily consists of two muscles, the **levator ani** and the **coccygeus.** Below (superficial to) the pelvic diaphragm, the urogenital diaphragm is formed by connective tissue membranes that are located within the infrapubic angle between the ischiopubic rami.

LEVATOR ANI. Of the two muscles that form the pelvic floor, the **levator ani** is the larger and the more important. The integrity of the pelvic floor depends on the appropriate function of the levator ani muscles. In females, these muscles are particularly vulnerable to damage during a strenuous delivery. When the muscles are damaged, support for the pelvic viscera is weakened. This may be followed by urinary incontinence and prolapse of the uterus. The levator ani muscles originate on the pelvic surface of the pubis and on the spine of the ischium. The fibers converge to insert on the coccyx. Some fibers insert on the levator ani muscle of the opposite side. The levator ani muscles are innervated by the pudendal nerve.

COCCYGEUS. The **coccygeus** is the smaller of the two muscles that form the pelvic diaphragm. From its origin on the spine of the ischium, the fibers fan out to form a triangular sheet that inserts on the sacrum and the coccyx. Branches of the pudendal nerve innervate the coccygeus muscle.

EXTRAPELVIC MUSCLES SEEN IN PELVIC SECTIONS

Numerous muscles, such as the gluteal and the thigh muscles, are external to the pelvis but are evident in sections through the pelvis. These muscles are associated with the hip joint and the movement of the lower extremity. Some of the more important and obvious extrapelvic muscles are named and described briefly in Table 4-1. They are discussed in more detail with the lower extremity.

TABLE 4-1. Extrapelvic Muscles

Muscle	Description
Gluteus maximus	The largest and most superficial muscle of the gluteal region; forms most of the mass of the buttocks
Gluteus medius	A thick, broad muscle of the gluteal region; located deep to the gluteus maximus but originating more superiorly, which results in the inferior one third's being covered by the gluteus maximus
Gluteus minimus	The smallest and deepest muscle of the gluteal region; underlies both the gluteus maximus and the gluteus medius
Gemelli	Two small muscle masses inferior to the piriformis muscle and deep to the lower part of the gluteus maximus; associated with the obturator internus tendon
Quadratus femoris	A short, flat, rectangular muscle that is located inferior to the gemelli and the extrapelvic portions of the obturator internus muscles
Tensor fasciae latae	A superficial muscle of the superior lateral thigh; overlies the superior lateral portion of the gluteus medius
Sartorius	A long, straplike muscle that courses obliquely and inferiorly across the anterior thigh; most superficial muscle of the anterior thigh
Rectus femoris	A large ropelike muscle mass that extends the length of the anterior thigh; one of the quadriceps femoris muscle group
Vastus lateralis	A large muscle that forms the lateral portion of the thigh; partially covered by the iliotibial fascia; one of the quadriceps femoris muscle group
Vastus medialis	A large muscle that forms the medial portion of the thigh; one of the quadriceps femoris muscle group
Vastus intermedius	An elongated muscle next to the shaft of the femur; deep to the rectus femoris, between the other two vastus muscles; one of the quadriceps femoris group
Pectineus	A flat muscle, medial to the iliopsoas, in the floor of the femoral triangle; apparent anterior to the pubic bone in transverse sections
Adductor longus	A superficial, flat muscle that extends obliquely from the pubis to the femur; medial to the pectineus
Adductor brevis	A flat muscle deep to the adductor longus on the medial aspect of the thigh
Adductor magnus	The largest of the adductor muscles in the medial compartment of the thigh; deep to the adductor longus and adductor brevis
Obturator externus	A relatively small, fan-shaped muscle that covers the obturator foramen; closely associated with the adductor muscles
Gracilis	A thin, straplike, superficial band of muscle that extends down the medial aspect of the thigh from the pubis to the tibia
Biceps femoris	A large, elongated muscle on the posterior and lateral aspects of the thigh; the superior portion is deep to the gluteus maximus; one of the hamstring muscles
Semitendinosus	A superficial muscle medial to the biceps femoris; the superior portion is deep to the gluteus maximus; one of the hamstring muscles
Semimembranosus	A fleshy muscle, deep and medial to the semitendinosus; one of the hamstring muscles

Vasculature of the Pelvis

At the level of the fourth lumbar vertebra, the abdominal aorta divides into the right and left **common iliac arteries.** These vessels descend to the pelvic brim, where they pass over the sacroiliac joint at the level of the disc between the fifth lumbar vertebra and the sacrum. Here, the common iliac arteries divide into the **external** and the **internal iliac arteries.** The external iliac artery follows around the pelvic brim, then passes under the inguinal ligament and becomes the femoral artery, which continues through the thigh. Only 4 cm long, the internal iliac artery is a short vessel that branches profusely. Parietal branches supply blood to the pelvic wall and the visceral branches supply blood to the pelvic organs. Blood is drained from the pelvis primarily by the internal iliac veins and their tributaries.

Nerve Supply to the Pelvis

Portions of the fourth and fifth lumbar nerves unite to form a thick, cordlike **lumbosacral trunk,** which descends obliquely over the sacroiliac joint and enters the pelvis. Within the pelvis, the junction of the lumbosacral trunk with the first four sacral nerves forms the **sacral plexus,** which lies on the piriformis muscle. Twelve nerves emerge from the sacral plexus. Five of these supply pelvic structures and the other seven go to the buttocks and the lower limb.

SCIATIC NERVE

One of the nerves of the sacral plexus, the **sciatic nerve,** is the largest nerve in the body. The sciatic nerve passes through the greater sciatic notch (foramen), enters the gluteal region at the lower border of the piriformis muscle, and then descends within the posterior compartment of the thigh. Its branches supply the flexor muscles of the thigh and all the muscles of the leg and foot. Because of its location, the sciatic nerve may be injured in dislocations and in fractures of the hip. It is also vulnerable to damage, such as when an individual is given an intramuscular injection into the buttock.

PUDENDAL NERVE

The **pudendal nerve,** another nerve of the sacral plexus, leaves the pelvis through the lesser sciatic notch (foramen) and supplies the perineum. It is important because, in addition to the perineum, it supplies the external anal sphincter and the sensory fibers of the external genitalia. This is the nerve that is anesthetized by a pudendal nerve block, which is used during childbirth and during surgical procedures on the female genitalia.

OBTURATOR NERVE

The **obturator nerve** does not arise from the sacral plexus but from a plexus of nerves in the abdomen. It enters the pelvis, runs along the lateral pelvic wall, and then leaves the pelvic cavity through the obturator foramen. The obturator nerve supplies the obturator externus and the adductor muscles of the thigh. It also sends branches to the hip and knee joints. Its proximity to the pelvic lymph nodes makes it vulnerable to injury during surgery for malignant disease in the pelvis. It may also be affected by pathologic changes in the ovary, which is near the nerve. The nerve supplies both the hip and knee joints; consequently, pain from the hip may be referred to the knee, making it difficult to locate the cause.

Viscera of the Pelvis

GASTROINTESTINAL ORGANS

Most of the length of the gastrointestinal tract is located in the abdomen. Loops of the mobile small intestine, especially the ileum, may extend into the pelvis. Portions of the colon are normally located within the false pelvis, and if the mobile transverse colon is particularly pendulous, it may extend down into the true pelvis. The mobility of the portions of the gastrointestinal tract that have a mesentery causes the extent of this tract in the pelvis to vary.

COLON. The **cecum** and the **ascending colon** are normally found within the greater or false pelvis on the right side. The **descending colon** on the left side becomes continuous with the **sigmoid colon** at the pelvic brim. In contrast to the descending colon, which is retroperitoneal, the sigmoid colon is surrounded by peritoneum and has a mesentery that allows considerable mobility. At its junction with the rectum, the sigmoid colon becomes fixed to the posterior pelvic wall.

RECTUM AND ANAL CANAL. The **rectum** begins near the middle of the sacrum and follows the curvature of the sacrum and the coccyx onto the pelvic floor. The levator ani muscle supports the rectum on the pelvic floor. The rectum penetrates the levator ani muscle and becomes the **anal canal.**

In the male, the peritoneum between the upper portion of the rectum and the urinary bladder forms the **rectovesical pouch** (see Fig. 4–2). Inferior to this, the rectum is related anteriorly to the bladder, the seminal vesicles, and the prostate, without the intervening peritoneum.

In the female, the peritoneum over the anterior surface of the rectum extends to the surface of the uterus, forming the **rectouterine pouch.** Inferior to the rectouterine pouch, the rectum is related anteriorly to the vagina (see Fig. 4–4).

URINARY ORGANS

URETERS. The upper half of each **ureter** is located in the abdomen, but the lower half enters the pelvis. The ureter crosses over the pelvic brim near the bifurcation of the common iliac artery. As it enters the pelvis, the ureter is anterior to the internal iliac artery and follows a course similar to that of the vessel. Retroperitoneally located and closely adherent to the peritoneum over it, the ureter descends along the lateral pelvic wall to a point near the ischial spine. There, it turns medially to enter the urinary bladder on its posterior surface.

URINARY BLADDER. The **urinary bladder** is a distensible, muscular organ, which, when empty, lies entirely within the true pelvis. As it fills, it extends into the abdomen. The muscle of the bladder wall is called the **detrusor muscle.** The urinary bladder rests on the pelvic floor posterior to the symphysis pubis. The superior surface is covered by peritoneum and is related to the sigmoid colon and to coils of the ileum. In the female, the peritoneum is reflected from the superior surface onto the anterior wall of the uterus, forming the **vesicouterine pouch** (see Fig. 4–4). In the male, the peritoneum is reflected from the superior surface of the bladder over the ductus deferens and the seminal vesicles onto the rectum, forming the **rectovesical pouch** (see Fig. 4–2). Held in place by peritoneal ligaments, the bladder is relatively movable, except in the inferior region called the neck, where it is firmly anchored. In the male, the neck rests on the prostate gland. In the female, the neck is attached to the pelvic diaphragm. Urine is conveyed from the bladder to the exterior through the urethra.

Internally, the lining of the bladder is thrown into irregular folds, which allow for expansion. The oblique openings of the two ureters, which convey urine from the kidneys to the bladder, are in the base of the bladder. These two openings, plus the **internal urethral orifice,** form a triangular region, called the **trigone,** in the base. The trigone is the region of the bladder that is most sensitive to pain.

FEMALE URETHRA. The **female urethra** is a short, muscular tube about 4 cm long that conveys urine from the bladder to the exterior. The external urethral orifice opens into the vestibule just anterior to the vagina.

MALE URETHRA. About five times as long as the female urethra, the **male urethra** opens to the exterior at the tip of the glans penis. For descriptive purposes, it may be divided into three regions (see Fig. 4–2). The **prostatic urethra** passes through the substance of the prostate gland. The ejaculatory duct and the ducts from the prostate gland open into this portion of the urethra. The **membranous urethra** penetrates the urogenital diaphragm to enter the penis. This is the shortest and narrowest portion of the urethra. The final portion is the **penile** or the **spongy urethra.** This is the longest part, which extends the full length of the corpus spongiosum of the penis. The ducts of the bulbourethral glands open into the proximal region of the spongy urethra.

MALE REPRODUCTIVE ORGANS

The male genital organs include the testes (singular, testis) within the scrotum, the epididymis, the ductus deferens, the ejaculatory duct, the seminal vesicles, the prostate, the bulbourethral glands, and the penis (Fig. 4–2). All of these are located within the pelvic cavity except for the testes and the penis. The penis is included with the

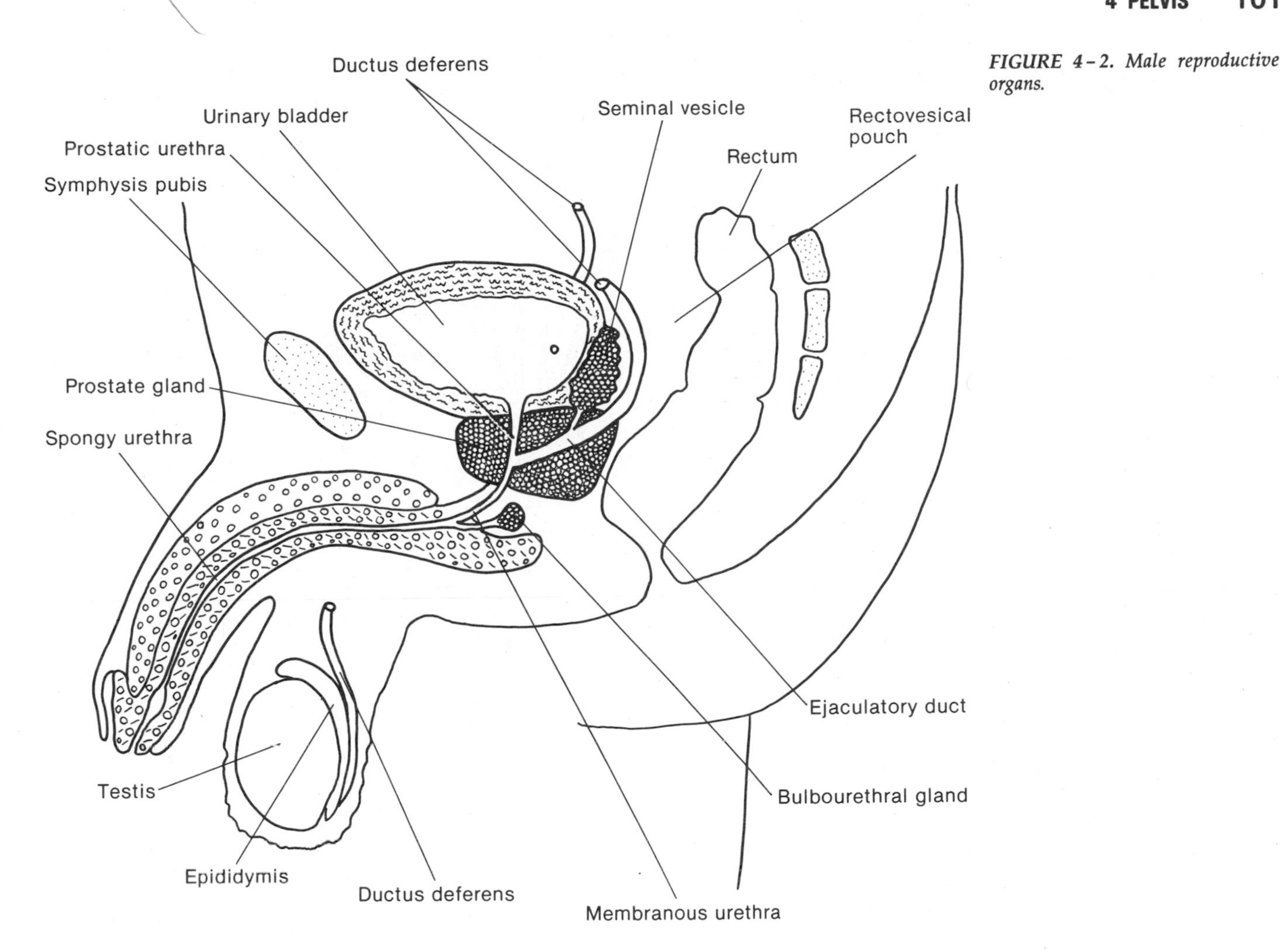

FIGURE 4-2. Male reproductive organs.

perineum as external genitalia. The testes are included in this section.

TESTES. The primary reproductive organs of the male are the paired **testes** (Fig. 4-3). These ovoid structures are suspended in a sac of skin called the **scrotum.** Each testis is covered by a tough fibrous connective tissue, the **tunica albuginea.** Extensions of the tunica albuginea project inwardly, dividing each testis into about 300 lobules. Each lobule contains from one to three tightly coiled **seminiferous tubules.** Spermatogenesis, the production of sperm, takes place within the seminiferous tubules.

In addition to the tunica albuginea, the testes are covered by a thin, serous sac, called the **tunica vaginalis,** which is derived from peritoneum. This is a serous membrane; therefore, it has two layers, a parietal layer and a visceral layer. The testes develop in the lumbar region of the abdomen and usually descend into the scrotum through the inguinal canal shortly before birth. An extension, or diverticulum, of peritoneum precedes the testes during their descent into the scrotum, and it is this peritoneum that becomes the tunica vaginalis. It is important that the testes descend into the scrotum, because spermatozoa cannot survive the warmer temperatures of the abdominal cavity.

EPIDIDYMIS. The **epididymis** is a flattened, tightly coiled, tubular structure on the posterior surface of each testis. As sperm are produced in the seminiferous tubules, they pass through efferent ductules, the ductus efferens, into the epididymis for maturation and storage. The epididymis is continuous with the ductus deferens, which transports sperm to the ejaculatory duct.

SCROTUM. The **scrotum** is a saclike structure that contains the testes along with their coverings, or tunics, and the epididymis. The scrotum consists of a layer of skin that covers a thin layer of connective tissue interspersed with smooth muscle called the **dartos muscle.** Contraction of the dartos muscle gives a wrinkled appearance to the scrotum. Internally, the scrotum is divided by a **median raphe,** or septum, into two compartments, which each contains a testis suspended by a **spermatic cord.**

DUCTUS DEFERENS. Each **ductus deferens** is a thick-walled, muscular tube that is a continuation of the epididymis. Beginning in the tail of the epididymis, at the inferior border of the testis, the ductus deferens ascends in the spermatic cord through the inguinal canal. As it enters the pelvis, it crosses over the external iliac vessels. From the inguinal canal, the ductus deferens descends retroperitoneally along the lateral wall of the pelvis, crosses the ureter, and then passes between the ureter

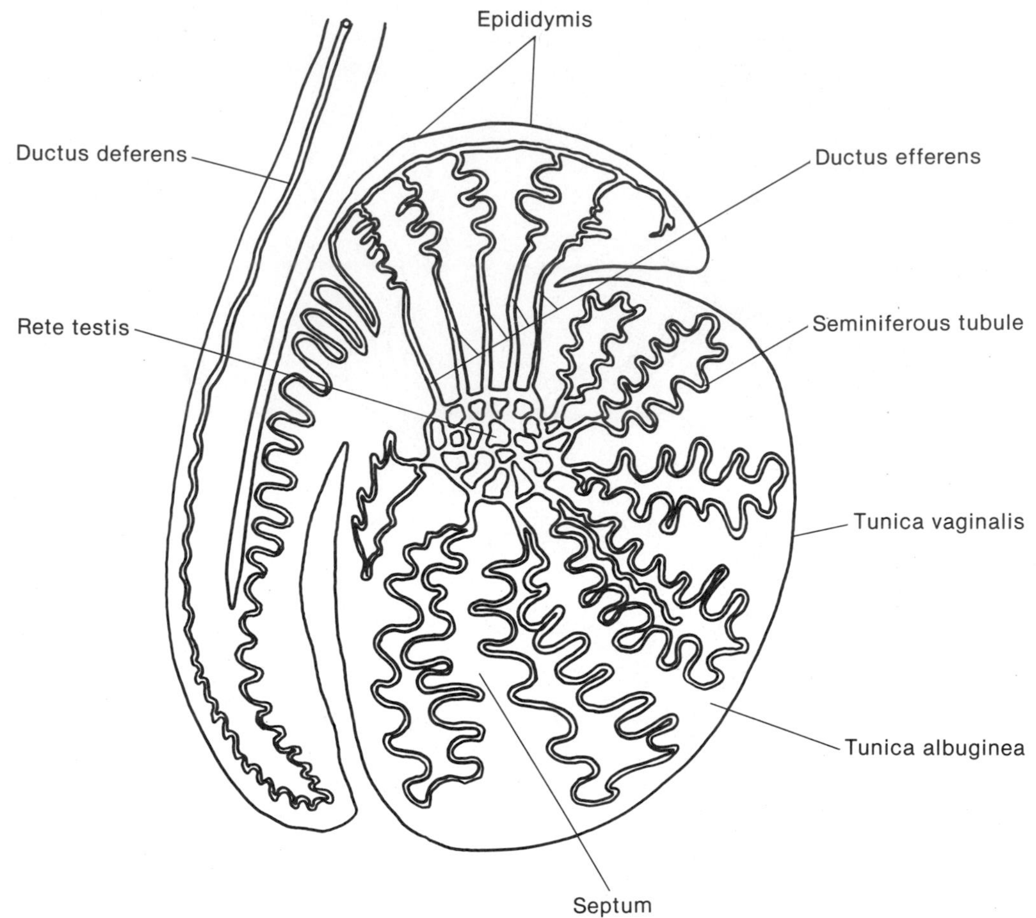

FIGURE 4-3. Sagittal section through the testis and the epididymis.

and the bladder. The ductus deferens descends posterior to the bladder and medial to the ureter and the seminal vesicles. The terminal portion, near the base of the bladder, is joined by the duct from the seminal vesicles to form the **ejaculatory duct,** which opens into the prostatic urethra.

SPERMATIC CORD. The **spermatic cord** extends from the testis to the inguinal canal. Structures passing to and from the testis make up the contents of the spermatic cord. This includes the ductus deferens, the testicular artery, the venous plexus (pampiniform plexus), the lymph vessels, the nerves, and some connective tissue. This bundle of structures is surrounded by connective tissue and muscle fibers to make a cord that suspends the testis in the scrotum. The **cremaster muscle,** derived from the internal oblique muscle of the abdomen, extends through the spermatic cord to the testis. The cremaster muscle functions with the dartos muscle to alter the position of the testis in order to maintain optimum temperature for the production and maturation of spermatozoa.

SEMINAL VESICLES. The **seminal vesicles** are paired accessory glands that consist of coiled tubes, which appear to be twisted to form small pouches, or vesicles. The glands are located between the posterior surface of the bladder and the rectum. The duct of each seminal vesicle joins with the associated ductus deferens to form an **ejaculatory duct,** which opens into the prostatic urethra. The secretion of the seminal vesicles has a high fructose content that provides an energy source for the sperm.

PROSTATE. The largest accessory gland of the male reproductive system is the **prostate.** It is composed partially of glandular parenchyma and partially of connective tissue stroma. Located inferior to the bladder, this chestnut-shaped gland surrounds the prostatic urethra. About 20 to 30 small prostatic ducts empty into the urethra. The secretion of the prostate gland aids in the motility and fertility of the sperm. The close relationship of the prostate to the urethra causes hypertrophy of the prostate to frequently interfere with the passage of urine. Subsequent problems related to the stasis of urine may then develop. Surgery may be indicated if urination becomes too difficult or impossible.

BULBOURETHRAL GLANDS. The two **bulbourethral glands,** also known as Cowper's glands, are about 1 cm in diameter. The glands, one on either side, are located posterior and lateral to the membranous urethra. Their relatively long ducts pass through the perineal membrane to empty into the proximal portion of the spongy urethra. The bulbourethral glands, in response to sexual stimulation, secrete a small amount of an alkaline, mucoid substance that neutralizes the acidity of the spongy urethra and lubricates the tip of the penis.

FEMALE REPRODUCTIVE ORGANS

The female reproductive organs within the pelvic cavity are the ovaries, the uterine tubes, the uterus, and a portion of the vagina.

To visualize the relationships of these organs and their peritoneal ligaments, it is necessary to understand the nature of the primary peritoneal fold in the female pelvis. The peritoneum is thrown upward in a fold over the midline uterus and then drapes from the uterine tubes, which extend laterally. To visualize this, imagine an individual, standing with outstretched arms with a sheet draped over the head and the arms. The sheet illustrates the peritoneum, the body represents the uterus, and the arms portray the uterine tubes. This large fold of peritoneum, called the **broad ligament,** divides the pelvic cavity into two compartments. The anteroinferior compartment contains the urinary bladder. The posterosuperior compartment contains the rectum.

OVARIES. A baby girl is born with about a quarter of a million primary oocytes, each capable of developing into a mature ovum. This multitude of potential ova is contained in a pair of ovaries. Each ovary is a small, solid, oval structure about 3 cm long that resembles an almond in size and shape. After menopause, the ovaries gradually become smaller.

Each ovary is located in an ovarian fossa, which is a shallow depression in the lateral wall of the pelvis on either side of the uterus. Three peritoneal ligaments loosely anchor each ovary in place. The **mesovarium** attaches the ovary to the posterior layer of the broad ligament of the uterus. The **ovarian ligament,** a cordlike thickening in the broad ligament, attaches the ovary to the lateral wall of the uterus. An extension of the broad ligament, the **suspensory ligament,** carries the ovarian vessels and attaches the ovary to the lateral pelvic wall. The position of the ovaries varies considerably, especially during pregnancy, when they are moved upward by the expanding uterus. Loops of intestine may also displace the ovaries.

The outer ovarian tissue is granulated in appearance because of the presence of the numerous ovarian follicles. Each month, an ovarian follicle matures into a graafian follicle. At ovulation, the mature follicle ruptures, releasing an ovum and some surrounding follicular cells into the peritoneal cavity.

UTERINE TUBES. The two slender uterine tubes, also called **fallopian tubes** (or oviducts), are about 10 or 12 cm long. They are in the upper border of the broad ligament, one on each side, and extend from the upper lateral angle of the uterus to the region of the ovary. The lumen of each tube is continuous with the cavity of the uterus. Near the ovary, the tube expands to form a funnel-shaped **infundibulum,** which is edged with fingerlike extensions, called **fimbriae.**

There is no direct connection between the ovary and the uterine tube that is associated with it; however, some of the fimbriae may make contact with the ovary. The ovum, or egg, is swept into the tube by a current that is set up in the peritoneal fluid by the beating motion of the fimbriae. Once inside, the ovum is propelled along by cilia that line the uterine tube and by the contraction of smooth muscle in the wall of the tube.

Fertilization, if it occurs, usually takes place in the uterine tube. The fertilized ovum, or zygote, continues the passage into the uterus for implantation and subsequent development. Occasionally, something interferes with the passage into the uterus, and implantation occurs in the uterine tube, resulting in an ectopic tubal pregnancy. The tube cannot expand sufficiently to accommodate the growing embryo; consequently, it ruptures in about 6 weeks. This frequently results in severe, life-threatening hemorrhage, which necessitates immediate surgery to control the bleeding.

The proximal end of each uterine tube opens into the uterine cavity, whereas the infundibular or distal end opens into the peritoneal cavity. The uterine tubes permit communication between the external environment and the peritoneal cavity, and they provide a pathway for pathogens to enter the abdominopelvic cavity.

UTERUS. A hollow, muscular, somewhat pear-shaped organ, the **uterus** receives the embryo that results from a fertilized egg and sustains the embryo's life during development. Although size and shape of the uterus change greatly during pregnancy, its typical, nonpregnant, premenopausal size is about 7 or 8 cm long and 5 cm across at its widest part. It is located in the anterior portion of the pelvic cavity, above the vagina, and it is usually bent forward over the upper surface of the urinary bladder. This is the **anteverted** position. Occasionally, the uterus tilts posteriorly in a **retroverted** position.

The wall of the uterus is relatively thick, and it is composed of three layers. The lining, called the **endometrium,** is a special type of mucous membrane that is covered with columnar epithelium and contains numerous glands, connective tissue, and blood vessels. The middle layer, or **myometrium,** is a thick layer of interlaced smooth muscle fibers. Contraction of the myometrium helps expel the fetus from the uterus during childbirth. A layer of serous peritoneum, called the **perimetrium,** covers the outside of the uterus. The perimetrium is continuous with the broad ligament.

The upper two thirds of the uterus, called the **body,** has a bulging superior surface known as the **fundus.** The **uterine tubes,** or oviducts, enter the uterus in this region at its broadest part. The lower one third of the uterus is a tubular **cervix,** which extends downward into the upper portion of the **vagina.** The opening of the cervix into the vagina is called the **external os.** In addition to the anterior tilt of the uterus as a whole **(anteversion),** the uterus is

FIGURE 4-4. Relationships of the uterus.

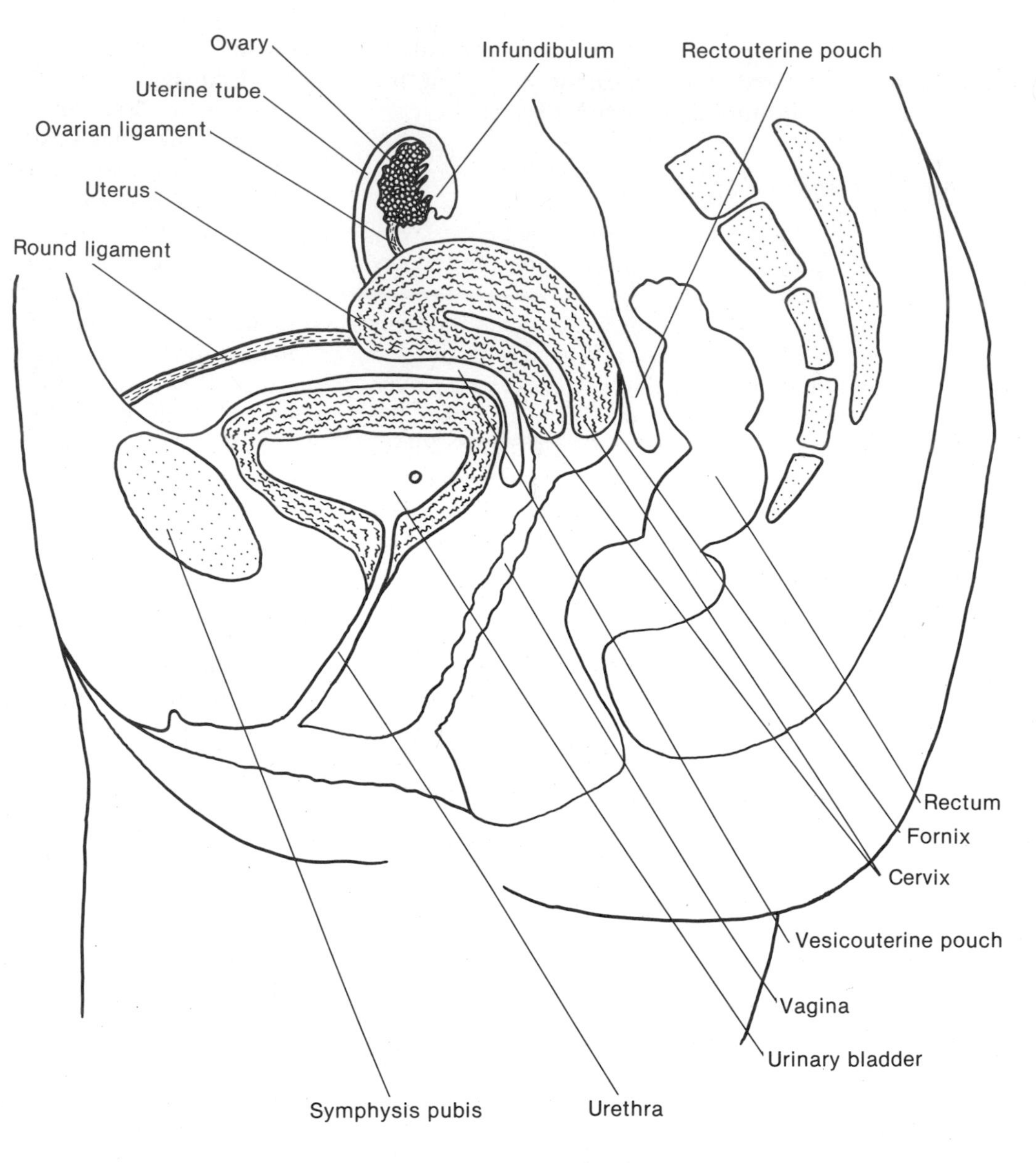

normally bent forward on its own axis; thus, the body is at an angle to the cervix. This is the normal **anteflexion** of the uterus. Figure 4-4 illustrates some of the relationships of the uterus.

Folds of peritoneum anchor and support the uterus in the pelvic cavity. Peritoneum is reflected from the superior surface of the bladder onto the uterus, forming the **vesicouterine pouch.** This pouch, or space, is usually empty, but it may contain a loop of small intestine. The peritoneum continues over the surface of the uterus as the perimetrium and then it is reflected onto the rectum. This forms the **rectouterine pouch,** or the pouch of Douglas.

Laterally, the peritoneum extends from the anterior and posterior uterine surfaces to the lateral pelvic wall. This peritoneal extension is the **broad ligament,** which not only supports the uterus but also encloses the uterine tubes. A pair of **round ligaments** extend from the lateral walls of the uterus, near the uterine tubes, to the lateral pelvic wall. The round ligament then follows a path similar to the vas deferens in the male as it passes through the inguinal canal and attaches to the subcutaneous tissue of the **labia majora.** The **uterosacral ligaments** extend from the uterus to attach on the sacrum. The **lateral cervical (cardinal) ligaments** extend from the lateral walls of the cervix to the pelvic floor and primarily stabilize the cervix. Even with the support of the various ligaments, the body of the uterus is relatively mobile and its principal support is the muscles of the pelvic floor and the pelvic viscera.

VAGINA. The vagina plays a key role in both the beginning and the end of the reproductive process. In addition to receiving the erect male penis during coitus, it functions as the birth canal during parturition, or childbirth. A muscular tube, 10 to 15 cm long, the vagina extends from the cervix of the uterus to the vestibule on the exterior of the body. It is situated between the urethra and the bladder anteriorly and the rectum posteriorly.

The cervix projects into the vagina at an angle; the anterior wall of the cervix is shorter than the posterior wall. Projections of the vagina form recesses or spaces around the cervix. These recesses are called **fornices** (singular, **fornix**). As mentioned earlier, the angle at which the cervix enters the vagina causes the posterior fornix to be longer than the anterior one. The posterior fornix is related to the rectouterine pouch of the peritoneum. As a consequence, instruments inserted into the vagina may penetrate the peritoneum of the rectouterine pouch, with subsequent hemorrhage and peritonitis.

The muscular wall of the vagina is lined with stratified squamous epithelium, which is thrown into transverse folds, called rugae, that allow for expansion during coitus and parturition.

Perineum and External Genitalia

The **perineum** is the region between the thighs, overlying (superficial to) the pelvic diaphragm. In anatomic position, it is a narrow region that extends from the pubic arch anteriorly, to the coccyx posteriorly. When the thighs are abducted, it is a diamond-shaped area, which is bounded laterally by the inferior pubic ramus, the ischial ramus, and the ischial tuberosity. For descriptive purposes, it is divided into a posterior **anal region** and an anterior **urogenital region** by drawing a line between the ischial tuberosities (Fig. 4–5). The posterior region contains the anus and the anterior urogenital region contains the external genitalia. In the female, the region between the vagina and the anus is called the **clinical perineum.** During childbirth, the clinical perineum may be surgically cut to avoid excessive stretching and tearing of the tissues as the fetal head emerges.

MUSCLES OF THE PERINEUM

The muscles of the perineum are illustrated in Figures 4–6 and 4–7. The muscles found in the anal region are the **levator ani** and the **sphincter ani.** The levator ani muscles have already been described as primary muscles of the pelvic diaphragm. The sphincter ani muscle surrounds the opening of the anal canal, the **anus.** The musculature of the urogenital region consists of the **bulbospongiosus,** the **ischiocavernosus,** and the **transversus perinei muscles.** These muscles are the same in both sexes; however, the arrangement differs. Figure 4–6 illustrates the arrangement in the male, and Figure 4–7 shows the arrangement in the female. The **transversus perinei muscles** are horizontal, arising on the ischial tuberosities and passing medially to insert on the central perineal tendon. In other words, they pass along the line that divides the perineum into the urogenital and the anal regions. The **ischiocavernosus muscles** also arise on the ischial tuberosities, but they pass forward and insert on the pubic arch and the crura of the penis in the male or the clitoris in the female. The **bulbospongiosus muscle** is in the median line of the urogenital region. In the female, the two parts of this muscle are separated by the urethra and the vagina. In the male, the fibers of the two muscles unite in the midline and encircle both the bulb and the adjacent corpus spongiosum of the penis.

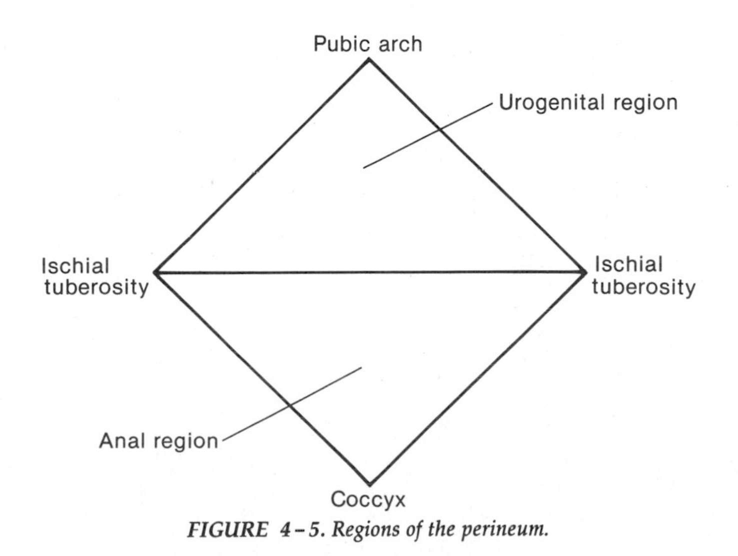

FIGURE 4–5. Regions of the perineum.

EXTERNAL MALE GENITALIA

External structures of the male reproductive system include the scrotum and the penis. The scrotum was discussed earlier with the testes. The penis is described here.

The **penis** is a copulatory organ that functions to introduce spermatozoa into the vagina of the female. It is cylindrical and is divided into a root and a body. Structurally, the penis consists of three cylindrical masses of erectile tissue, each surrounded by a connective tissue tunica albuginea (Fig. 4–8). The two dorsal cylinders are the **corpora cavernosa;** the smaller, midventral cylinder that encircles the spongy urethra is the **corpus spongiosum.** Distally, the corpus spongiosum expands to form the **glans penis.**

The **root** of the penis is the attached portion, which consists of a bulb and two crura. The **bulb** is the expanded proximal end of the corpus spongiosum. It is anchored to the tissue of the urogenital diaphragm in the pelvic floor and is enclosed by the bulbospongiosus muscle of the perineum. The two **crura** are the tapered proximal ends of the corpora cavernosa, which diverge and attach to the ischiopubic rami. The ischiocavernosus muscle of the perineum envelops the crura.

The **body** of the penis is the free part that is pendulous in the flaccid condition. The three cylindrical columns of erectile tissue are surrounded by connective tissue fascia and by skin. Facing anteriorly when flaccid, the dorsum of the penis is continuous with the anterior abdominal wall. The ventral or urethral aspect faces posteriorly in the flaccid condition and anteriorly when erect. The **glans penis,** at the distal end of the body, is corpus spongiosum with the opening for the urethra. The skin that covers the body of the penis continues over the glans penis as the **prepuce.** The prepuce, or foreskin, is often removed shortly after birth in a surgical procedure called circumcision.

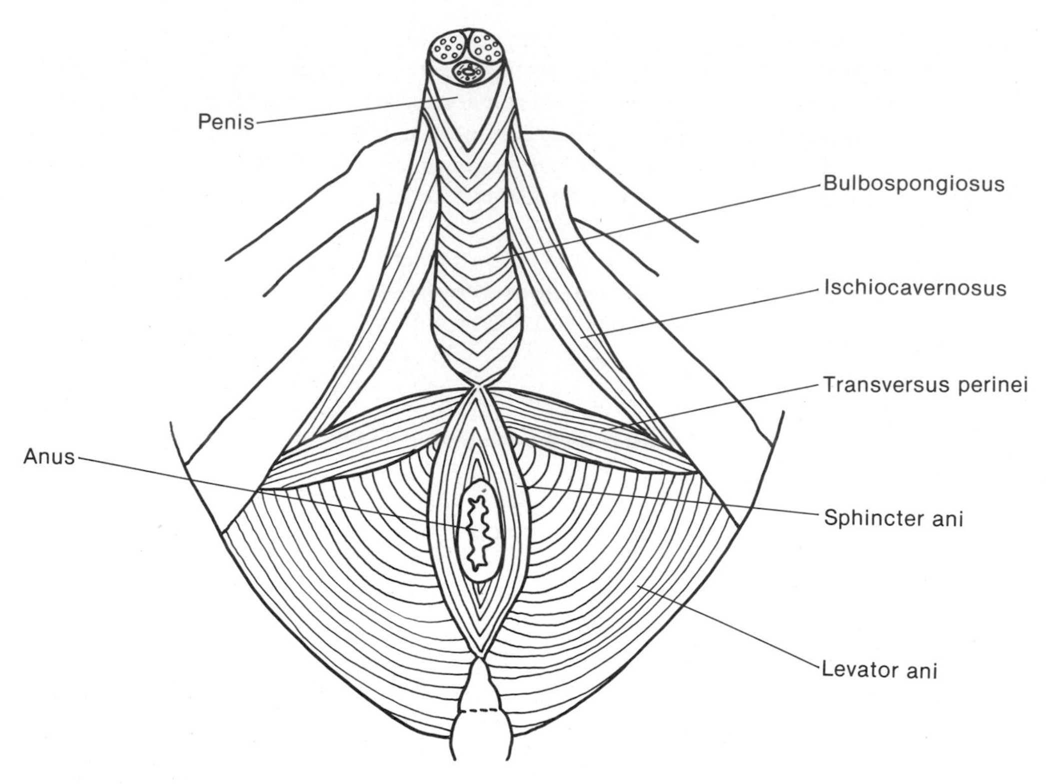

FIGURE 4-6. Muscles in the male perineum.

During sexual stimulation, parasympathetic reflexes cause dilation of the arteries that supply the penis, and this fills the sinuses of the erectile tissue with blood. At the same time, the pressure of the dilated arteries and filled sinuses compresses the veins leaving the penis so that the blood is retained. These vascular changes result in an erection. The penis returns to its flaccid state when the arteries constrict and pressure on the veins is reduced.

EXTERNAL FEMALE GENITALIA

The external accessory structures of the female genital system are closely associated with the perineum. The term **vulva,** or **pudendum,** is a collective term that refers to all the female external genitalia. These include the mons pubis, the labia majora, the labia minora, the vestibule, the clitoris, and the vestibular glands.

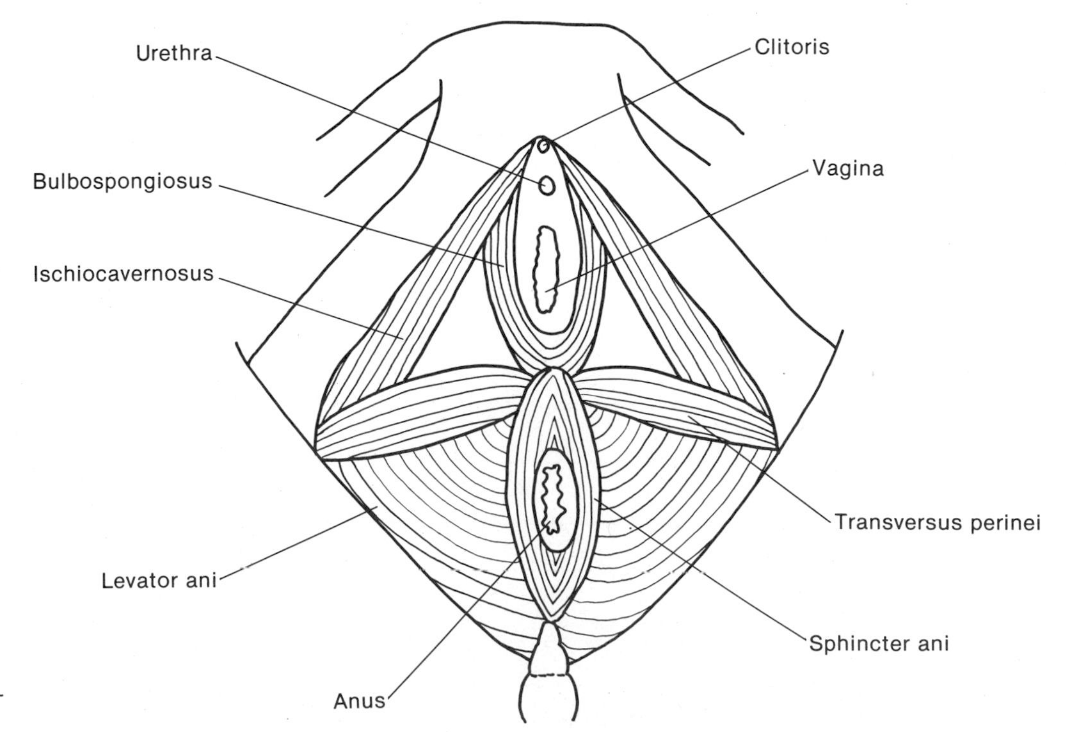

FIGURE 4-7. Muscles in the female perineum.

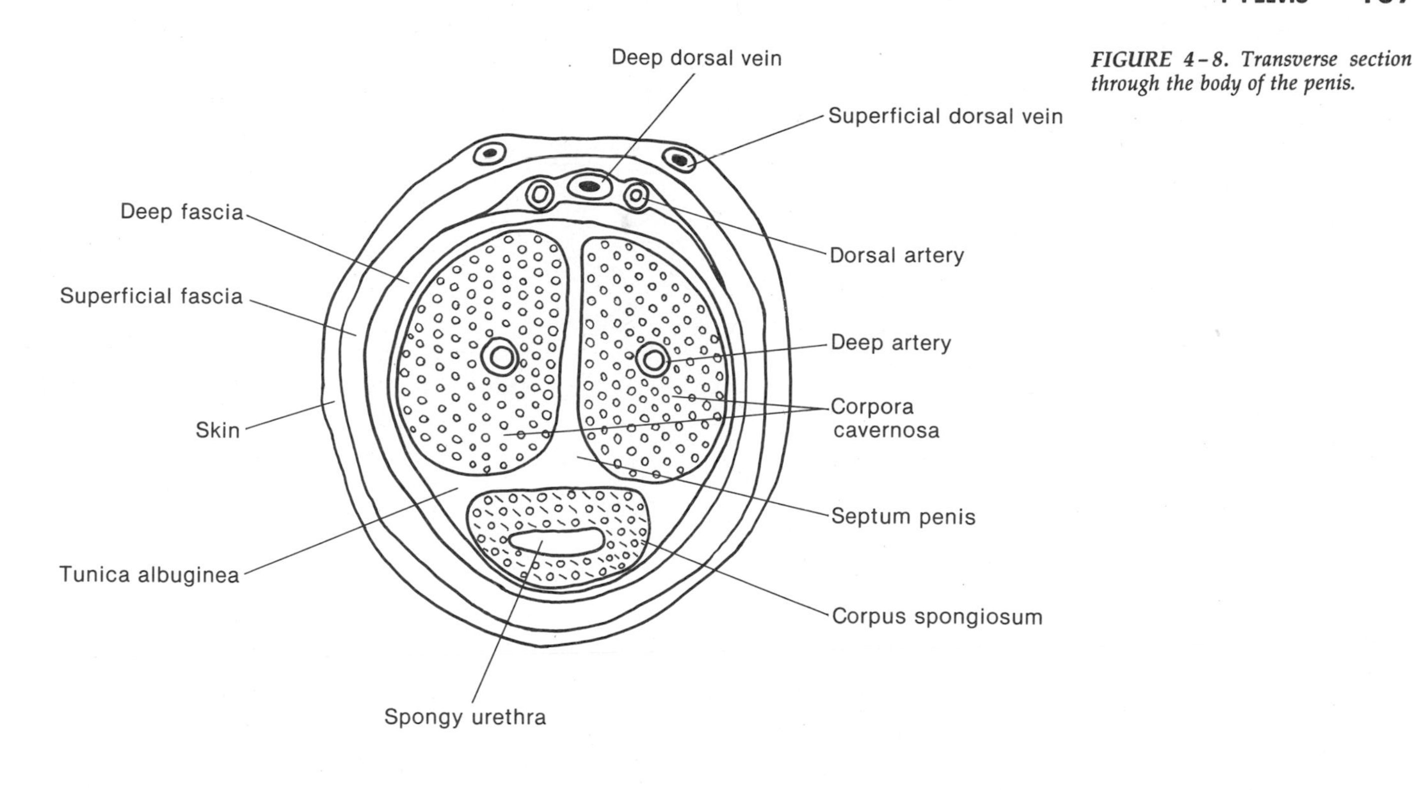

FIGURE 4-8. Transverse section through the body of the penis.

MONS PUBIS. A subcutaneous pad of fatty tissue that is covered with skin forms a rounded elevation anterior to the symphysis pubis. This is the **mons pubis.** During puberty, the mons pubis becomes covered with coarse pubic hairs.

LABIA MAJORA. Passing posteriorly from the mons pubis, the **labia majora** are large folds of skin, filled with subcutaneous fat. The skin on the lateral surface has sweat glands and sebaceous or oil glands. Also, after puberty, it is covered by pubic hair. Embryologically, the labia majora are homologous with the scrotum in the male. This means they have a similar structure and are derived from the same undifferentiated tissue. The labia majora are the lateral margins of the vulva.

LABIA MINORA. The **labia minora** are two thin delicate folds of skin located between the labia majora. Although devoid of fat and hair, the labia minora are richly supplied with blood vessels, nerves, and sebaceous glands. They have few sweat glands. Enclosing the vestibule, the labia minora lie on either side of the urethral and vaginal openings. Anteriorly, the folds of the labia minora unite to form a prepuce over the clitoris.

CLITORIS. The **clitoris,** homologous to the penis in the male, is located inferior to the mons pubis and between the anterior ends of the labia minora. Although only 2 to 3 cm in length, it is composed of two columns of erectile corpora cavernosa and is anchored to the os coxae by two crura. Like the male penis, it is capable of enlargement when stimulated. The clitoris is not associated with the urethra and has no corpus spongiosum.

VESTIBULE. The **vestibule** is the narrow cleft between the two labia minora. The clitoris is at the anterior end of the vestibule. Posterior to the clitoris, the urethra opens into the anterior portion of the vestibule. The vaginal orifice is posterior to the urethra. Openings for the **paraurethral glands** are located on either side of the external urethral orifice. These glands are homologous to the prostate gland of the male and secrete mucus for lubrication. The **greater vestibular glands,** also called Bartholin's glands, are located within the vestibule on either side of the vaginal orifice and also secrete a mucoid substance for lubrication. Normally, the vestibular glands are not palpable, but some become greatly enlarged and painful when infected. These glands are homologous to the bulbourethral glands of the male.

SECTIONAL ANATOMY OF THE PELVIS

For convenience, this discussion of the sectional anatomy of the pelvis includes the false pelvis. In other words, transverse sections begin with the iliac crest. In these more superior sections of the false pelvis, the organs seen are familiar because they are continuous with those studied in the abdomen.

Sections of the Male Pelvis

TRANSVERSE SECTION THROUGH THE SACROILIAC JOINT

A line through the most superior point of the two iliac crests usually intersects the body of the fourth lumbar vertebra. This is the level of the aortic bifurcation into the **right** and the **left common iliac arteries.** Inferior to this, at level L-5, the two **common iliac veins** join to form the

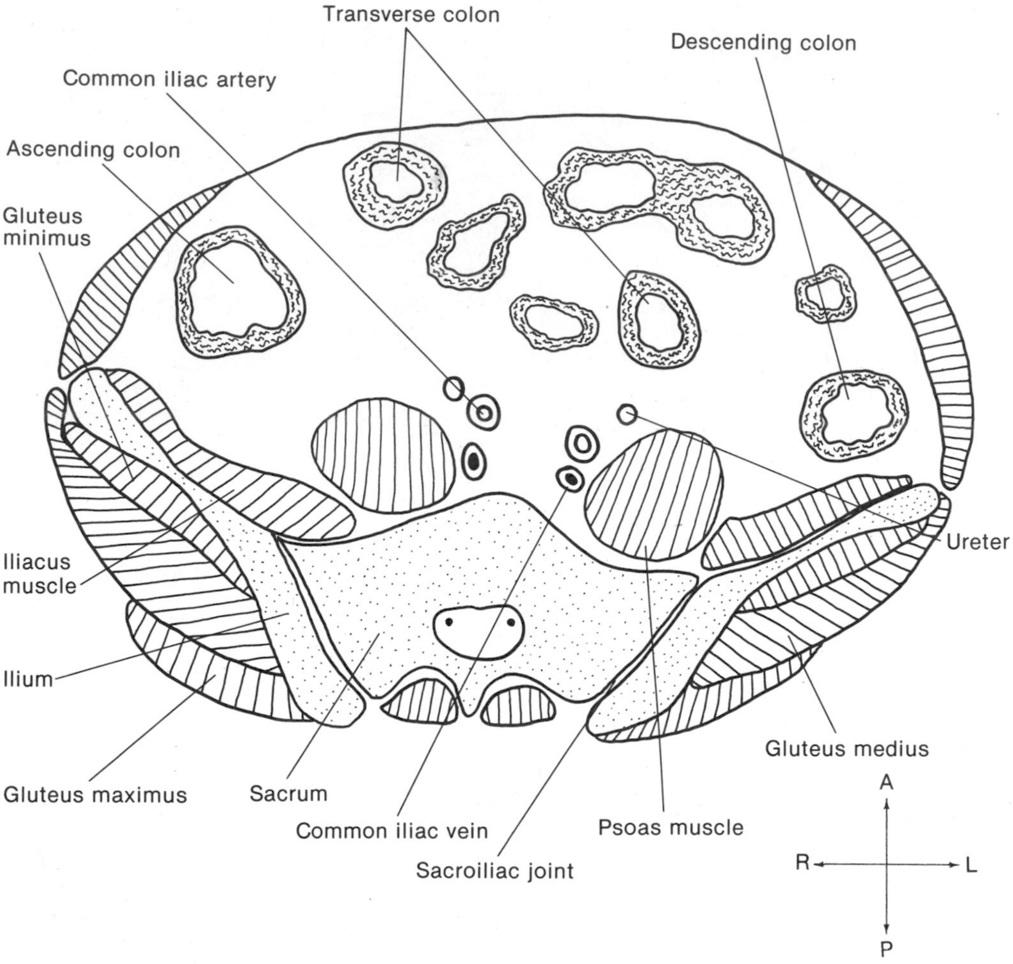

FIGURE 4-9. Transverse section through the sacroiliac joint.

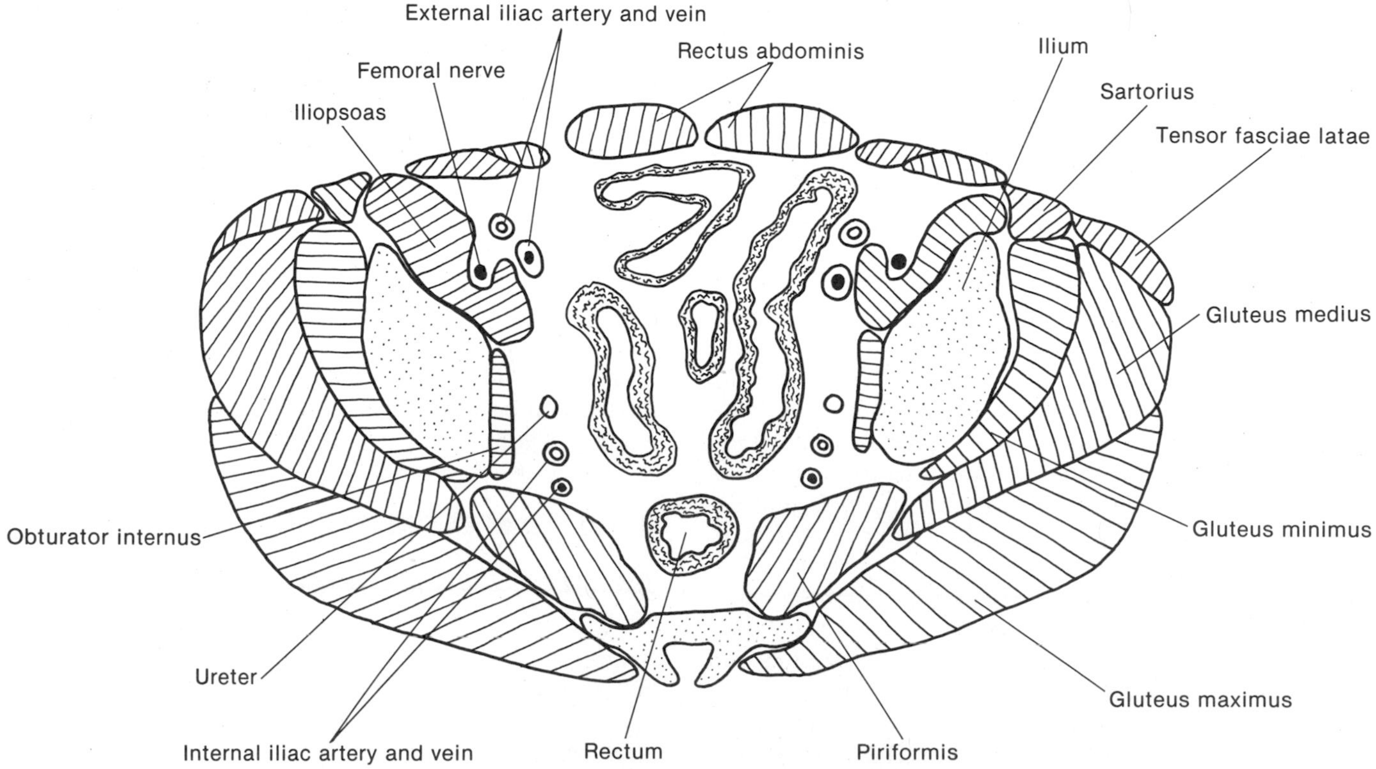

FIGURE 4-10. Transverse section through the lower part of the sacrum.

inferior vena cava. Figure 4–9 illustrates a section through the sacroiliac joint. Common iliac arteries and common iliac veins are present at this level. The lower portions of the **ascending colon** and the **cecum** are located in the right iliac fossa. Whether the ascending colon or the cecum is present depends on individual variation and the level of the section. The smaller **descending colon** is seen in the left iliac fossa.

The variable nature of the **transverse colon** makes it impossible to state specifically when it will or will not be seen. It drapes inferiorly and anteriorly between the right and the left colic flexures. In some individuals, it is very pendulous and extends down into the pelvis; transverse sections through the false pelvis of these individuals show two regions of the transverse colon. One region is on the right side as the transverse colon descends, and the other is on the left as it ascends toward the left colic (splenic) flexure. More commonly, sections through the sacroiliac joint are inferior to the transverse colon.

As seen in sections of the abdomen, the **psoas muscles** are on either side of the vertebral column. The **ureters** descend anterior to the psoas muscle. The **iliacus muscle** lines the iliac fossa, and the **gluteal muscles** originate on the lateral surface of the ilium (Fig. 4–9).

TRANSVERSE SECTION THROUGH THE LOWER PART OF THE SACRUM

The relationships of structures, primarily muscles, in transverse sections through the lower part of the sacrum are illustrated in Figure 4–10. The fibers of the psoas and iliacus muscles are merged at this level, forming the **iliopsoas muscle,** which is apparent in Figure 4–10. Observe the close relationship of the **femoral nerve** to this muscle. All three gluteal muscles are evident lateral and posterior to the ilium. The **gluteus maximus** is the most superficial, and the **gluteus minimis** is the deepest, next to the ilium. The **piriformis muscle** is located in the **greater sciatic foramen** (notch) between the ilium and the sacrum. The **obturator internus** muscle, which lines the cavity of the true pelvis, is medial to the ilium. The common iliac vessels have usually bifurcated at this level, with the **external iliac vessels** associated with the iliopsoas muscles as they proceed to the upper thigh region. The **ureter** is more closely associated with the **internal iliac vessels** in the true pelvis.

TRANSVERSE SECTION THROUGH THE SEMINAL VESICLES

Figure 4–11 illustrates the relationships of the **ureters,** the **seminal vesicles,** and the **ductus deferens.** The ureters are shown as they penetrate the wall of the bladder. The **seminal vesicles** are glandular structures between the rectum and the urinary bladder. The **ductus deferens** are medial to both the ureters and the seminal vesicles. The space between the bladder and the rectum is the **rectovesical space,** a peritoneal cul-de-sac in the male. Anteriorly, between the iliopsoas and the pectineus muscles, the **femoral triangle** contains the femoral artery, vein, and nerve. The **spermatic cord,** anterior and medial to the margin of the pectineus muscle, contains the testicular artery, the ductus deferens, and the pampiniform venous plexus.

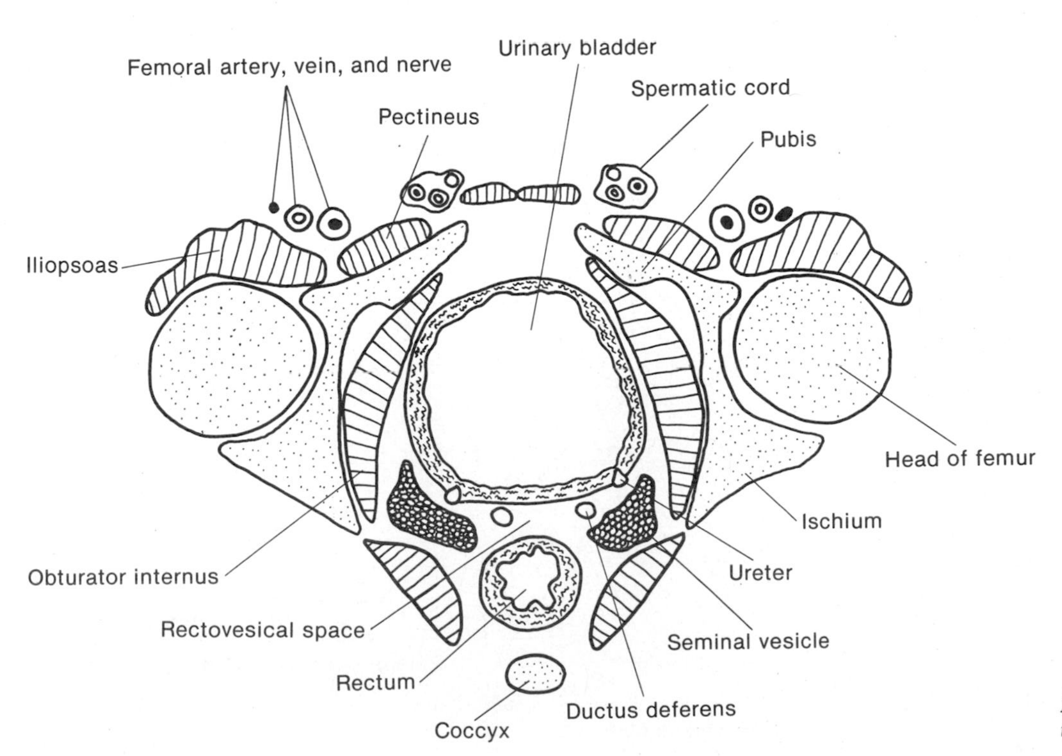

FIGURE 4–11. Transverse section through the seminal vesicles.

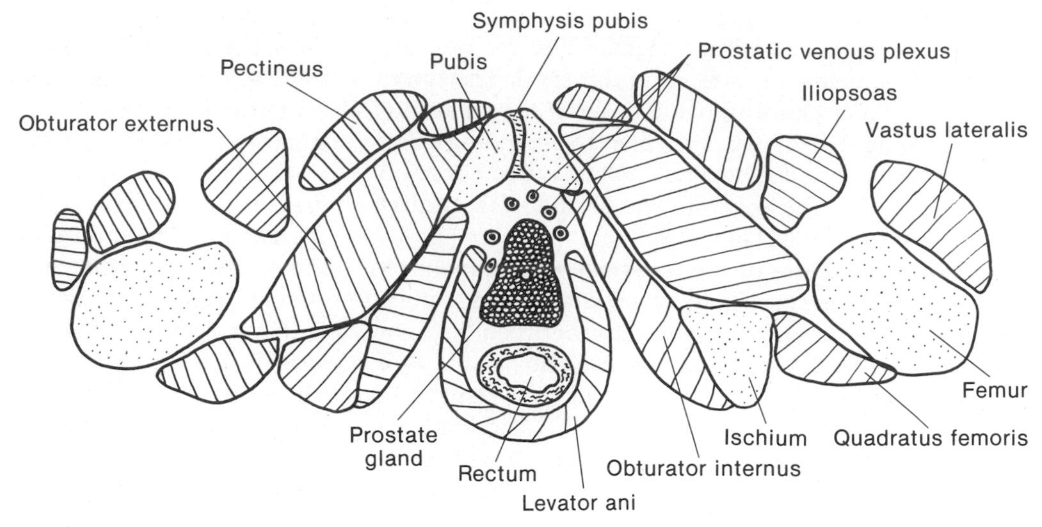

FIGURE 4-12. Transverse section through the prostate gland.

TRANSVERSE SECTION THROUGH THE PROSTATE GLAND

The **prostate gland** is inferior to the bladder and posterior to the symphysis pubis, and it surrounds the **prostatic urethra** (Fig. 4-12). A **prostatic venous plexus** surrounds the gland. The **rectum** is posterior to the prostate. The prostate rests on the curved pelvic floor; thus a portion of the floor, specifically the **levator ani muscle,** is evident. The pelvic walls are lined with the **obturator internus muscle.**

TRANSVERSE SECTION THROUGH THE ROOT OF THE PENIS

Figure 4-13 illustrates the root of the penis, which consists of a bulb and two crura. The **bulb** portion of the root is the **corpus spongiosum,** which surrounds the urethra. The **transversus perinei muscle** extends horizontally between the two ischial rami, with the bulb anterior, and the anal canal posterior, to the muscle. The **bulbospongiosus muscle** is associated with the bulb of the penis. Lateral to the bulb, the two **crura** of the **corpora cavernosa** diverge and attach to the ischial rami. The **ischiocavernosus muscle** is related to the crura.

MIDSAGITTAL SECTION OF THE MALE PELVIS

Figure 4-14 illustrates a midsagittal section of the male pelvis. Posteriorly, this shows how the **rectum** follows the curvature of the **sacrum.** The peritoneum continues down the posterior wall and then curves over the **seminal vesicles** and the **urinary bladder.** The seminal vesicles are located between the bladder and the rectum, just inferior to the peritoneal **rectovesical pouch.** Posterior to the symphysis pubis, the urinary bladder rests on the pelvic floor. The duct from the seminal vesicles joins the ductus

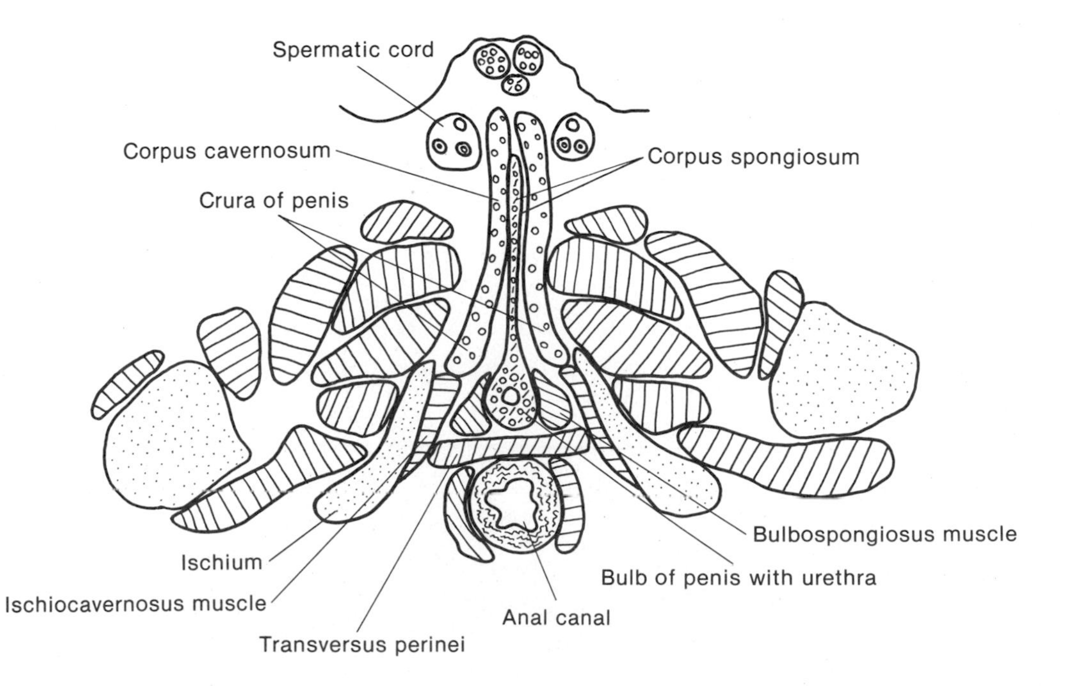

FIGURE 4-13. Transverse section through the root of the penis.

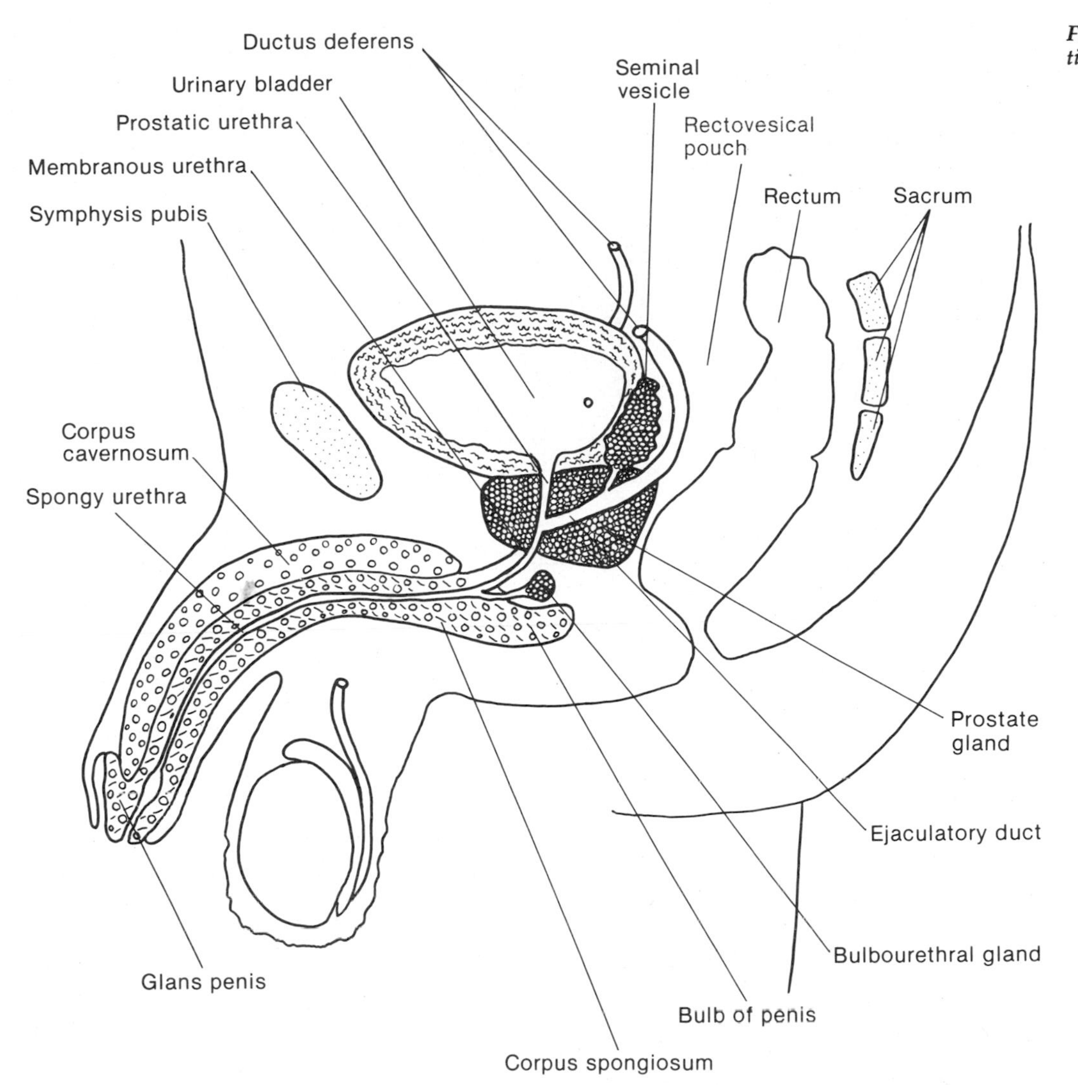

FIGURE 4-14. Midsagittal section of the male pelvis.

deferens, and they form the ejaculatory duct, which penetrates the **prostate gland** to empty into the **prostatic urethra.** The urethra continues through the urogenital diaphragm as the **membranous urethra** and then enters the corpus spongiosum of the penis to become the **penile** or **spongy urethra.** The duct from the bulbourethral gland empties into the spongy urethra.

CORONAL SECTION THROUGH THE PROSTATE GLAND AND THE ROOT OF THE PENIS

Figures 4-15 *A* and *B* illustrate a coronal section of the male pelvis through the prostate gland and the root of the penis. **Obturator internus muscles** line the pelvic wall and fill the space of the obturator foramen. **Levator ani muscles** form the hammock-shaped pelvic floor, and the **transversus perinei muscle** extends between the two ischial tuberosities. The **prostate gland,** inferior to the **urinary bladder,** rests on the pelvic floor. It encircles the **prostatic urethra.** The urethra continues through the muscle and the fascia of the **urogenital diaphragm** as the **membranous urethra,** and then it penetrates the corpus spongiosum in the bulb of the penis to become the **spongy urethra.**

The components of the root of the penis, the bulb and the two crura, in the perineal space are also illustrated in Figure 4-15. The **bulb,** which consists of the **corpus spongiosum,** is surrounded by the **bulbospongiosus muscle** and is anchored to the perineal membrane. It encircles the spongy urethra. The two **crura** of the **corpus cavernosum** are surrounded by the **ischiocavernosus muscle** and are anchored to the ischial tuberosities.

Sections of the Female Pelvis

TRANSVERSE SECTION THROUGH THE UTERUS

Figure 4-16 illustrates a transverse section through the **uterus.** This section shows the **broad ligament** that extends from the uterus to the lateral pelvic wall. The **ovaries** are attached to the posterior portion of the broad ligament by the mesovarium, and the **uterine tube,** is in the upper margin of the broad ligament. The uterine tube

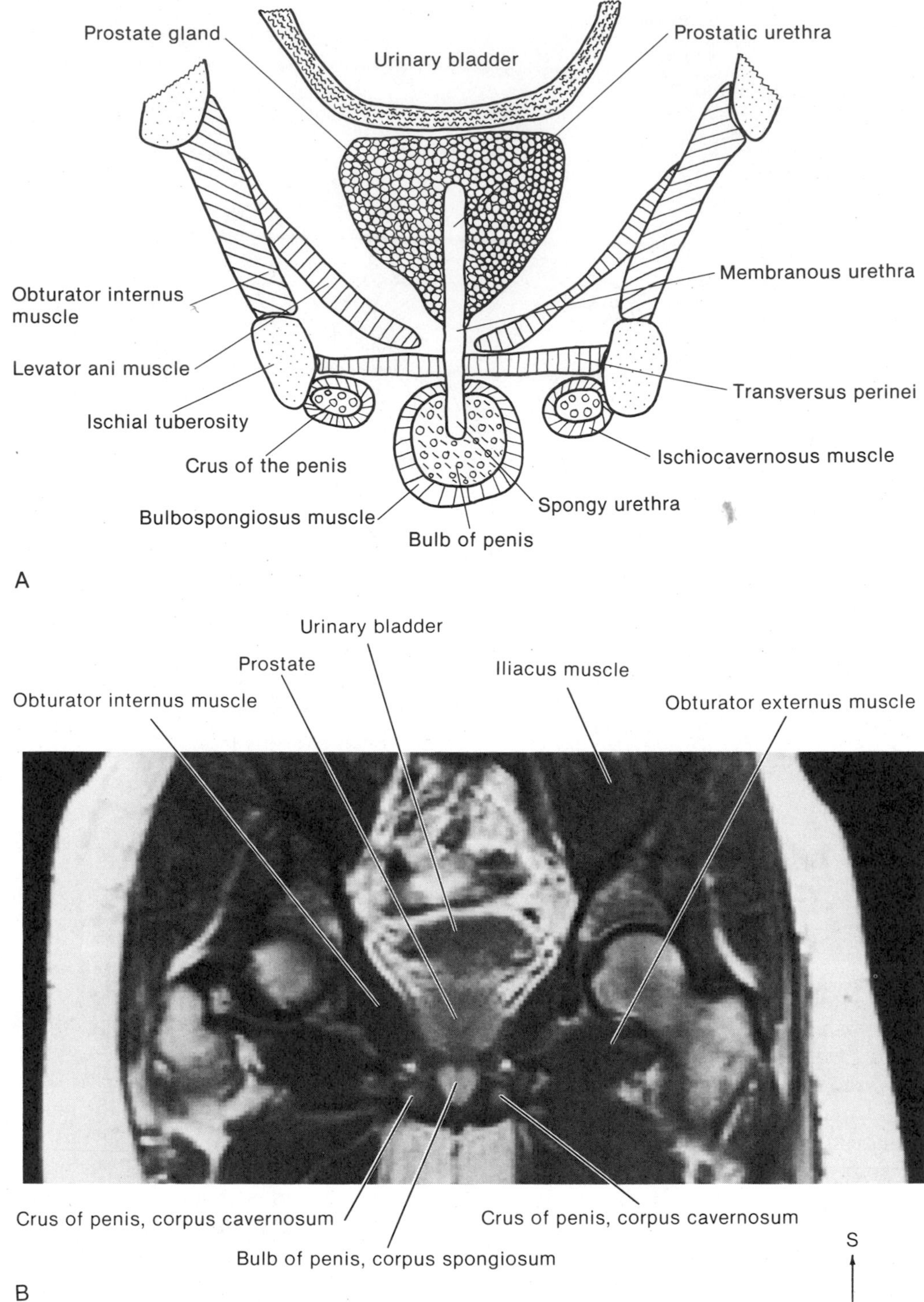

FIGURE 4-15. A, Coronal section through the prostate gland and the root of the penis. B, Coronal magnetic resonance image through a similar region.

may be difficult to see. The space between the uterus and the rectum is called the **rectouterine pouch** (pouch of Douglas) and may contain loops of bowel. The **obturator internus** muscle originates on the inner surface of the ilium and covers most of the lateral wall of the true pelvis. The musculature at this level is similar to that in the male.

TRANSVERSE SECTION THROUGH THE URINARY BLADDER

Figure 4-17 illustrates a transverse section through the superior surface of the urinary bladder. This shows the **ureters** situated posterior to the bladder as they are about

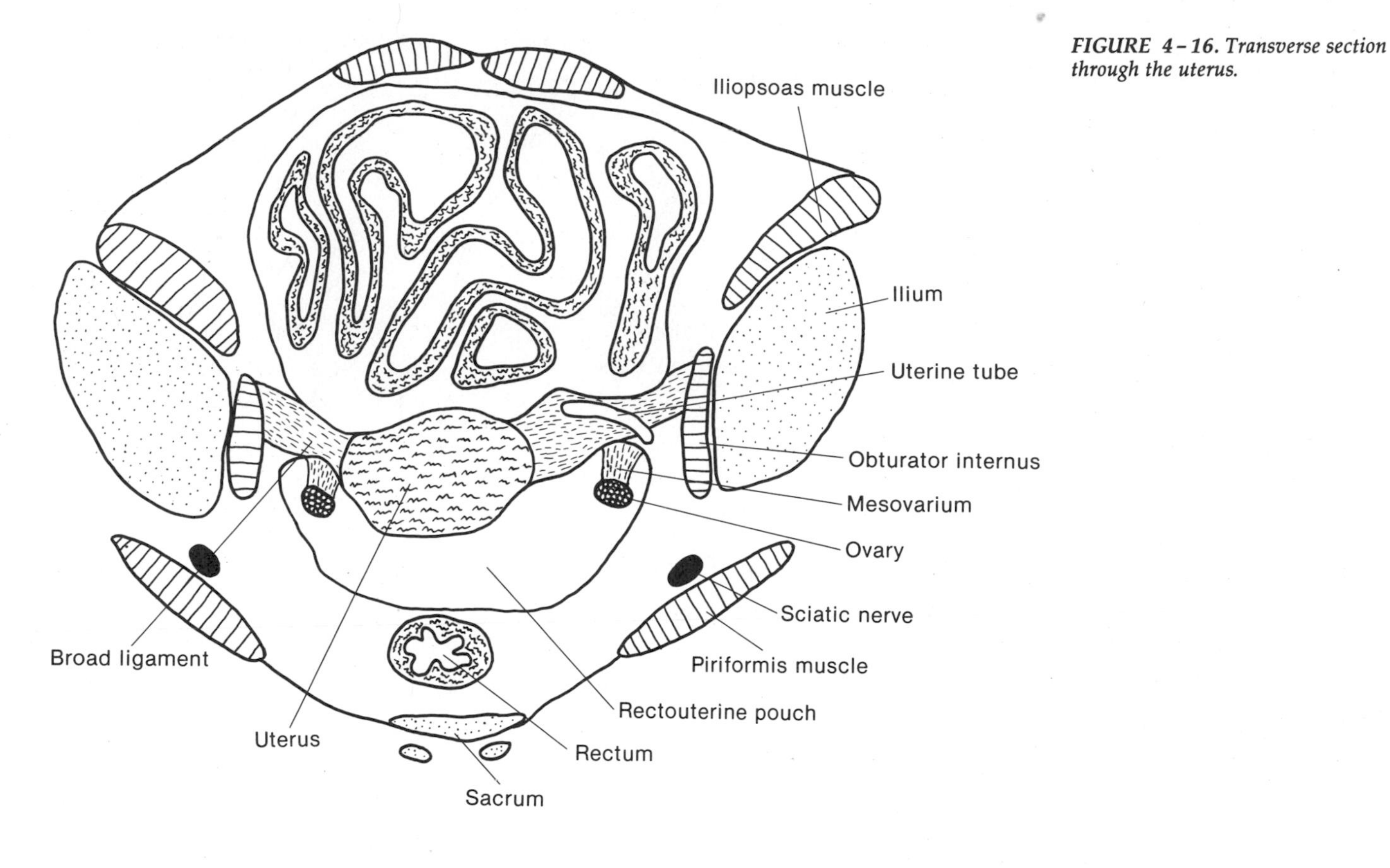

FIGURE 4–16. Transverse section through the uterus.

to penetrate the bladder wall. The **cervix** is interposed between the bladder and the rectum. The **obturator internus** and the **levator ani** muscles form the lateral walls and the floor of the pelvic cavity. Anteriorly, the **pectineus muscle** originates on the pubis.

MIDSAGITTAL SECTION THROUGH THE FEMALE PELVIS

Figure 4–18 illustrates a midsagittal section through the female pelvis. Posteriorly, the **rectum** follows the curvature of the sacrum. Anteriorly, the **urinary bladder** is in a

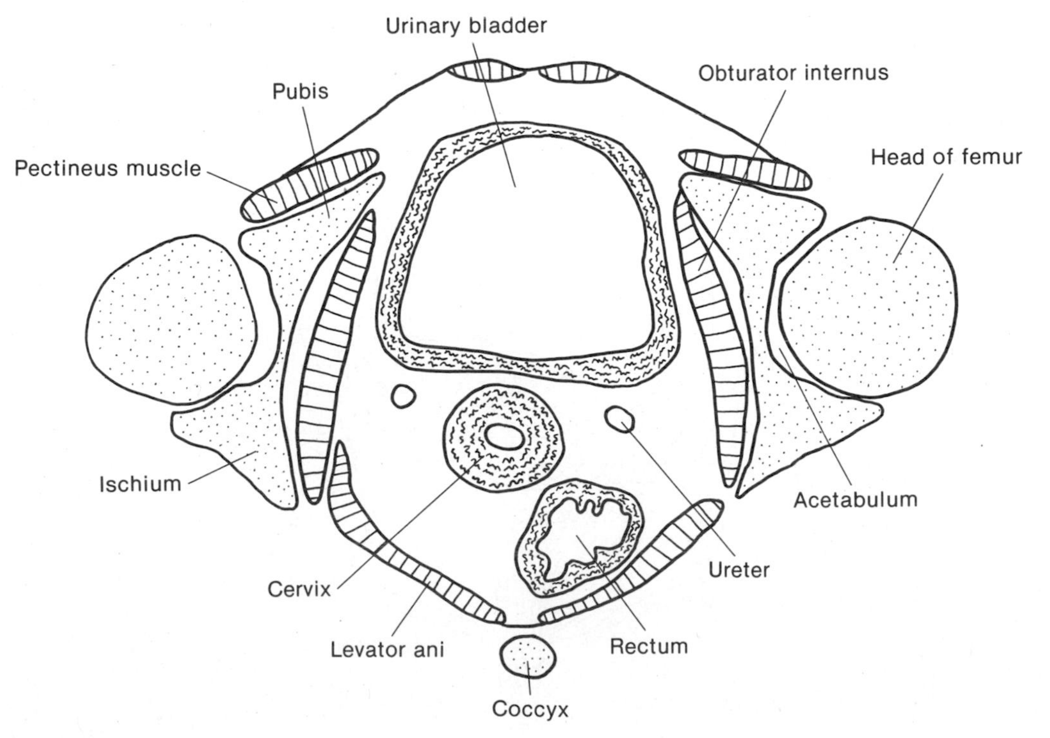

FIGURE 4–17. Transverse section through the urinary bladder.

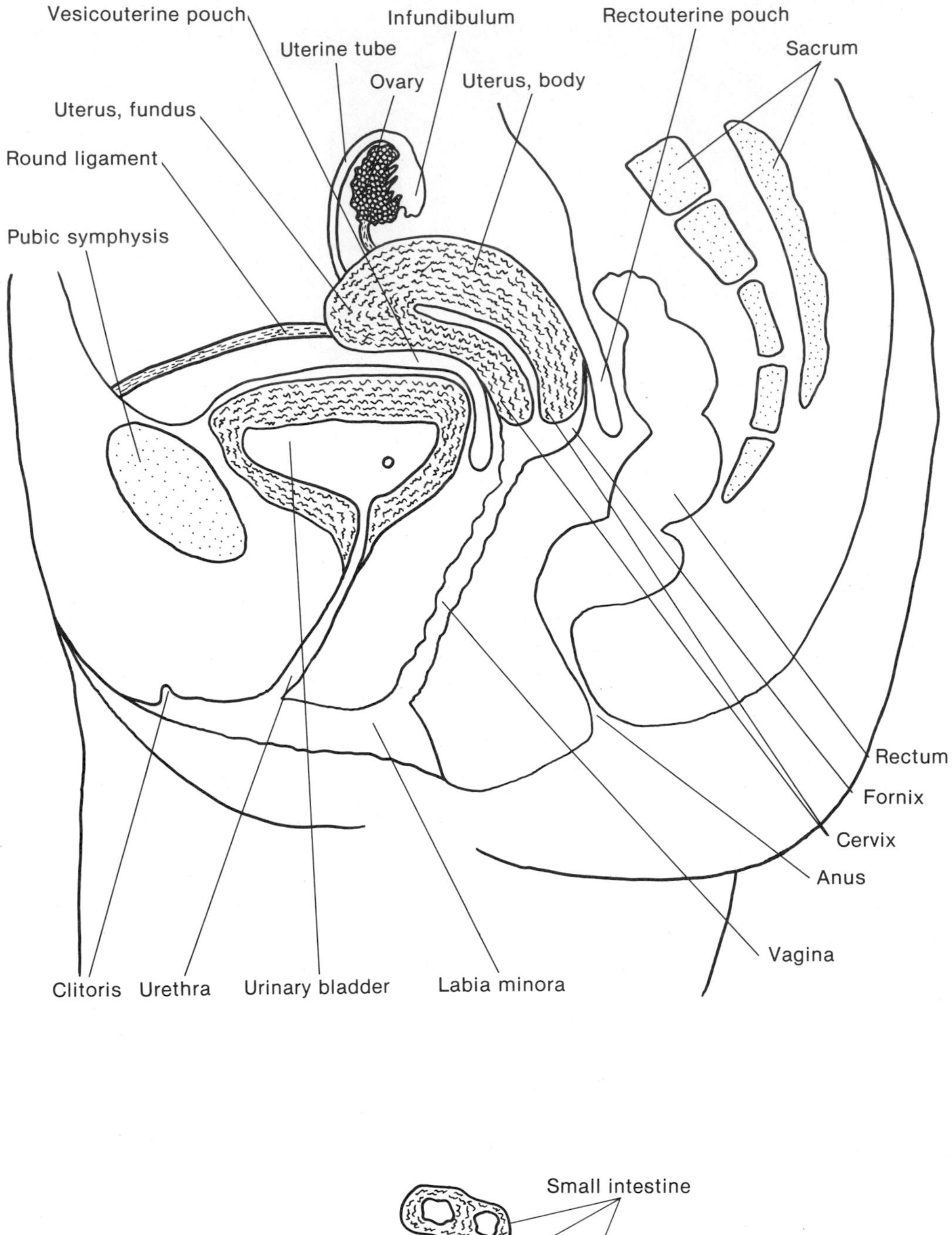

FIGURE 4–18. Midsagittal section through the female pelvis.

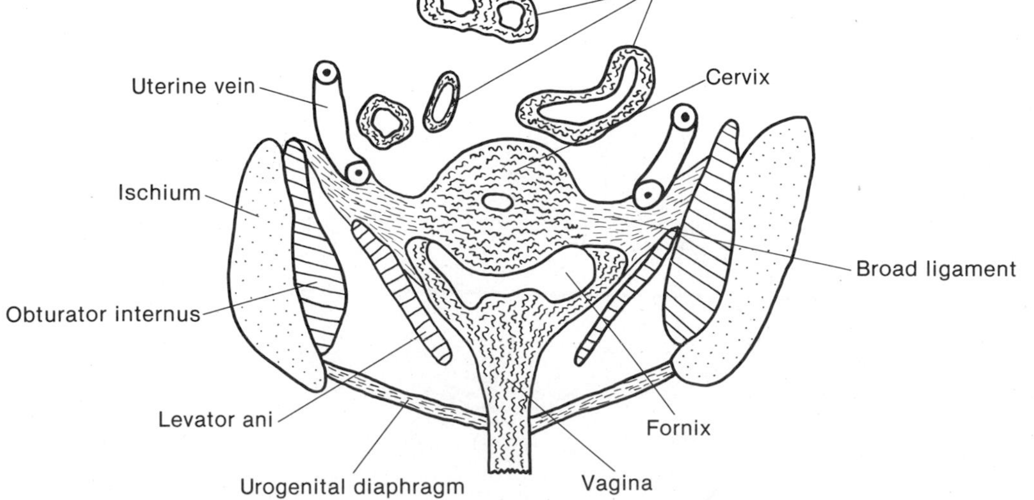

FIGURE 4–19. Coronal section through the cervix and the vagina.

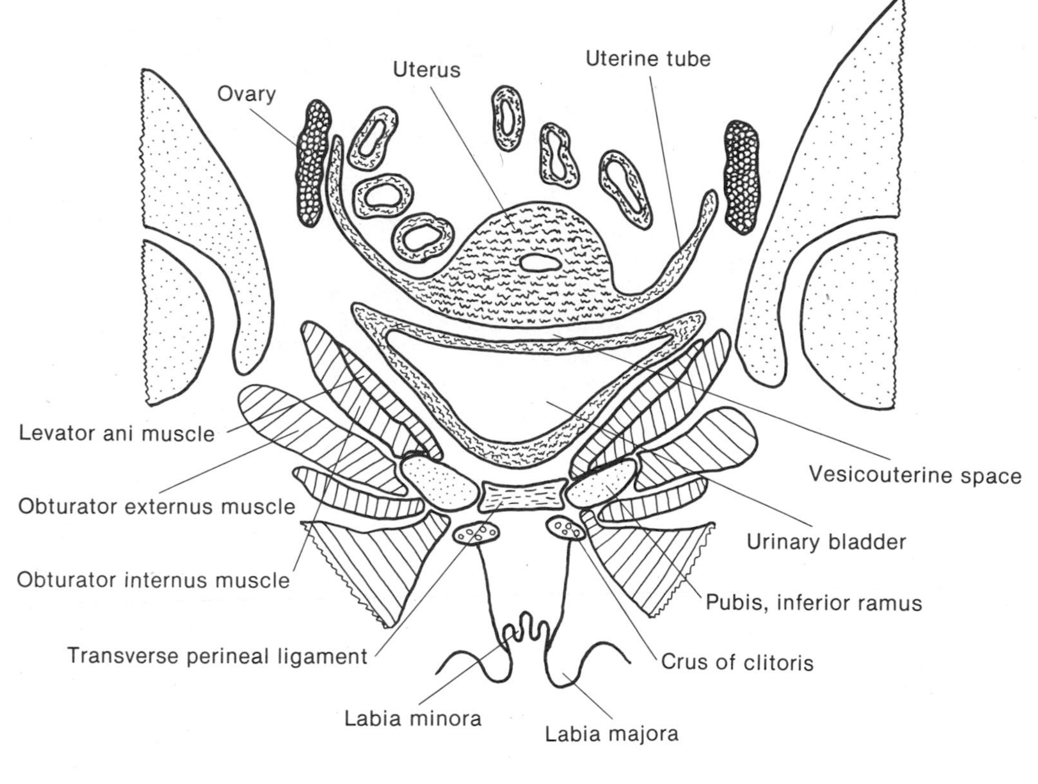

FIGURE 4-20. Coronal section through the body of the uterus and the urinary bladder.

position immediately posterior to the symphysis pubis. The **fundus** and the **body** of the uterus are anteverted over the superior surface of the bladder and are anteflexed with the **cervix.** The peritoneum forms two cul-de-sacs that are related to the uterus; these cul-de-sacs are nicely illustrated in midsagittal sections. Posteriorly, the **rectouterine pouch** extends between the rectum and the uterus; then, as the peritoneum continues from the fundus of the uterus to the superior surface of the bladder, it forms the **vesicouterine pouch.** The **vagina** slants posteriorly as it ascends to the cervix of the uterus, where the vagina surrounds the cervix to form the **fornices.** The **posterior fornix** is longer than the anterior fornix. The **urethra** extends from the bladder and through the urogenital diaphragm and opens into the **vestibule,** anterior to the vaginal orifice and posterior to the **clitoris.** The **labia minora** form the lateral margins of the vestibule.

CORONAL SECTION THROUGH THE CERVIX AND THE VAGINA

Figure 4-19 illustrates a coronal section through the posterior region of the female pelvis. In normal anatomic relationships, the most posterior part of the **uterus** is the **cervix,** which is the portion shown in this illustration. The **posterior fornix** is evident, since it is the most posterior part of the vagina. The obturator internus muscles form the lateral walls of the pelvic cavity.

CORONAL SECTION THROUGH THE BODY OF THE UTERUS

Figure 4-20 illustrates a coronal section through the anterior part of the pelvis, and shows the **body** of the uterus and the **urinary bladder. Uterine tubes** extend from the uterus toward the lateral pelvic wall. The urinary bladder is inferior to the uterus and is separated from it by the **vesicouterine** space. The **obturator internus** muscles form the lateral walls of the cavity, and the hammocklike **levator ani** muscles form the pelvic floor. Two **crura** of the clitoris, consisting of corpus cavernosum, are associated with the ischiopubic ramus. The **labia majora,** which are homologous to the scrotum in the male, enclose the smaller **labia minora,** which form the lateral margins of the vestibule.

REVIEW QUESTIONS

1. The term "pelvis" refers to
   a. the cavity enclosed by the sacrum and os coxae.
   b. the region of the body inferior to the abdomen where the lower limb articulates with the trunk.
   c. the coccyx, the sacrum, the ilium, the ischium, and the pubis.
   d. two of the above.
   e. all of the above.
2. **True** or **False:** In the child, the os coxa consists of an ilium, an ischium, and a pubis that are separated by hyaline cartilage.

3. **True** or **False:** The true pelvis is limited by the iliac crest superiorly and by the pelvic outlet inferiorly.

4. The muscle that passes through the greater sciatic notch and closes off that opening is the
   a. psoas.
   b. obturator internus.
   c. pectineus.
   d. piriformis.
   e. levator ani.

5. Which of the following is **NOT** a true statement?
   a. Both the pelvic diaphragm and the urogenital diaphragm contribute to the pelvic floor.
   b. The urogenital diaphragm is principally connective tissue and the coccygeus muscle.
   c. The levator ani is the principal muscle of the pelvic diaphragm.
   d. The urogenital diaphragm is superficial to the pelvic diaphragm.

6. The ____________ nerve originates in the sacral plexus and supplies the perineum.

7. **True** or **False:** The rectum is related anteriorly to the urinary bladder in both the male and the female. In addition to the urinary bladder, the rectum is related anteriorly to the seminal vesicles in the male or to the vagina in the female.

8. In the male, the neck of the urinary bladder rests on the ____________. In the female it rests on the ____________.

9. Given the following numeric code, which of the sequences best represents the pathway for sperm: (1) ejaculatory duct, (2) seminiferous tubules, (3) ductus deferens, (4) urethra, (5) epididymis?
   a. 1, 2, 3, 4, 5
   b. 2, 3, 1, 5, 4
   c. 2, 1, 5, 3, 4
   d. 2, 5, 3, 1, 4
   e. 5, 2, 1, 3, 4

10. The ____________ secrete(s) a fluid with a high fructose content that provides an energy source for sperm.

11. Which of the following is **NOT** a true statement?
    a. Ovaries are normally located near the lateral pelvic walls.
    b. Mesovarium, suspensory, and ovarian ligaments help keep the ovaries in place.
    c. Ovaries are usually displaced inferiorly during pregnancy.
    d. Loops of small intestine may displace the ovaries.

12. The double layer of peritoneum that extends laterally from the surface of the uterus is the ____________.

13. The uterus is normally
    a. anteverted and anteflexed.
    b. anteverted and dorsiflexed.
    c. retroverted and anteflexed.
    d. retroverted and dorsiflexed.

14. The muscle in the midline of the urogenital region of the perineum is the ____________.

15. Which of the following is **NOT** a true statement?
    a. There are two dorsal columns of corpus spongiosum in the body of the penis.
    b. The expanded distal end of the ventral column is the glans penis.
    c. The ventral column expands proximally to form the bulb.
    d. The crura of the penis are associated with the ischiocavernosus muscle.

16. Given the following numeric code, which of the sequences best describes the locations of the structures from anterior to posterior: (1) vaginal orifice, (2) mons pubis, (3) rectum, (4) clitoris, (5) urethral orifice?
    a. 4, 2, 1, 5, 3
    b. 2, 4, 5, 1, 3
    c. 2, 4, 1, 5, 3
    d. 3, 1, 5, 4, 2
    e. 4, 2, 5, 1, 3

17. Which of the following are true statements?
    (1) The ascending colon is in the right iliac fossa.
    (2) The sciatic nerve is associated with the iliopsoas muscle.
    (3) The psoas muscle is more medial than is the iliacus.
    (4) The obturator internus muscle is lateral to the ilium.
    (5) The external iliac vessels are associated with the iliopsoas muscle, and the internal iliac vessels are associated with the obturator internus muscle.
    a. 1, 2, 5
    b. 2, 4, 5
    c. 2, 4
    d. 1, 3, 5
    e. 1, 3

18. The glandular structures between the rectum and the urinary bladder in the male are the ____________.

19. The peritoneal cul-de-sac between the uterus and urinary bladder is the ____________.

20. The recesses of the vagina around the cervix are called the ____________.

21. Identify the indicated structures on Figure *A*.

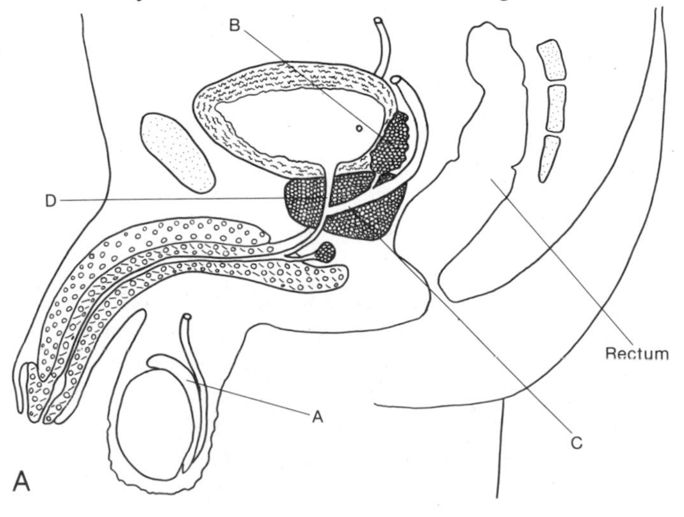

22. Identify the indicated structures on Figure *B*.

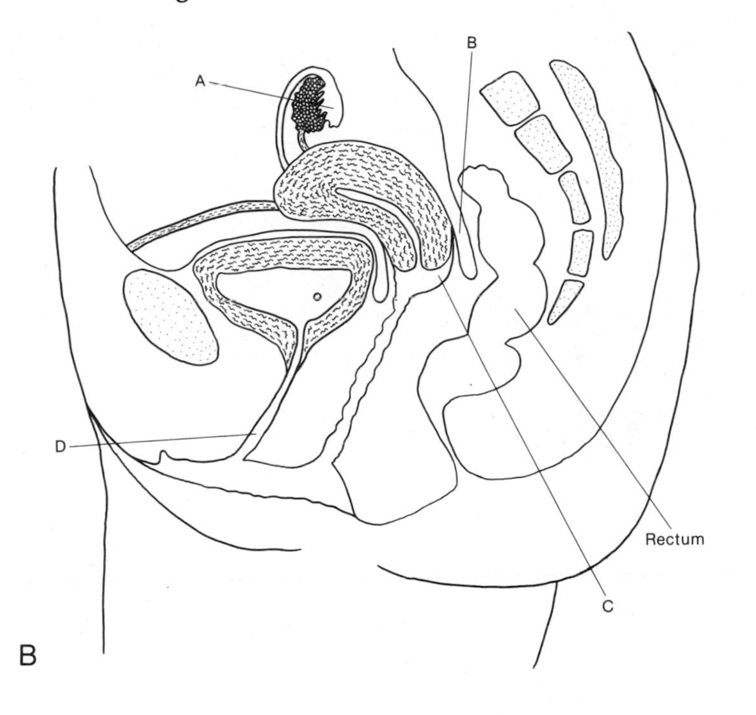

CHAPTER 5

EXTREMITIES AND ARTICULATIONS

OBJECTIVES

Upon completion of this chapter, the student should be able to do the following:

- Identify the bones that make up the pectoral girdle.
- Describe the location and the functions of three groups of muscles that are associated with the attachment of the pectoral girdle and the upper extremity to the trunk of the body.
- Describe the boundaries and contents of the axilla.
- Identify the skeletal, muscular, vascular, and neural components of the arm.
- Describe the boundaries and the contents of the cubital fossa.
- Identify the skeletal, muscular, vascular, and neural components of the forearm.
- Discuss the structure and the significance of the carpal tunnel.
- Describe the structure of the shoulder joint, and discuss the anatomic relationships of its components.
- Describe the structure of the elbow joint, and discuss the anatomic relationships of its components.
- Identify the structural components of the upper extremity in transverse sections.
- Name the muscles in the gluteal region, and describe their anatomic relationships with other body structures.
- Identify the skeletal and the muscular components of the thigh.
- Describe the location, boundaries, and contents of the femoral triangle.
- Identify the skeletal and muscular components of the leg.
- Describe the location, boundaries, and contents of the popliteal fossa.
- Describe the structure of the hip (coxal) joint, and discuss the anatomic relationships of its components.
- Describe the structure of the knee joint, and discuss the anatomic relationships of its components.
- Describe the structure of the ankle joint, and discuss the anatomic relationships of its components.
- Identify the structural components of the lower extremity in transverse sections.
- Identify the structural components of the articulations associated with the upper and lower extremities in transverse, sagittal, and coronal planes.

UPPER EXTREMITY

The upper extremity consists of the arm, forearm, wrist, hand, and fingers. There is a single bone, the humerus, in the arm. Two bones, the radius on the lateral side and the ulna on the medial side, form the framework of the forearm. The wrist consists of eight small bones that are collectively called carpals. The hand has five bones that are known as the metacarpals, and distal to these, there are fourteen phalanges that form the fingers. These bones are covered with muscle, fascia, and skin.

Attachment of the Upper Extremity to the Trunk

The scapula and the clavicle make up the **pectoral girdle,** which provides the connection between the upper extremity and the axial skeleton. Muscles anchor the bones of the upper extremity and the pectoral girdle to the trunk of the body. These muscles can be divided into three groups.

One group extends from the trunk to the scapula. The muscles in this group can, in appropriate combinations, move the scapula upward, downward, forward, backward, clockwise, or counterclockwise. These actions assist in movement of the shoulder.

A second group of muscles extends between the scapula and the humerus. These muscles move the arm at the glenohumeral joint. The muscles and tendons that extend over the shoulder strengthen and stabilize the joint.

A third group attaches the humerus to the trunk. This group, which includes the pectoralis major and the latissimus dorsi, functions to adduct the arm. Table 5–1 summarizes the muscles associated with the trunk, the scapula, and the humerus.

Axilla

The space at the junction of the arm and the thorax, between the upper limb and the chest wall, is called the **axilla.** The anterior wall of the axilla is formed by the **pectoralis major** and the **pectoralis minor** muscles. Predominant structures in the posterior wall are the **scapula** and the **subscapularis** muscle. Medially, the axilla is delineated by the **ribs,** the **intercostal muscles,** and the **serratus anterior muscle.** The narrow lateral wall is formed by the head of the **humerus;** specifically, it is formed by the intertubercular groove, where the anterior and posterior walls converge. The boundaries of the axilla are illustrated in Figure 5–1.

The axilla functions as a passageway for vessels and nerves that pass between the root of the neck and the arm. The vessels in this region include the axillary artery and vein, together with their branches. The nerves, which are all branches of the brachial plexus, innervate the arm. The axilla contains numerous lymph nodes, which are drained by axillary lymph vessels that pass through this region. The lymph nodes in the area are of particular significance because of their frequent involvement in breast cancer.

TABLE 5-1. Muscles Associated With the Trunk, the Scapula, and the Humerus

Muscle	Origin	Insertion	Action	Innervation
Extend From Trunk to Scapula				
Trapezius	Thoracic vertebrae	Spine of scapula	Adduct scapula	Accessory (XI)
Rhomboids	Cervical and thoracic vertebrae	Medial border and spine of scapula	Adduct scapula	Dorsal scapular
Levator scapulae	Cervical vertebrae	Medial border of scapula	Elevate scapula	Dorsal scapular
Pectoralis minor	Third to fifth ribs	Coracoid process of scapula	Pull scapula inferiorly	Medial pectoral
Serratus anterior	First eight ribs	Medial border of scapula	Rotate scapula	Long thoracic nerve
Extend from Scapula to Humerus				
Deltoid	Acromion and spine of scapula; clavicle	Deltoid tuberosity of humerus	Abduct arm	Axillary
Supraspinatus	Supraspinous fossa	Greater tubercle of humerus	Abduct arm	Suprascapular
Subscapularis	Subscapular fossa	Lesser tubercle of humerus	Medial rotation of arm	Subscapular
Infraspinatus	Infraspinous fossa	Greater tubercle of humerus	Lateral rotation of arm	Suprascapular
Teres minor	Lateral margin of scapula	Greater tubercle of humerus	Lateral rotation of arm	Axillary
Teres major	Superior lateral margin of scapula	Intertubercular groove of humerus	Adduct arm	Subscapular
Coracobrachialis	Coracoid process of scapula	Shaft of humerus	Adduct arm	Musculocutaneous
Extend from Trunk to Humerus				
Pectoralis major	Clavicle, sternum, costal cartilages	Intertubercular groove of humerus	Adduct and medially rotate humerus	Medial and lateral pectoral
Latissimus dorsi	Thoracic and lumbar vertebrae; crest of ilium	Intertubercular groove of humerus	Adduct and medially rotate humerus	Thoracodorsal

General Anatomy of the Arm

The region from the shoulder to the elbow is the arm, or brachium. The only bone in the arm is the humerus. This is the longest bone in the upper extremity. The muscles of the arm are arranged in anterior and posterior compartments that are separated by fascia. The muscles of the arm are summarized in Table 5-2.

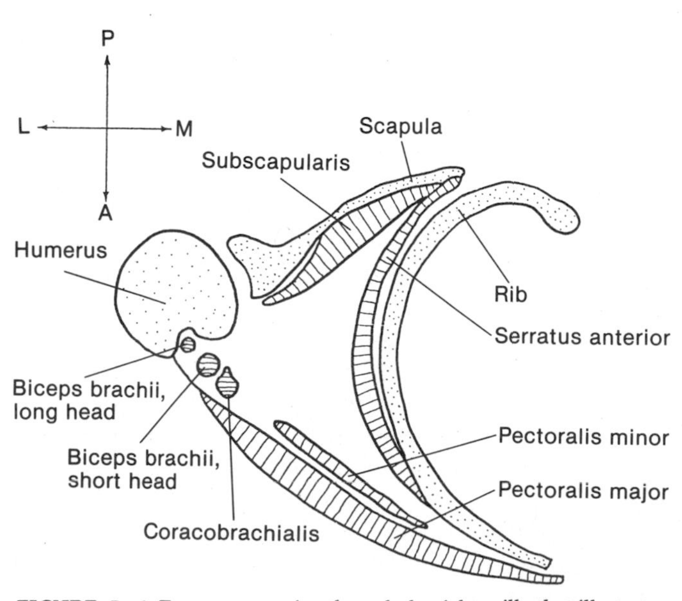

FIGURE 5-1. Transverse section through the right axilla that illustrates boundaries.

MUSCLES LOCATED IN THE ARM

ANTERIOR MUSCLE COMPARTMENT. The anterior muscle compartment consists of three muscles that act as flexors and are innervated by the musculocutaneous branch of the brachial plexus.

Biceps Brachii. The largest of the muscles in the anterior compartment is the **biceps brachii.** As the name implies, the biceps brachii has two heads of origin. The short head originates on the coracoid process of the scapula. The long head originates from a tubercle just above the glenoid fossa (supraglenoid tubercle). The two heads merge, forming a single muscle belly that inserts on the radial tuberosity and on the ulna by way of an aponeurosis. The biceps brachii is a flexor of the forearm at the elbow and a supinator of the forearm. The muscle originates superior to the shoulder and passes over the joint; thus, it assists in stabilizing and strengthening that joint. The biceps brachii also acts as a flexor of the arm at the shoulder.

Coracobrachialis. The **coracobrachialis** is a short muscle on the medial surface of the superior part of the arm. It has a common origin with the short head of the biceps brachii on the coracoid process of the scapula, and it inserts on the medial side of the humerus near the midpoint of the shaft. The coracobrachialis along with the biceps brachii acts as a weak flexor and adductor of the shoulder. This muscle is visible only in sections through the upper part of the arm. The coracobrachialis is the only muscle that, although predominantly located in the arm, acts on the shoulder joint.

TABLE 5-2. Muscles Located in the Arm

Muscle	Origin	Insertion	Action	Innervation
Anterior Compartment				
Biceps brachii	Short head: coracoid process Long head: supraglenoid tubercle	Radius and ulna	Flex and supinate forearm	Musculocutaneous
Coracobrachialis	Coracoid process	Medial humerus	Flex and adduct shoulder	Musculocutaneous
Brachialis	Distal humerus	Ulna	Flex forearm	Musculocutaneous
Posterior Compartment				
Triceps brachii	Long head: infraglenoid tubercle Lateral head: proximal shaft of humerus Medial head: distal shaft of humerus	Olecranon process of ulna	Extend arm at elbow and stabilize shoulder	Radial

Brachialis. The third muscle of the anterior compartment is the **brachialis.** This is a deep muscle, underlying the biceps brachii. It has an extensive origin along the anterior surface of the distal half of the humerus, and it terminates on the coronoid process of the ulna. The brachialis is a strong flexor of the forearm at the elbow joint. It is seen only in sections through the lower part of the arm.

POSTERIOR MUSCLE COMPARTMENT. The posterior compartment of the arm is occupied by a single, large muscle, the **triceps brachii.** As the name implies, this muscle has three heads of origin. The long head originates via a tendon from the infraglenoid tubercle. The lateral head attaches to the proximal shaft of the humerus on the posterior surface. The medial head is deep to both the long and the lateral heads. Its origin is on the posterior surface of the shaft of the humerus distal to the origin of the lateral head. All three heads merge, forming a single muscle belly, which inserts via a single tendon on the olecranon process of the ulna. There is an olecranon bursa between the tendon and the olecranon process. The triceps brachii is a powerful extensor of the elbow. The long head spans the shoulder joint; consequently, it also helps to stabilize that joint. The radial nerve innervates the triceps brachii.

VASCULATURE OF THE ARM

The primary arterial blood supply to the arm is the **brachial artery** and its branches. The brachial artery begins at the inferior border of the teres major muscle, as a continuation of the axillary artery, and it ends in the cubital fossa, where it divides into the radial and ulnar arteries. The vessel is superficial throughout its length, and it runs its course in the fascia of the medial intermuscular septum that divides the muscles into the anterior and posterior compartments. In the septum, it is associated with the basilic vein, the median nerve, and the ulnar nerve. It gives off numerous branches, which supply the muscles of the arm.

One, or possibly two, **brachial veins** accompany the brachial artery. These deep veins ascend through the arm to continue as the axillary vein. In addition to the deep brachial vein, there are two important superficial veins in the arm. The **cephalic vein** is in the superficial fascia, anterolateral to the biceps brachii muscle. As it courses superiorly, it passes between the deltoid and the pectoralis major muscles and empties into the axillary vein. The **basilic vein** is in the superficial fascia on the medial side of the arm. About one third of the way up the arm from the elbow, the basilic vein passes deep to the superficial fascia and continues upward to merge with the brachial vein, forming the axillary vein. Both the superficial cephalic and basilic veins are frequently visible through the skin.

NERVES IN THE ARM

The major nerves traversing the arm are the **musculocutaneous,** the **median,** the **ulnar,** and the **radial.** A fifth nerve, the **axillary,** supplies the skin over the upper part of the arm. The nerves are all terminal branches of the **brachial plexus.** Both the median and the ulnar nerves descend the arm without giving off branches. They supply the elbow joint and the forearm. In the uppermost part of the arm, the median nerve may be either lateral or anterior to the brachial artery. About midway down the arm, the nerve crosses over the vessel to the medial side. The ulnar nerve is situated near the brachial artery in the upper half of the arm. It then penetrates the intermuscular septum and descends through the arm just anterior to the medial head of the triceps brachii. The musculocutaneous and radial nerves give off branches to supply the muscles of the arm. The musculocutaneous nerve descends between the biceps brachii and the brachialis muscles. As it descends, it gives off branches that innervate those two muscles and the coracobrachialis muscle. The radial nerve enters the arm on the medial side of the humerus and then curves around the bone, in the radial groove, descending in the intermuscular septum on the lateral side.

Transverse Sections of the Arm

SECTION THROUGH THE PROXIMAL ARM

Figure 5–2 illustrates a transverse section through the upper region of the arm. The **deltoid** muscle is superficial on the lateral side of the humerus and the long head of the **biceps brachii** is anterior. The **cephalic vein** is in the superficial fascia, anterior to these two muscles. At upper levels, such as the level in Figure 5–2, the **coracobrachialis muscle** is adjacent to the long head of the biceps brachii. The coracobrachialis has a common origin with the short head of the biceps, and at this level, the two may be indistinguishable. The **musculocutaneous nerve** enters the arm by penetrating the coracobrachialis muscle; in Figure 5–2, this nerve is apparent between the coracobrachialis and the biceps brachii. The posterior compartment contains the lateral head and the long head of the **triceps brachii.** A very small portion of the medial head is evident on the posterior surface of the humerus. Several vessels and nerves are apparent in the fascia of the medial intermuscular septum. The artery of note is the **brachial artery,** with the **median nerve** located anterior to it, and the **ulnar nerve** located posterior to it. The deep vein is the **brachial vein,** and the **basilic vein** is more superficial.

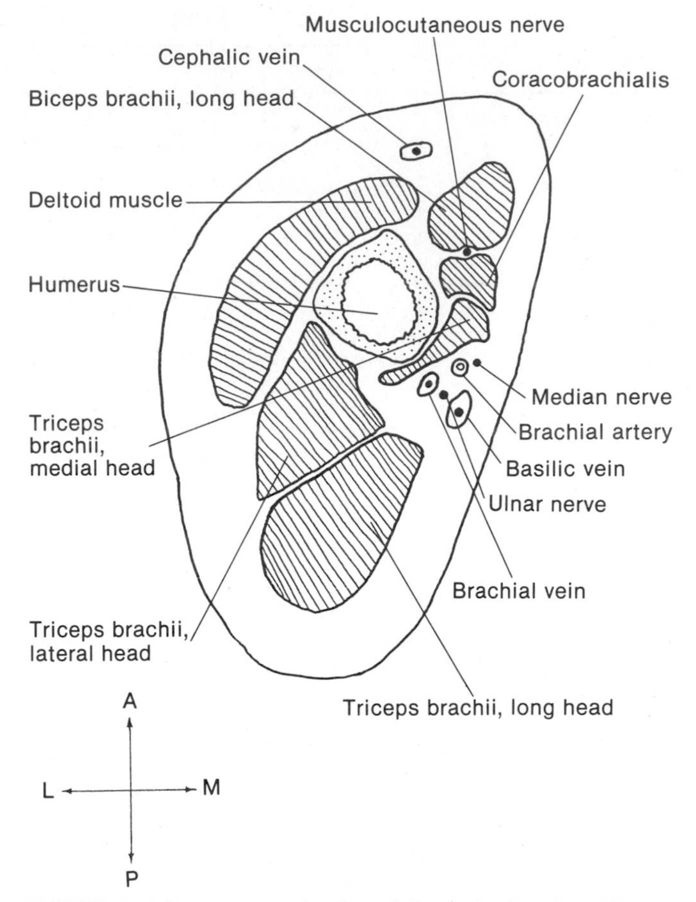

FIGURE 5–2. Transverse section through the proximal portion of the arm.

SECTION THROUGH THE DISTAL ARM

In the distal half of the arm, as illustrated in Figure 5–3, the anterior muscle compartment includes the **biceps**

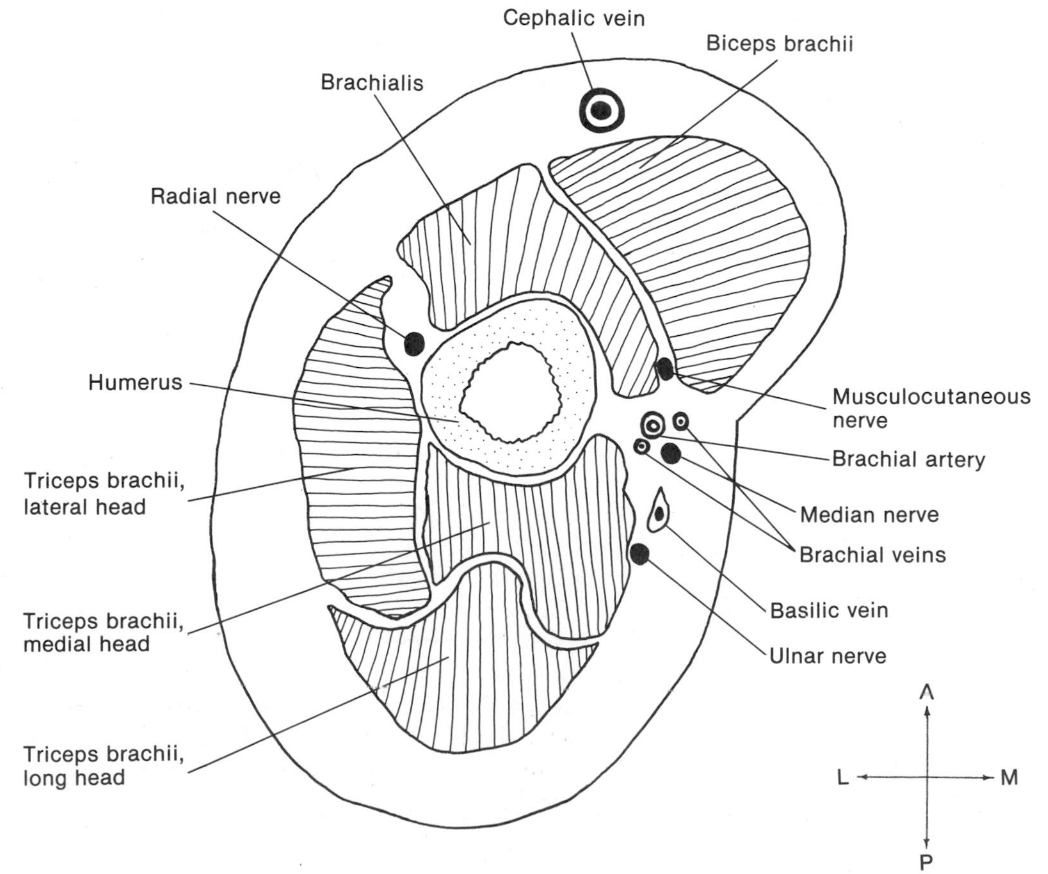

FIGURE 5–3. Transverse section through the distal portion of the arm. Note the anterior and the posterior muscle compartments.

brachii and the **brachialis** muscles. The brachialis is lateral and deep to the biceps brachii. The coracobrachialis muscle, seen in the proximal arm, is not evident in the lower sections, because this short muscle inserts near the middle of the shaft. The posterior compartment continues to contain the **triceps brachii.** The **radial nerve** is now lateral to the shaft of the humerus, between the triceps brachii and the brachialis muscles. At this level, the **median nerve** has crossed over the **brachial artery** and is now medial to it. The **ulnar nerve** is closely associated with the triceps brachii on the medial side.

Cubital Fossa

The cubital fossa is a triangular area on the anterior side of the elbow joint. It contains vessels and nerves that pass from the arm to the forearm. The sides of the fossa are formed by the **brachioradialis muscle** laterally and the **pronator teres muscle,** medially. Both these muscles follow a somewhat oblique course; thus they meet to form the apex of a triangle. The base of the triangle is an imaginary line between the lateral and medial epicondyles of the humerus. The floor is formed by the **brachialis** and the **supinator muscles.** The roof consists of deep fascia and a triangular sheet of tendon called the **bicipital aponeurosis.** The contents of the fossa from medial to lateral include the **median nerve,** the **brachial artery,** with its accompanying veins, the **tendon of the biceps brachii,** and the **radial nerve.** In the distal part of the fossa, near the apex, the brachial artery branches into the radial and ulnar arteries, and the radial nerve divides into the superficial radial and posterior interosseous branches. All of these structures are embedded in fatty connective tissue within the fossa. The superficial fascia, overlying the cubital fossa, contains superficial blood vessels and nerves. One of the most significant of these is the **median cubital vein,** which connects the basilic and cephalic veins and is frequently used for venipuncture. A transverse section through the cubital fossa is illustrated in Figure 5–4.

General Anatomy of the Forearm

The forearm, or antebrachium, extends from the elbow to the wrist. The skeleton of the forearm consists of two bones, the **radius** and the **ulna.** In anatomic position, the bones are parallel, with the radius on the lateral side and the ulna on the medial side. An interosseous membrane connects the two bones and also separates the muscles into the anterior flexor and the posterior extensor compartments. Rotation of the proximal and distal radioulnar joints allows the hand to function in either the pronated or the supinated position.

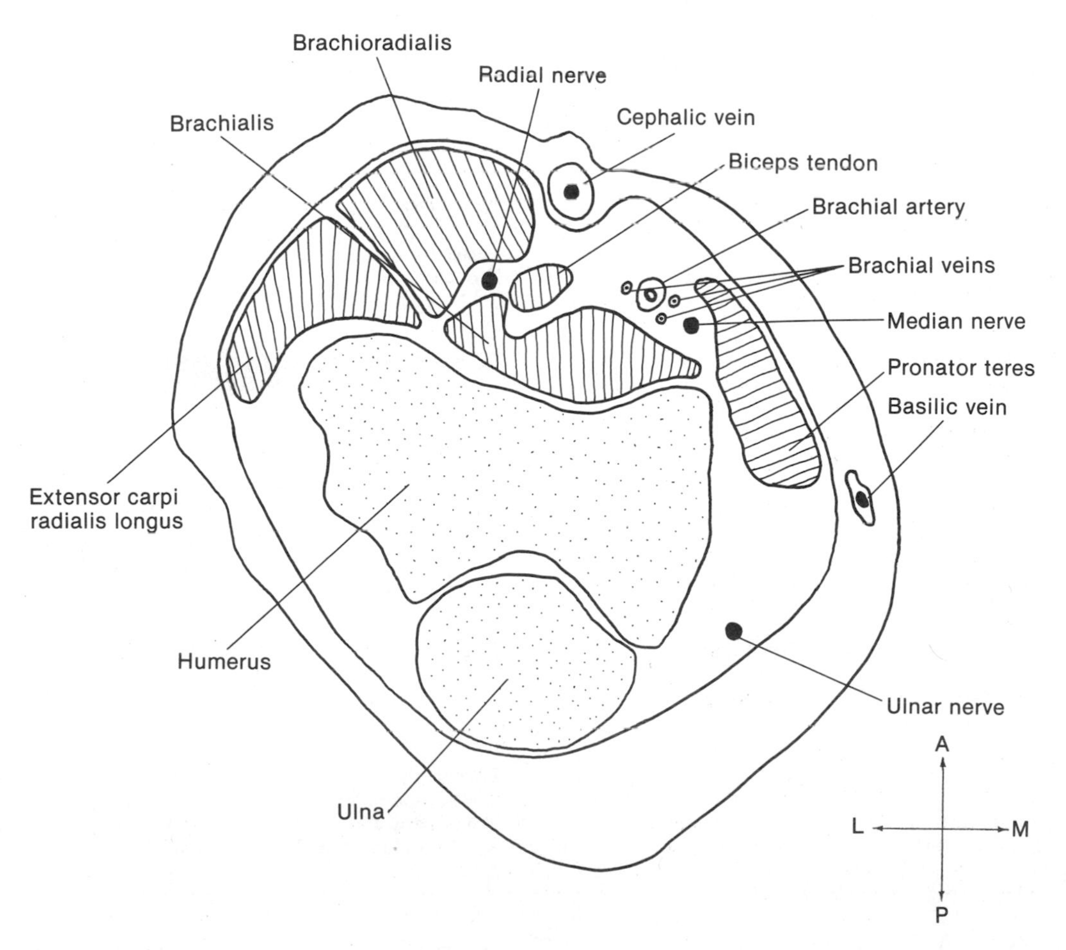

FIGURE 5–4. Transverse section through the cubital fossa showing boundaries and contents of fossa.

MUSCLES IN THE FOREARM

The muscles of the forearm, in general, act on the wrist, the hand, and the digits (Table 5–3). Exceptions to this include the brachioradialis, which flexes the elbow joint, and the pronator and supinator muscles. The radius and ulna, with an interosseous membrane between them, form a dividing line, which separates the muscles into anterior and posterior compartments.

ANTERIOR MUSCLE COMPARTMENT. The anterior muscle compartment contains the **flexor/pronator** group of muscles. Superficial muscles in this group arise from the medial epicondyle of the humerus by a common flexor origin. The superficial muscles cross the elbow joint and are anterior to it; consequently, they act as weak flexors of the elbow, in addition to their function with the wrist or hand. All of the superficial muscles, except the flexor carpi ulnaris, are innervated by the median nerve. The flexor carpi ulnaris is innervated by the ulnar nerve.

The deep muscles of the anterior compartment originate from the anterior surfaces of the radius and the ulna. The deep muscles include two flexor muscles and a pronator. These are innervated by the anterior interosseous branch of the median nerve. A portion of the flexor digitorum profundus is supplied by the ulnar nerve.

POSTERIOR MUSCLE COMPARTMENT. The posterior muscle compartment contains the **extensor/supinator** group of muscles. In addition, the brachioradialis muscle is considered to be part of this compartment. Four of the superficial extensors arise from a common origin on the lateral epicondyle of the humerus.

The brachioradialis and the extensor carpi radialis longus arise from the lateral supracondylar ridge of the humerus rather than from the common extensor origin with the other superficial muscles. The brachioradialis is a lateral muscle, which inserts on the styloid process of the radius, and it flexes the elbow joint rather than acting on the wrist or hand. Four of the five deep muscles are extensors. The fifth muscle, the supinator, acts on the forearm, rather than on the hand and digits. The radial nerve, and its branches innervate all the muscles of the posterior compartment.

TABLE 5–3. Muscles Located in the Forearm

Muscle	Origin	Insertion	Action	Innervation
Anterior Compartment				
Superficial Muscles				
Flexor carpi radialis	Medial epicondyle	Metacarpals, second and third	Flexes hand; weak flexion of the elbow	Median
Flexor carpi ulnaris	Medial epicondyle	Carpals, pisiform and hamate	Flexes hand; weak flexion of the elbow	Ulnar
Palmaris longus	Medial epicondyle	Flexor retinaculum	Flexes hand; weak flexion of the elbow	Median
Pronator teres	Medial epicondyle	Radius, lateral surface	Pronates hand	Median
Flexor digitorum superficialis	Medial epicondyle	Middle phalanges of fingers	Flexes fingers	Median
Deep Muscles				
Flexor digitorum profundus	Ulna	Distal phalanges of fingers	Flexes fingers	Median and ulnar
Flexor pollicis longus	Radius	Distal phalanx of thumb	Flexes thumb	Median
Pronator quadratus	Distal ulna	Distal radius	Pronates hand	Median
Posterior Compartment				
Superficial Muscles				
Extensor carpi ulnaris	Lateral epicondyle (common extensor origin)	Fifth metacarpal	Extends hand	Radial
Extensor carpi radialis brevis	Lateral epicondyle (common extensor origin)	Third metacarpal	Extends hand	Radial
Extensor digiti minimi	Lateral epicondyle (common extensor origin)	Proximal phalanx of fifth finger	Extends fifth finger	Radial
Extensor digitorum	Lateral epicondyle (common extensor origin)	Phalanges	Extends fingers	Radial
Extensor carpi radialis longus	Lateral supracondylar ridge	Second metacarpal	Extends hand	Radial
Brachioradialis	Lateral supracondylar ridge	Styloid of radius	Flexes forearm	Radial
Deep Muscles				
Extensor pollicis brevis	Radius	Proximal phalanx of thumb	Extends thumb	Radial
Extensor pollicis longus	Ulna	Distal phalanx of thumb	Extends thumb	Radial
Extensor indicis	Ulna	Phalanx of index finger	Extends index finger	Radial
Abductor pollicis longus	Radius and ulna	First metacarpal	Extends and abducts thumb	Radial
Supinator	Lateral epicondyle	Proximal radius	Rotates radius to supinate forearm	Radial

VASCULATURE OF THE FOREARM

The brachial artery in the arm divides into the **radial** and the **ulnar arteries** in the cubital fossa. The radial artery is the smaller of the two vessels. The radial artery courses distally deep to the brachioradialis muscle. Near the wrist, it becomes more superficial and can be palpated against the anterior surface of the radius. The radial artery enters the palm, terminating in the deep palmar arch. Along its course, the radial artery gives off branches to nearby muscles.

The ulnar artery continues distally between the superficial and deep muscle layers on the medial side of the anterior compartment. Near its origin in the cubital fossa, the ulnar artery gives off a **common interosseous branch.** This branch immediately divides into the **anterior** and **posterior interosseous arteries.** The anterior branch courses distally along the anterior surface of the interosseous membrane. The posterior interosseous artery enters the posterior compartment and supplies the muscles in that region. The ulnar artery terminates in superficial and deep palmar arches. Veins accompany most of the arteries.

NERVES IN THE FOREARM

The largest nerve in the forearm is the median nerve. It begins in the axilla as a branch of the brachial plexus and descends through the arm without giving off branches. In the cubital region, the median nerve crosses over the ulnar artery and then descends through the forearm, lateral to the ulnar artery, between the superficial and deep muscle layers in the anterior compartment. Near the wrist, the nerve becomes superficial. The median nerve supplies all of the superficial muscles in the anterior compartment, except the flexor carpi ulnaris. The anterior interosseous branch of the median nerve supplies most of the deep muscles of the anterior compartment.

The **ulnar nerve** is on the medial side of the anterior compartment. It supplies the flexor carpi ulnaris muscle.

The **radial nerve** descends along the lateral side of the arm and supplies muscles of the posterior compartment. In the region of the elbow, the radial nerve divides into the superficial and the deep branches. The superficial branch continues distally under the brachioradialis muscle and innervates the brachioradialis and the flexor carpi radialis muscles. The deep branch is the posterior interosseous nerve, which descends along the posterior surface of the interosseous membrane. This nerve supplies all the muscles of the posterior compartment, with the exception of the two muscles supplied by the superficial branch of the radial nerve.

Transverse Sections of the Forearm

SECTION THROUGH THE PROXIMAL FOREARM

Transverse sections through the upper part of the forearm show the **radius** and the **ulna,** with the **interosseous membrane** between them as illustrated in Figure 5–5. These structures separate the muscles into anterior flexor and posterior extensor compartments. Within the anterior compartments, the **ulnar artery** and **nerve** are on the medial side, the **median nerve** is centrally located, and the **radial artery** is lateral and near the surface. The superficial branch of the **radial nerve** is adjacent to the brachioradialis muscle. The deep branch of the radial nerve is more posteriorly located. The medial **basilic** and the lateral **cephalic veins** are situated in the superficial fascia.

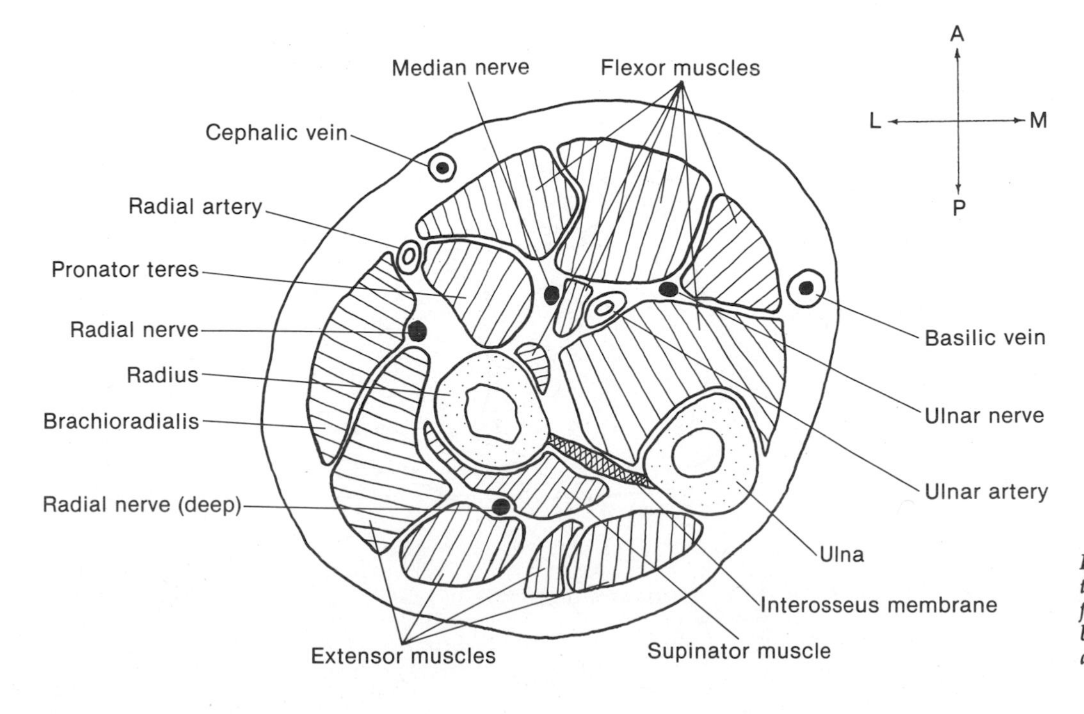

FIGURE 5–5. Transverse section through the proximal portion of the forearm showing interosseus membrane, anterior flexor muscle group, and posterior extensor muscle group.

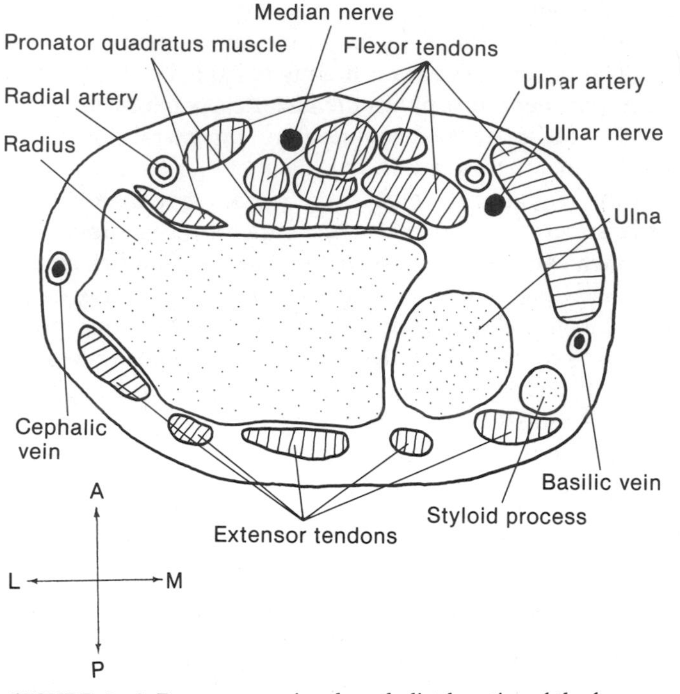

FIGURE 5-6. Transverse section through distal portion of the forearm showing numerous tendons that pass from the forearm to the wrist and hand. Note styloid process of ulna on medial side.

SECTION THROUGH THE DISTAL FOREARM

Transverse sections through the most distal part of the forearm show little musculature, because this region consists primarily of tendons going to the hand and fingers. A transverse section through the distal radius and ulna is illustrated in Figure 5-6. The large distal **radius** dominates this section. It articulates with the small **ulnar head** at the inferior radioulnar joint. The **styloid process** of the ulna is medial to the ulnar head. The space posterior to the osseous components is filled with tendons of the extensor muscles and fascia. Tendons of the flexor muscles, together with arteries and nerves, fill the anterior space. The **ulnar artery** and **nerve** are anterior to the ulnar head. The **radial artery** is close and just anterior to the lateral side of the radius. The **median nerve** is close to the surface, anterior to the midpoint of the radius. The superficial **cephalic** and **basilic veins** are on the lateral and medial sides, respectively.

Carpal Tunnel

The carpal bones in the wrist are tightly bound together by ligaments in such a way that they form an anterior depression, or concavity, called the carpal groove. A fibrous connective tissue sheet, called the **flexor retinaculum,** bridges over the carpal groove, making it into a carpal tunnel. On the medial side, the flexor retinaculum is anchored to the hook of the hamate bone. On the lateral side it attaches to the trapezium bone.

The carpal tunnel, illustrated in Figure 5-7, is completely filled with flexor tendons as they pass from the forearm to the hand and digits. In addition to the tendons, the **median nerve** is just beneath the flexor retinaculum on the lateral side. At times, the tendons may compress the nerve leading to "carpal tunnel syndrome." The **ulnar artery** and **nerve** are more medially located and are superficial to the retinaculum.

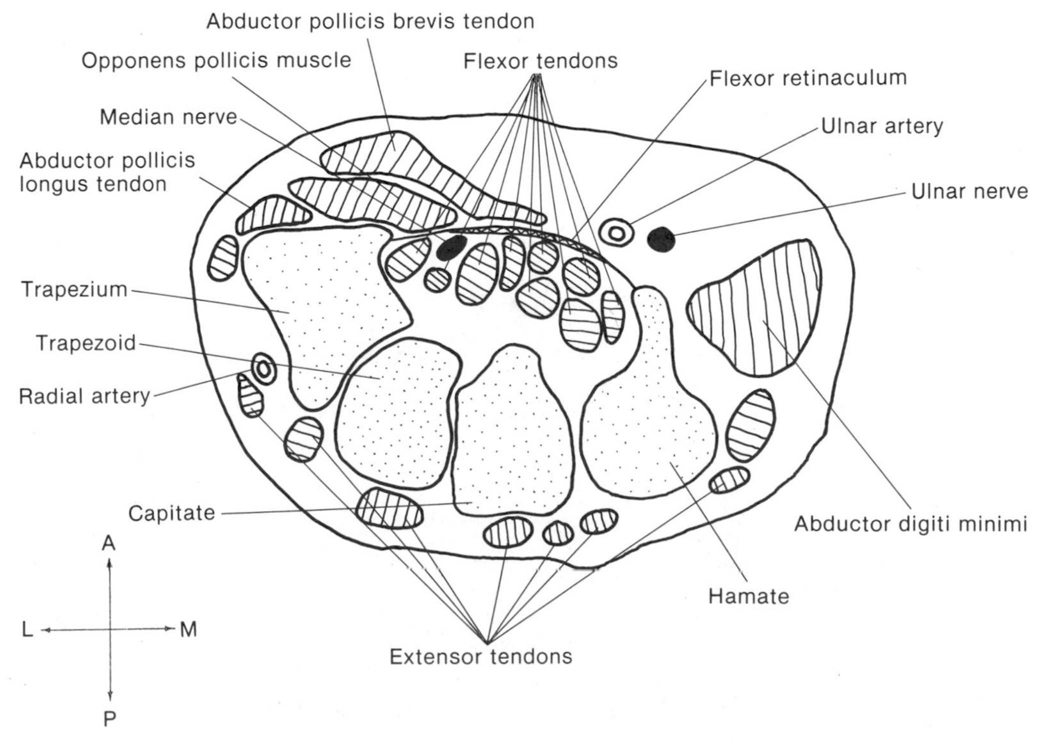

FIGURE 5-7. Transverse section through the carpal tunnel in wrist. Note relationship of median nerve to tendons just deep to the flexor retinaculum.

ARTICULATIONS ASSOCIATED WITH THE UPPER EXTREMITY

An articulation, or joint, is where two bones come together. There are numerous joints in the body that allow very little, if any, movement. For example, the sutures in the skull are joints, but they permit no movement. The symphysis pubis and intervertebral discs permit limited movement because the fibrocartilage in these joints is somewhat flexible. Other joints, such as the shoulder, allow a wide range of motion. These are synovial joints, which are more complex in structure than other types of joints. Synovial joints are principally found in the appendicular skeleton because this is the part of the skeleton involved in movement. The other, less movable joints, are more common in the axial skeleton, where they contribute rigidity to form and structure.

Synovial joints are characterized by a **fibrous joint capsule** that is lined with a **synovial membrane.** The synovial membrane that lines the fibrous layer secretes **synovial fluid,** which helps lubricate the joint. The ends of the bones comprising the joint are covered with a layer of hyaline cartilage called the **articular cartilage.** The synovial membrane lines all aspects of the joint, except over the articular cartilage. Figure 5–8 illustrates the structure of a "typical" synovial joint. In addition to the components common to all synovial joints, some have additional features, such as articular discs and intracapsular ligaments. Frequently, the fibrous capsule itself is thickened in places, forming a type of ligament.

It is impossible and unnecessary in this chapter to describe all the synovial joints found in the body. The few presented here have been selected because of their interest and/or importance in imaging.

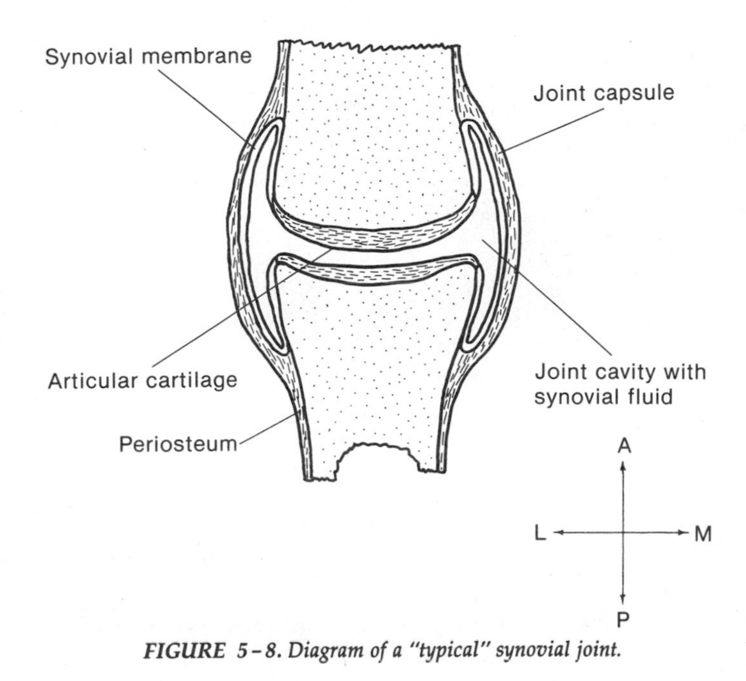

FIGURE 5–8. Diagram of a "typical" synovial joint.

Description of the Shoulder Joint

The shoulder joint is a ball-and-socket joint in which the rounded head of the humerus articulates with the shallow concavity of the glenoid fossa of the scapula. It is given several names, reflecting the osseous components of the joint. It may be called the humeral, the glenohumeral, or the humeroscapular joint. The shoulder joint offers a wide range of motion, but this is at the expense of stability. In other words, stability has been sacrificed for mobility. Although three ligaments help to support the joint, most of the support for the joint comes from the strong muscle tendons that pass over it. For this reason, the shoulder is easily dislocated in young children before muscular strength is developed.

ARTICULAR SURFACES AND THE JOINT CAPSULE

The shallow **glenoid fossa** is deepened by a fibrocartilage rim called the **glenoid labrum.** The joint capsule extends from the glenoid labrum to the anatomic neck of the humerus. The capsule is somewhat thin and loose, which contributes to the flexibility of the joint. There is an arch over the joint to protect it from above and to help prevent superior displacement. The arch is formed by the **acromion** and the **coracoid processes of the scapula** and by the **coracoacromial ligament** between them. The **deltoid muscle** covers the joint. Features of the shoulder are illustrated in Figure 5–9.

CAPSULAR LIGAMENTS OF THE SHOULDER

Three ligaments help to reinforce the joint capsule (see Fig. 5–9*A*). The **transverse humeral ligament** thickens the joint capsule between the greater and the lesser tubercles of the humerus. This ligament holds the tendon from the long head of the biceps in place. The **coracohumeral ligament** strengthens the superior part of the capsule. This ligament extends from the coracoid process of the scapula to the anatomic neck of the humerus, near the greater tubercle. The **glenohumeral ligament** consists of three slight thickenings on the anterior side of the capsule. These thickenings extend from the margin of the glenoid fossa to the anatomic neck and lesser tubercle of the humerus. The glenohumeral ligaments may be indistinct or absent entirely.

MUSCULAR SUPPORT FOR THE SHOULDER

The primary support for the shoulder joint comes from the muscles surrounding it. Four of these muscles, the **supraspinatus,** the **infraspinatus,** the **subscapularis,** and the **teres minor,** are collectively known as the rotator cuff muscles. These muscles, associated with the scapula, pull the head of the humerus upward and medially into the glenoid fossa. The tendon of the long head of the biceps brachii muscle also helps hold the humeral head in place. This tendon attaches to the supraglenoid tubercle

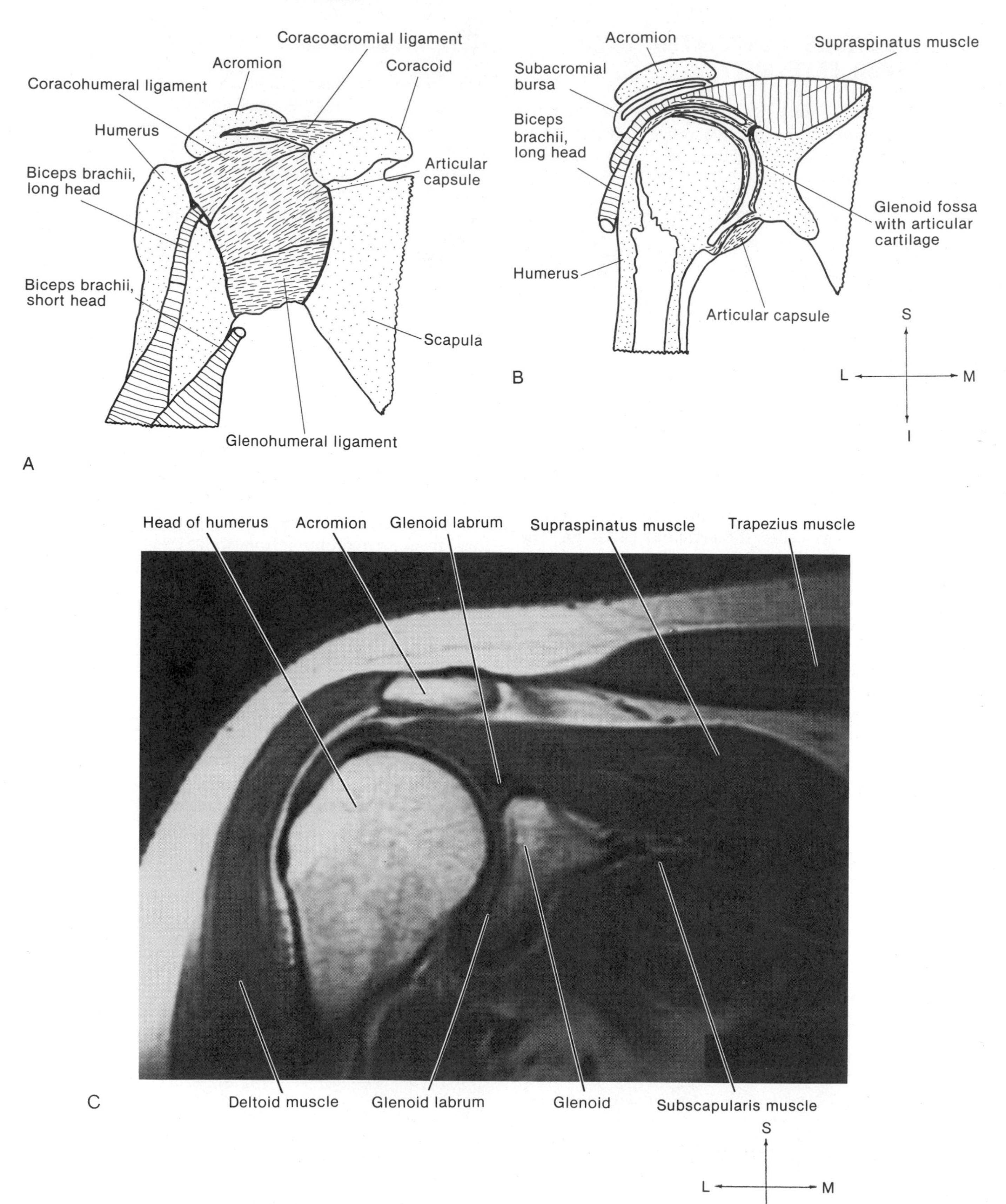

FIGURE 5-9. This figure illustrates three views of the humeroscapular joint. A, *Anterior surface view, showing joint capsule and reinforcing ligaments;* B, *Coronal view, illustrating interior of joint cavity;* C, *Coronal magnetic resonance image.*

of the scapula, passes over the head of the humerus within the joint capsule, and descends along the intertubercular groove. There is little support for the shoulder joint inferiorly; consequently, most dislocations are in that direction.

BURSAE

Several bursae are associated with the shoulder. Bursae are synovial membrane sacs filled with synovial fluid. They are found where tendons cross bones, ligaments, or other tendons. Bursae act as cushions to reduce the friction between the moving parts. Four of the shoulder bursae are the subdeltoid, the subacromial, the subscapularis, and the subcoracoid.

Sectional Anatomy of the Shoulder Joint

Actually, there are relatively few components to look for when examining the sectional anatomy of the shoulder. The osseous elements are the humerus and the scapula, with its acromion, spine, coracoid, and glenoid. Superiorly, the clavicle is seen as it contributes to the coracoclavicular articulation. The muscular elements are the rotator cuff muscles and the tendons of the biceps brachii. The deltoid muscle covers the joint anteriorly, laterally, and posteriorly. Representative sections showing the osseous and muscular components are illustrated.

TRANSVERSE SECTION THROUGH THE HUMERAL HEAD

Figure 5 – 10 illustrates a transverse section through the superior portion of the head of the humerus. The **coracoid of the scapula** and the **clavicle** are close together, separated only by the coracoclavicular ligament. The other osseous component, the **spine of the scapula,** is situated more posteriorly. The muscular components at this level are the **infraspinatus,** the **supraspinatus,** and the **deltoid** muscles. The supraspinatus muscle occupies the space between the coracoid and the spine, with its tendon extending to the greater tubercle of the humeral head. The infraspinatus is evident posterior to the spine, actually in the infraspinous fossa. The tendon of this muscle also extends to the greater tubercle of the humeral head, but it is more posterior. The deltoid encloses the joint on the anterior, lateral, and posterior sides.

TRANSVERSE SECTION THROUGH THE GLENOID

In transverse sections through the glenoid, the **glenoid labrum** appears at the edges of the glenoid. The **tendon for the supraspinatus muscle** occupies the space be-

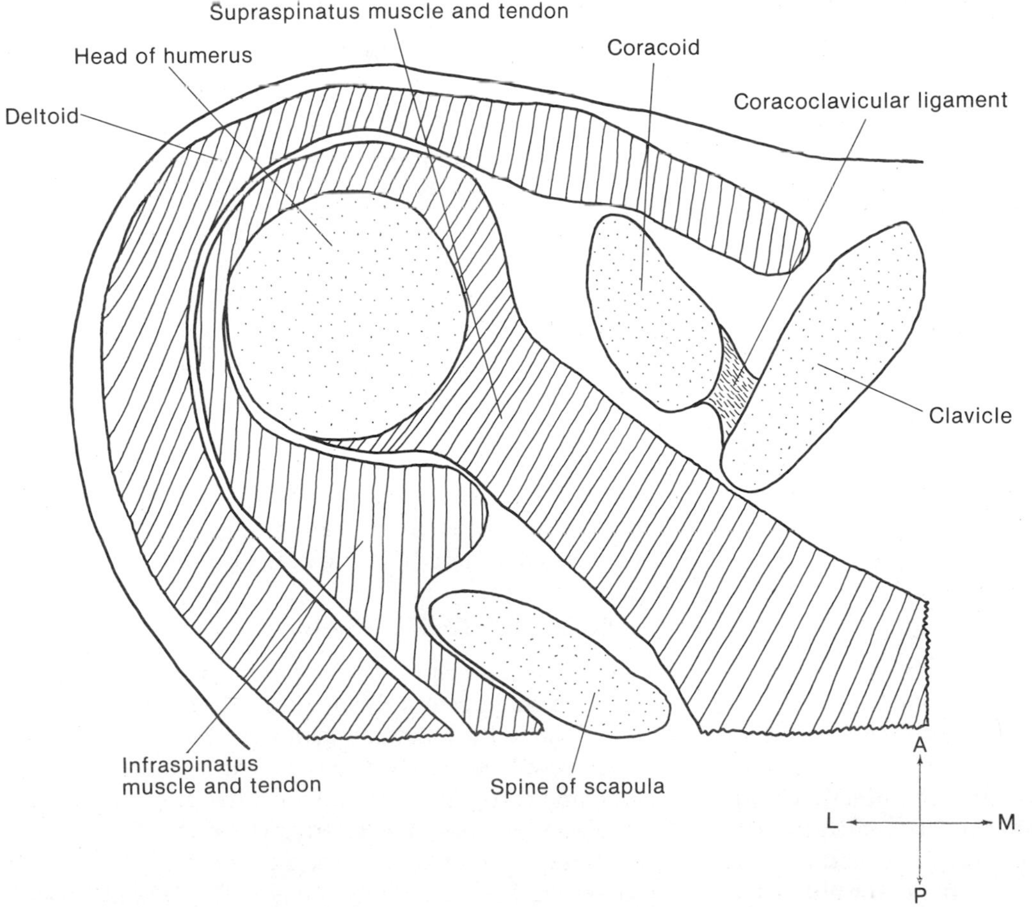

FIGURE 5 - 10. Transverse section of shoulder joint through humeral head. Note relationships of musculoskeletal components.

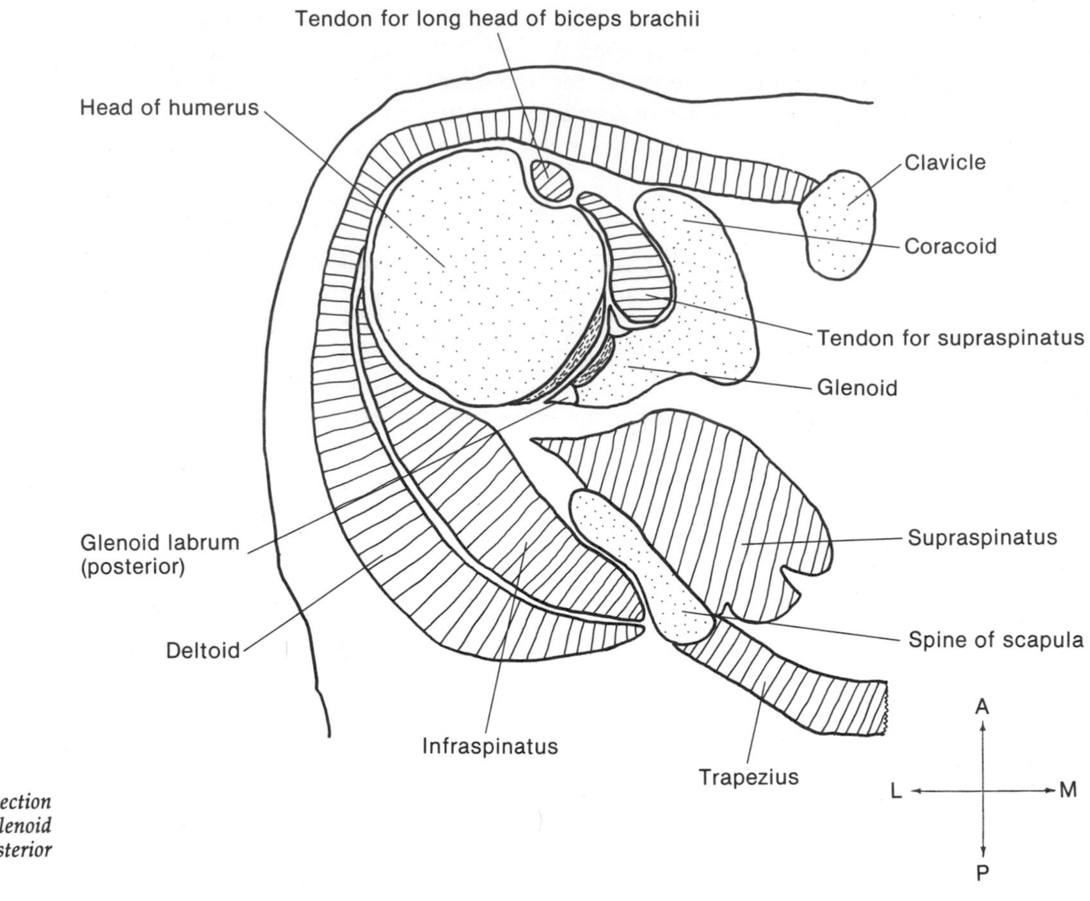

FIGURE 5-11. Transverse section of shoulder joint through glenoid fossa. Note anterior and posterior portions of glenoid labrum.

tween the **coracoid** and the **head of the humerus.** The tendon for the long head of the **biceps brachii** is in the bicipital groove of the humerus. The **supraspinatus** and **infraspinatus muscles** are associated with the spine of the scapula. The **deltoid muscle** continues to enclose the joint anteriorly, laterally, and posteriorly. These features are illustrated in Figure 5-11.

Description of the Elbow Joint

The joint capsule at the elbow encloses three separate articulations. Two of the articulations are uniaxial hinge joints that allow flexion and extension of the forearm. The third is a pivot joint that permits pronation and supination of the forearm.

HUMEROULNAR ARTICULATION

On the medial side, the **trochlear notch of the ulna** articulates with the **trochlea of the humerus,** forming the humeroulnar joint. This is a uniaxial hinge joint that permits flexion and extension. On the distal posterior surface of the humerus, there is a deep concavity called the **olecranon fossa,** which accommodates the **olecranon process of the ulna.** On the distal anterior surface of the humerus, there is a shallow depression called the **coronoid fossa,** which accommodates the **coronoid process of the ulna.**

HUMERORADIAL ARTICULATION

The humeroradial articulation is on the lateral side. The articulating surfaces are the **head of the radius** and the **capitulum of the humerus.** This is also a uniaxial hinge joint that permits flexion and extension. The humeroulnar and the humeroradial articulations make up what is commonly called the elbow joint.

RADIOULNAR ARTICULATION

The third articulation, enclosed within the elbow joint capsule, is between the **head of the radius** and the **radial notch of the ulna.** This is the proximal radioulnar joint. It is a pivot joint, which allows rotation of the radius. The **annular ligament** wraps around the head of the radius and attaches to the anterior and posterior margins of the radial notch to help hold the radial head in place.

JOINT CAPSULE OF THE ELBOW

The fibrous joint capsule of the elbow is relatively weak anteriorly and posteriorly, but it is strengthened laterally and medially by collateral ligaments. Proximally, the fibrous capsule is attached to the superior margins of the radial and coronoid fossae on the anterior surface and to the olecranon fossa posteriorly. The distal attachments are to the margins of the trochlear notch, the coronoid process, and the annular ligament. The **lateral (radial) collateral** and **medial (ulnar) collateral ligaments** strengthen the joint capsule on the sides. Stability of the joint depends on these two ligaments.

MUSCULAR ACTION ON THE ELBOW

The primary flexor of the forearm at the elbow is the **brachialis muscle.** The **biceps brachii** is also an important flexor when there is resistance to the movement. Flexion is limited by the presence of the collateral ligaments, the tension in the antagonistic muscles, and the opposing surfaces of the arm and forearm. The principal extensor muscle acting on the elbow is the **triceps brachii.** Extension is limited by the collateral ligaments, the tension in the antagonistic muscles, and the olecranon process of the ulna in the olecranon fossa of the humerus.

Blood supply to the elbow joint is from anastomoses of branches of the **brachial,** the **radial,** and the **ulnar vessels.**

Sectional Anatomy of the Elbow Joint

The musculoskeletal features of the elbow are probably best illustrated by sagittal sections through the humeroulnar and humeroradial articulations and by a transverse section through the radioulnar articulation.

SAGITTAL SECTION THROUGH THE HUMERUS AND THE ULNA

Figure 5–12 illustrates the medial portion of the elbow joint. In this region, the **trochlear notch** of the ulna articulates with the **trochlea** of the humerus. The powerful flexor muscle, the **brachialis,** is on the anterior surface of the joint. The **biceps brachii,** which also acts as a flexor, is superficial to the brachialis. These muscles are opposed by the **triceps brachii,** a powerful extensor, that is located on the posterior surface of the humerus. These muscles insert on the ulna.

SAGITTAL SECTION THROUGH THE HUMERUS AND THE RADIUS

In sagittal sections through the lateral portion of the elbow (Fig. 5–13), the **radius** and the **capitulum of the humerus** are evident. The **triceps brachii** and **brachialis** are less apparent, because they insert on the ulna and only their association with the humerus is evident in lateral sections. More of the muscles associated with the forearm are visible in this section.

TRANSVERSE SECTION THROUGH THE RADIUS AND THE ULNA

Figure 5–14 illustrates a transverse section through the **head of the radius** as it articulates in the **radial notch of**

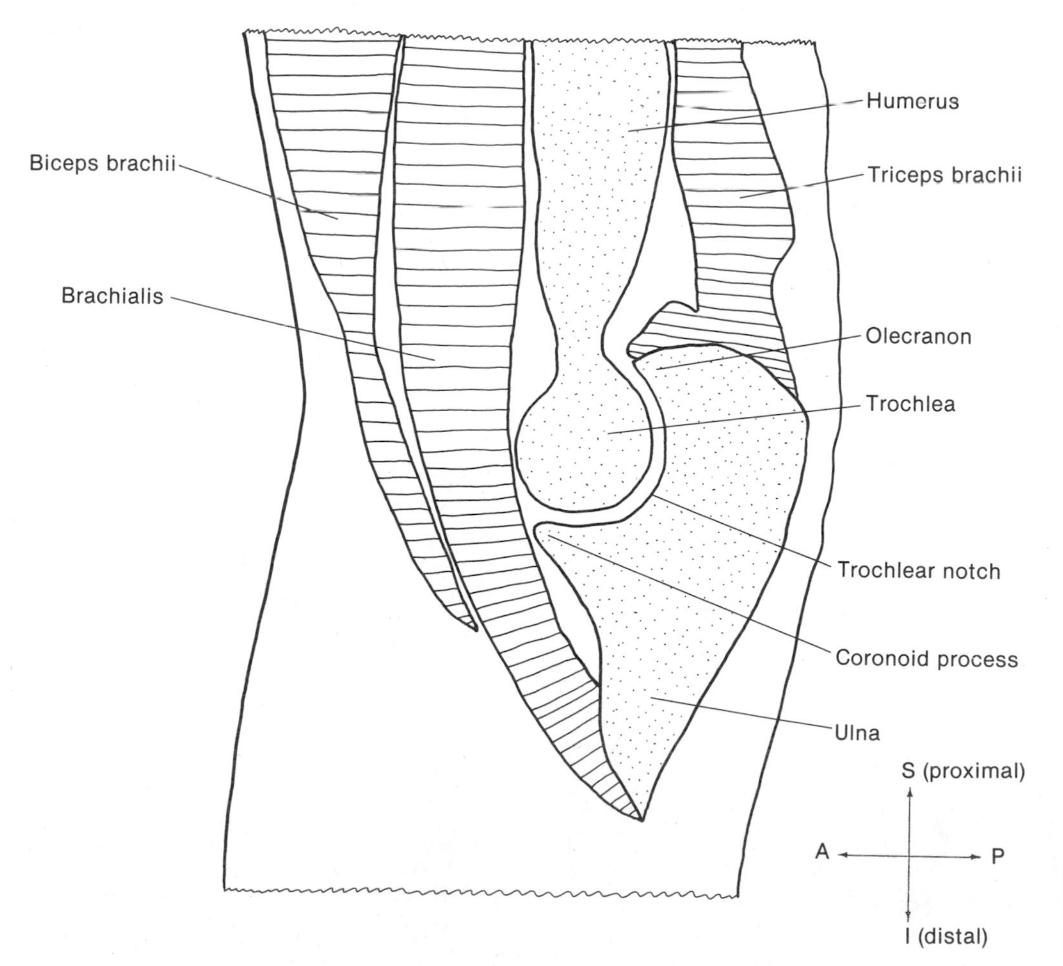

FIGURE 5–12. Sagittal section through medial portion of the elbow, showing articulation of humerus and ulna.

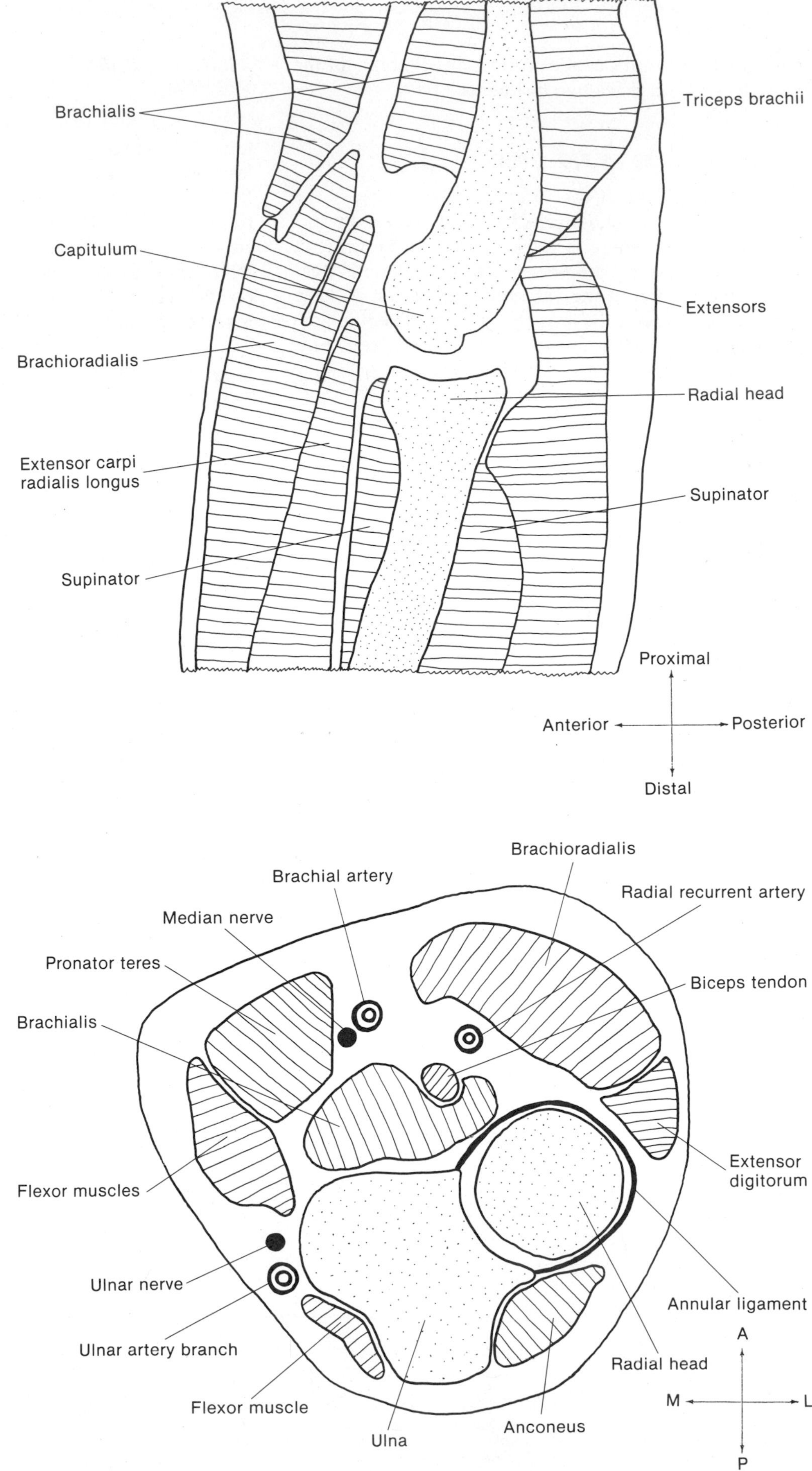

FIGURE 5-13. Sagittal section through lateral portion of the elbow, showing articulation of humerus and radius.

FIGURE 5-14. Transverse section through the proximal portion of the forearm, showing relationship of the head of the radius with the ulna. Note the annular ligament and the extensor and flexor muscle groups.

the ulna. This is a pivot joint that allows rotation of the forearm. The **annular ligament** holds the radial head in place. The **ulnar nerve** and a branch of the **ulnar artery** are near the surface on the medial side. The extensor and flexor muscles are quite apparent at this level.

LOWER EXTREMITY

The lower extremity can be divided into three regions: the thigh, the leg, and the foot. The two os coxae make up the pelvic girdle, which provides the connection between the lower extremity and the axial skeleton. The gluteal region consists of the muscles that overlie the posterior portion of the pelvic girdle. The thigh is the superior portion of the lower extremity, which articulates with the pelvic girdle. The single bone in the thigh is the femur. The intermediate portion of the lower extremity is the leg. The bones of the leg are the tibia and the fibula. The leg articulates with the thigh and the foot by hinge joints at the knee and ankle. The foot constitutes the most distal part and includes the tarsals, metatarsals, and phalanges. The lower extremity supports the weight of the body; thus, some freedom of movement in the joints has been sacrificed to acquire strength and stability.

Gluteal Region

The gluteal, or buttock region, is bounded superiorly by the iliac crest, and inferiorly by the lower margin of the gluteus maximus muscle, which is marked by a crease or groove just below the gluteal fold. The region is continuous with the lower trunk above and the posterior surface of the thigh below. The muscles of the gluteal region are summarized in Table 5–4.

TENSOR FASCIAE LATAE. The **tensor fasciae latae** is the most lateral muscle in the gluteal region. It originates on the iliac crest and inserts on the iliotibial tract, which continues down the side of the leg and attaches to the lateral condyle of the tibia. The tensor fasciae latae flexes the thigh, but it also helps to extend the leg at the knee by putting tension on the fascia lata. The muscle is innervated by the superior gluteal nerve.

GLUTEUS MAXIMUS, MEDIUS, AND MINIMUS. The three gluteal muscles account for the bulk and contour of the region. These muscles act on the hip joint to move the thigh. The largest and most superficial of the gluteal muscles is the **gluteus maximus.** This muscle has an extensive origin on the ilium, the sacrum, and the coccyx. The fibers descend obliquely to insert on the gluteal tuberosity of the femur and the iliotibial tract. The gluteus maximus is a powerful extensor of the thigh. The **gluteus medius** lies deep to the gluteus maximus; however, a portion of the medius usually extends more superiorly than the maximus. The gluteus medius extends from its origin on the ilium to its insertion on the greater trochanter of the femur. The **gluteus minimus** is the smallest and the deepest of the gluteal muscles. It also extends from the external surface of the ilium to the greater trochanter of the femur, but its points of attachment are less extensive than those of gluteus medius. Both the medius and the minimus muscles abduct and medially rotate the thigh.

PIRIFORMIS. The **piriformis muscle** serves as a landmark for structures that enter the gluteal region deep to the gluteus maximus. These structures are described as entering the region either above or below the piriformis

TABLE 5–4. Muscles Located in the Gluteal Region

Muscle	Origin	Insertion	Action	Innervation
Tensor fasciae latae	Iliac crest	Iliotibial tract	Flexes thigh; tenses fascia lata	Superior gluteal
Gluteus maximus	Ilium, sacrum, coccyx	Gluteal tuberosity; iliotibial tract	Extends thigh	Inferior gluteal
Gluteus medius	Ilium	Greater trochanter	Abducts and medially rotates thigh	Superior gluteal
Gluteus minimus	Ilium	Greater trochanter	Abducts and medially rotates thigh	Superior gluteal
Piriformis	Sacrum	Greater trochanter	Rotates thigh laterally	Branches from S_1 and S_2
Obturator internus	Margin of obturator foramen	Greater trochanter	Rotates thigh laterally	L_5 and S_1
Gemelli, superior and inferior	Ischium; margin of lesser sciatic notch	Obturator internus tendon; greater trochanter	Rotates thigh laterally	Sacral plexus
Quadratus femoris	Ischial tuberosity	Greater trochanter and shaft of femur	Rotates thigh laterally	L_4, L_5, S_1
Obturator externus	Margin of obturator foramen	Trochanteric fossa	Rotates thigh laterally	L_5 and S_1

muscle. The superior gluteal artery and nerve both arise in the pelvis; the artery arises from the internal iliac artery, and the nerve arises from the sacral plexus. They then pass through the greater sciatic notch and enter the gluteal region superior to the piriformis. The piriformis muscle itself originates on the anterior surface of the sacrum, passes through the greater sciatic notch, and then inserts on the greater trochanter of the femur. The **sciatic nerve,** the largest nerve in the region, enters the gluteal region through the greater sciatic notch, just inferior to the piriformis. Muscles inferior to the piriformis are the obturator internus, gemelli, and quadratus femoris. The inferior gluteal artery and nerve also enter the region inferior to the piriformis.

OBTURATOR INTERNUS. The **obturator internus** muscle is an intrapelvic muscle, but its tendon passes through the lesser sciatic notch and becomes extrapelvic in the gluteal region. The muscle originates around the margin of the obturator foramen and covers the space of the foramen. It passes through the deep gluteal region and inserts on the greater trochanter of the femur. The obturator internus is one of the deep lateral rotators of the thigh.

GEMELLI. The **superior** and **inferior gemelli muscles** are closely associated with the obturator internus muscle and often obscure the extrapelvic portion of the obturator internus. Some anatomists consider the three muscles to be three parts of one muscle. Both the superior and inferior gemelli muscles originate from the ischium, around the margin of the lesser sciatic notch, and insert on the greater trochanter of the femur with the obturator internus tendon. The gaster, or belly, of the superior gemellus is superior to the obturator internus tendon, whereas the inferior gemellus is inferior to the tendon. Both gemelli muscles are deep lateral rotators of the thigh.

QUADRATUS FEMORIS. The **quadratus femoris** is the most inferior muscle in the deep gluteal region. The muscle originates on the ischial tuberosity, inserts on the greater trochanter and shaft of the femur, and with the other deep gluteal muscles, laterally rotates the thigh.

OBTURATOR EXTERNUS. The **obturator externus muscle** is often described with the adductor group of thigh muscles because of its location. It is included with the deep gluteal muscles because it is closely related to them functionally. The obturator externus originates around the margin of the obturator foramen and inserts on the trochanteric fossa of the femur. It covers the space of the obturator foramen on the exterior side. The obturator externus laterally rotates the thigh.

General Anatomy of the Thigh

The thigh is the most proximal portion of the lower extremity. The gluteal region is an intermediate area on the posterior surface between the trunk and the thigh. There is only one bone in the thigh, the femur, which is the longest bone in the body. Anterior, posterior, and medial muscle compartments surround the bone. A neurovascular bundle accompanies each muscle compartment. The muscles are invested by deep fascia (fascia lata), which also projects inward to create intermuscular septa.

On the lateral side of the thigh, the deep fascia (fascia lata) is thickened to form the iliotibial tract, which extends from the ilium to the tibia. The tensor fasciae latae and gluteus maximus muscles are attached to the upper part of the iliotibial tract.

Superficial fascia and skin complete the coverings of the thigh. The superficial fascia contains the great (long) saphenous vein and its tributaries. There are numerous lymph nodes in the superficial fascia of the inguinal region near the great (long) saphenous vein. The great (long) saphenous vein penetrates the fascia lata to drain into the femoral vein.

OSSEOUS COMPONENT OF THE THIGH

The only bone in the thigh is the **femur,** which is the longest and heaviest bone in the body. The length of an individual's femur is approximately one fourth of the person's height. The proximal end of the femur consists of a rounded head that fits into the acetabulum of the os coxa, a slender neck that projects laterally away from the head, and the greater and lesser trochanters, which provide points of attachment for muscles. The distal end of the femur, at the knee joint, is broadened with two large condyles for articulation with the tibia in the leg. In between the two ends of the bone, there is a long, fairly smooth shaft. A rough ridge on the posterior surface, the linea aspera, provides attachment for muscles. Superiorly, the linea aspera expands and becomes the gluteal tuberosity.

MUSCLES LOCATED IN THE THIGH

Some muscles in the thigh region have origins on the pelvic girdle and insertions on the femur. These muscles move the thigh by acting on the hip joint. Other muscles that are seen in sections of the thigh originate on the femur and insert on the tibia or fibula of the leg. These muscles move the leg by acting on the knee joint. Another muscle group extends from the pelvic girdle to the leg and acts on both the hip and the knee joints.

The muscles of the thigh are divided into three compartments by intermuscular septa of deep fascia. The muscles within each compartment have similar functions, and each compartment has its own neurovascular bundle. The muscles of the thigh are summarized in Table 5–5.

ANTERIOR MUSCLE COMPARTMENT. The largest muscle mass in the anterior muscle compartment of

TABLE 5-5. Muscles Located in the Thigh

Muscle	Origin	Insertion	Action	Innervation
Anterior Compartment				
Iliopsoas	Lumbar vertebrae and iliac fossa	Lesser trochanter of femur	Flex thigh	Femoral
Sartorius	Anterior superior iliac spine	Superior medial tibia	Flex thigh and leg	Femoral
Quadriceps femoris				
Rectus femoris	Anterior superior iliac spine	Tibial tuberosity	Extend leg and flex thigh	Femoral
Vastus lateralis	Lateral shaft of femur	Tibial tuberosity	Extend leg	Femoral
Vastus medialis	Medial shaft of femur	Tibial tuberosity	Extend leg	Femoral
Vastus intermedius	Anterior shaft of femur	Tibial tuberosity	Extend leg	Femoral
Medial (Adductor) Compartment				
Pectineus	Pubis	Femur	Adduct thigh	Obturator and femoral
Adductor longus	Pubis	Linea aspera of femur	Adduct thigh	Obturator
Gracilis	Pubis	Superior medial tibia	Adduct thigh	Obturator
Adductor brevis	Pubis	Linea aspera of femur	Adduct thigh	Obturator
Adductor magnus	Pubis and ischium	Linea aspera and other aspects of femur	Adduct thigh	Obturator and sciatic
Posterior (Hamstring) Compartment				
Biceps femoris	Ischial tuberosity	Fibula	Flex leg at knee; extend thigh at hip	Sciatic
Semitendinosus	Ischial tuberosity	Medial surface of tibia	Flex leg at knee; extend thigh at hip	Sciatic
Semimembranosus	Ischial tuberosity	Medial condyle of tibia	Flex leg at knee; extend thigh at hip	Sciatic

the thigh consists of the extensor muscle group, the **quadriceps femoris.** The compartment also contains the terminal portion of the **iliopsoas** from the posterior abdominopelvic wall and the **sartorius.**

Iliopsoas. The **iliopsoas** inserts on the lesser trochanter of the femur and flexes the thigh. This muscle is discussed in more detail with the musculature of the posterior abdominal wall.

Sartorius. The **sartorius** is a long, straplike muscle that is superficial to the other anterior thigh muscles. It is the longest muscle in the body. The sartorius originates on the anterior superior iliac spine, courses obliquely across the anterior part of the thigh, and then inserts on the upper part of the medial surface of the tibia. The sartorius extends over the hip and the knee; thus, it has an effect on both joints. Contraction of the sartorius causes the thigh to be flexed at the hip and the leg to be flexed at the knee. The muscle is innervated by the femoral nerve.

Quadriceps Femoris. The **quadriceps femoris** is a powerful extensor group of muscles that occupies the major portion of the anterior compartment. As the name implies, there are four muscles in the group, namely, the **rectus femoris,** the **vastus lateralis,** the **vastus medialis,** and the **vastus intermedius.** These muscles cover the front and the sides of the femur. The rectus femoris originates on the anterior superior iliac spine with fibers that run straight down the thigh. The three vasti muscles originate on the shaft of the femur. The vastus lateralis is on the lateral side, the vastus medialis on the medial side, and the vastus intermedius between the other two on the anterior portion of the shaft. The three vasti muscles make a groove, or trough, that holds the rectus femoris. The four muscles of the quadriceps femoris have a common tendinous insertion on the superior part of the patella. The tendon continues over the patella as the patellar ligament and finally attaches to the tibial tuberosity. All four parts of the quadriceps femoris act as extensors of the leg at the knee joint and are used in climbing, walking, running, and rising from a chair. The rectus femoris extends over the hip joint; thus, it also affects that joint by flexing the thigh. The quadriceps femoris muscles are innervated by the femoral nerve.

MEDIAL MUSCLE COMPARTMENT. The main action of the muscles in the medial compartment is to adduct the thigh. All of the muscles in this compartment originate on either the pubis or the ischium and insert on the femur, except the gracilis, which inserts on the tibia. The muscles in the medial compartment are innervated by the obturator nerve with the exception of portions of the pectineus and adductor magnus, which are supplied by the femoral and sciatic nerves, respectively. The muscles are arranged in three levels or layers. The superficial or anterior layer consists of the **pectineus,** the **adductor longus,** and the **gracilis.**

Pectineus. The **pectineus** is a rectangular muscle just medial to the iliopsoas in the floor of the femoral triangle. It has its origin on the pectineal line of the pubis, and it inserts on the pectineal line of the femur.

Adductor Longus. The **adductor longus** is medial to the pectineus. It originates on the body of the pubis and inserts on the middle part of the linea aspera of the femur.

Gracilis. The most medial muscle of the anterior layer is the **gracilis.** It is also the longest muscle of the group, since it extends down the medial side of the thigh from the pubis bone to the upper part of the medial surface of the tibia.

Adductor Brevis and Adductor Magnus. The middle or intermediate layer of the middle compartment is occupied by the **adductor brevis.** This muscle also originates on the pubic bone, but it inserts on the upper part of the linea aspera of the femur. The deep or posterior layer consists of the **adductor magnus.** The adductor magnus originates on the inferior pubic ramus and the ischium and has an extensive insertion on the linea aspera and on other aspects of the femur.

POSTERIOR MUSCLE COMPARTMENT. The muscular contents of the posterior compartment consist of the hamstring muscle group. These three muscles, the **biceps femoris,** the **semitendinosus,** and the **semimembranosus,** all arise from the ischial tuberosity. The **biceps femoris** is the most lateral of the three muscles as it descends to insert on the upper part of the fibula. The middle muscle is the **semitendinosus,** which inserts on the upper part of the medial surface of the tibial shaft. The most medial of the three muscles is the **semimembranosus,** which inserts on the posteromedial surface of the medial condyle of the tibia. Near their origin, on the ischial tuberosity, the tendons of the biceps femoris and the semitendinosus overlie the tendon of the semimembranosus. The hamstring muscles extend over the hip and the knee joints and, consequently, have an effect on both. These muscles extend the thigh at the hip and flex the leg at the knee. All three of the hamstring muscles are innervated by the sciatic nerve.

FEMORAL TRIANGLE

The femoral triangle is in the upper, medial part of the anterior muscle compartment of the thigh. The base of the triangle (superior margin) is formed by the inguinal ligament. The lateral margin is the medial border of the sartorius muscle. The lateral border of the adductor longus muscle forms the medial margin. The iliopsoas and pectineus muscles make up the floor of the triangle. The central structures that pass through the femoral triangle are the femoral artery and vein, which are enclosed in a femoral sheath made of tough connective tissue. The artery is lateral to the vein within the sheath. Fat and lymphoid tissue, also within the sheath, are medial to the artery and vein. The femoral nerve is not enclosed within the sheath, but it passes through the femoral triangle lateral to the sheath. Figure 5–15 illustrates the components of the femoral triangle.

A subsartorial canal links the apex of the femoral trian-

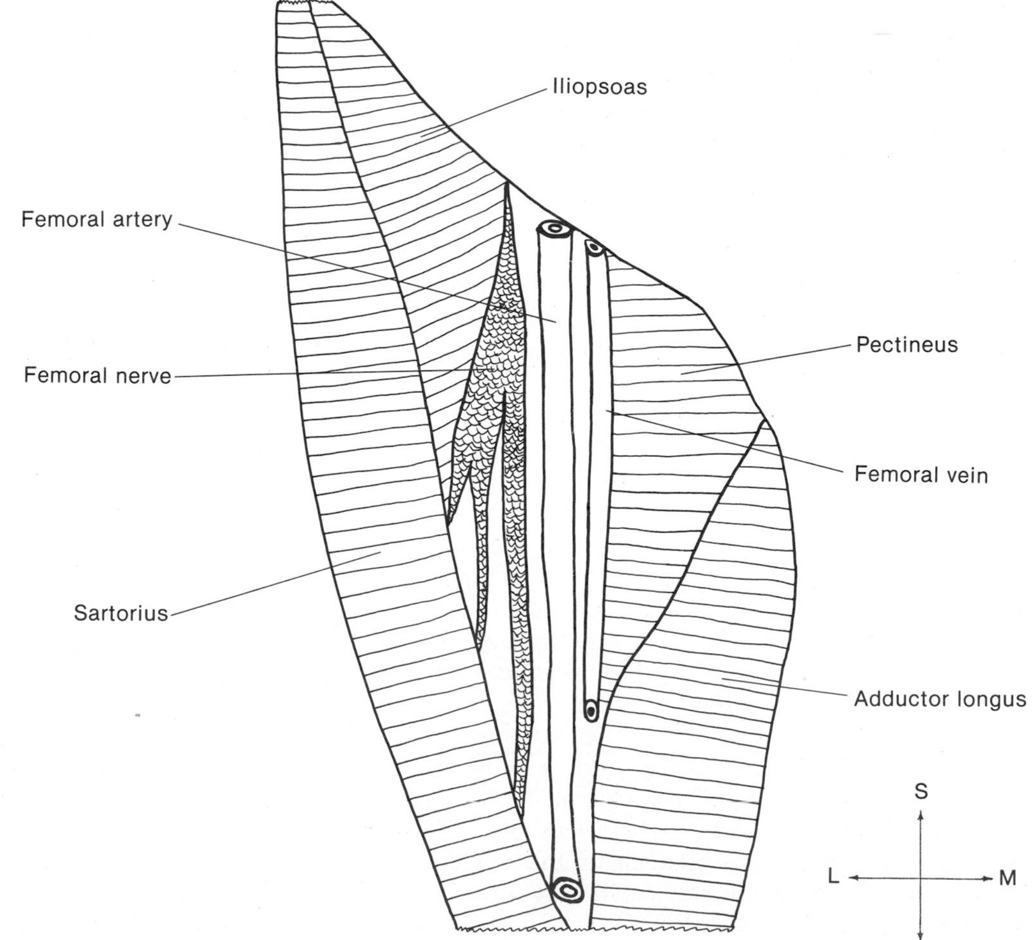

FIGURE 5–15. Surface view of femoral triangle, showing boundaries and contents. Note sequence of femoral artery, vein, and nerve within triangle.

gle to the popliteal fossa. The canal is really an intermuscular groove beneath the sartorius muscle. The subsartorial canal contains connective tissue in addition to the femoral artery and vein, which are continuing their descent to the popliteal fossa.

Transverse Sections of the Thigh

TRANSVERSE SECTION THROUGH THE PROXIMAL FEMUR

This section, illustrated in Figure 5-16, shows the three muscle compartments of the thigh. The anterior compartment contains the **quadriceps femoris muscle** group and the **sartorius.** Just deep to the sartorius, the **femoral artery** and **vein** descend in the subsartorial canal. The **great saphenous vein** is in the subcutaneous tissue near the medial boundary of this compartment. The medial compartment contains the **adductor longus, brevis,** and **magnus muscles,** as well as the **gracilis,** which is more superficial than the adductors. The posterior compartment contains the **hamstrings.** At superior levels of the thigh, the **gluteus maximus** may be evident before it inserts on the iliotibial tract and gluteal tuberosity of the femur. The **sciatic nerve** supplies the posterior compartment.

TRANSVERSE SECTION THROUGH THE DISTAL FEMUR

The adductor muscles in the medial compartment are noticeably absent in sections through the distal femur because these muscles insert at higher levels. Only muscles that extend over the knee joint are evident. The three **vastus muscles** with the quadriceps tendon occupy the anterior compartment. In proximal regions of the femur, the sartorius is superficial in the anterior compartment. As the sartorius descends, it crosses obliquely over the thigh to insert on the medial tibia. In sections through the distal thigh, the **sartorius** is next to the **gracilis** on the medial side. The three muscles of the **hamstrings** occupy the posterior compartment. A transverse section through the distal femur is illustrated in Figure 5-17.

General Anatomy of the Leg

OSSEOUS COMPONENTS OF THE LEG

The leg is the portion of the lower extremity that is between the knee and the foot. Its framework consists of two bones, the **tibia** and the **fibula.** The **tibia,** on the medial side, is the larger of the two bones and is the weight-bearing bone. The tibia articulates proximally with the condyles of the femur at the knee and distally with the talus, which is one of the tarsal bones. The distal end of the tibia is the medial malleolus. This is near the surface and can be palpated as a subcutaneous bump on the medial side of the ankle. The **fibula,** on the lateral side, is a long slender bone that functions primarily as an attachment for muscles and to lend stability. The distal end of the fibula is the lateral malleolus, which can be palpated as a subcutaneous bump on the lateral side of

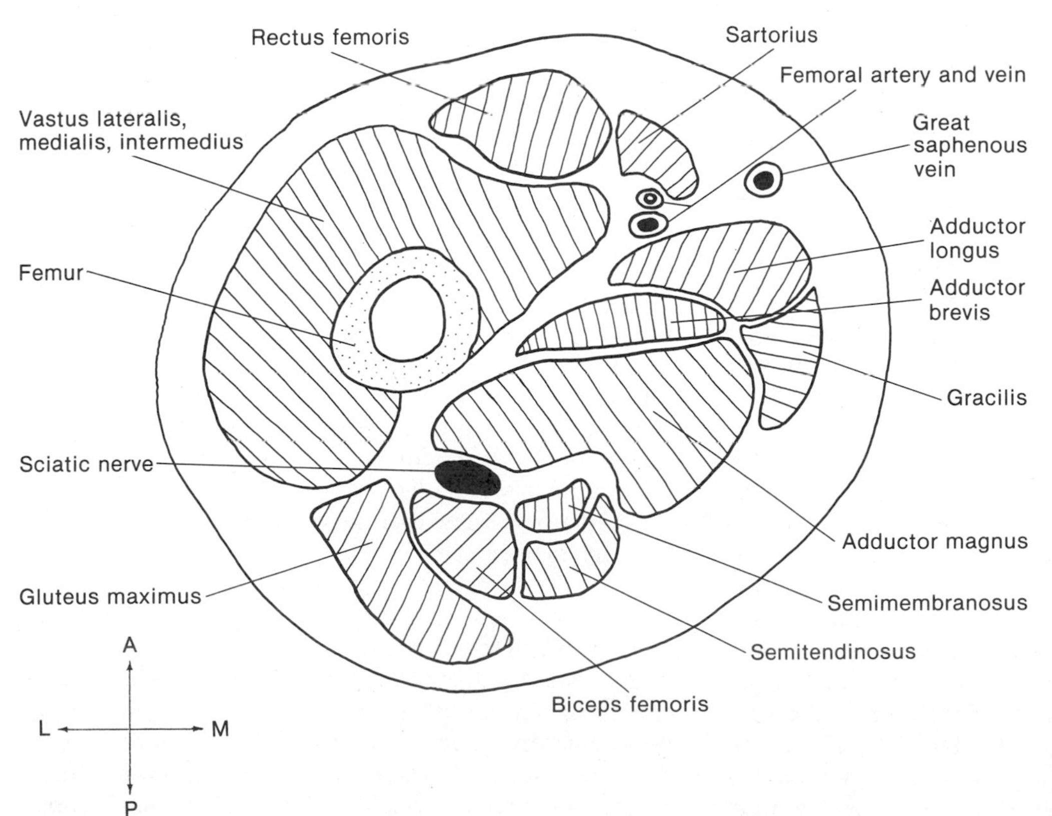

FIGURE 5-16. Transverse section through proximal portion of the femur, illustrating the anterior, medial, and posterior muscle groups. Note sciatic nerve.

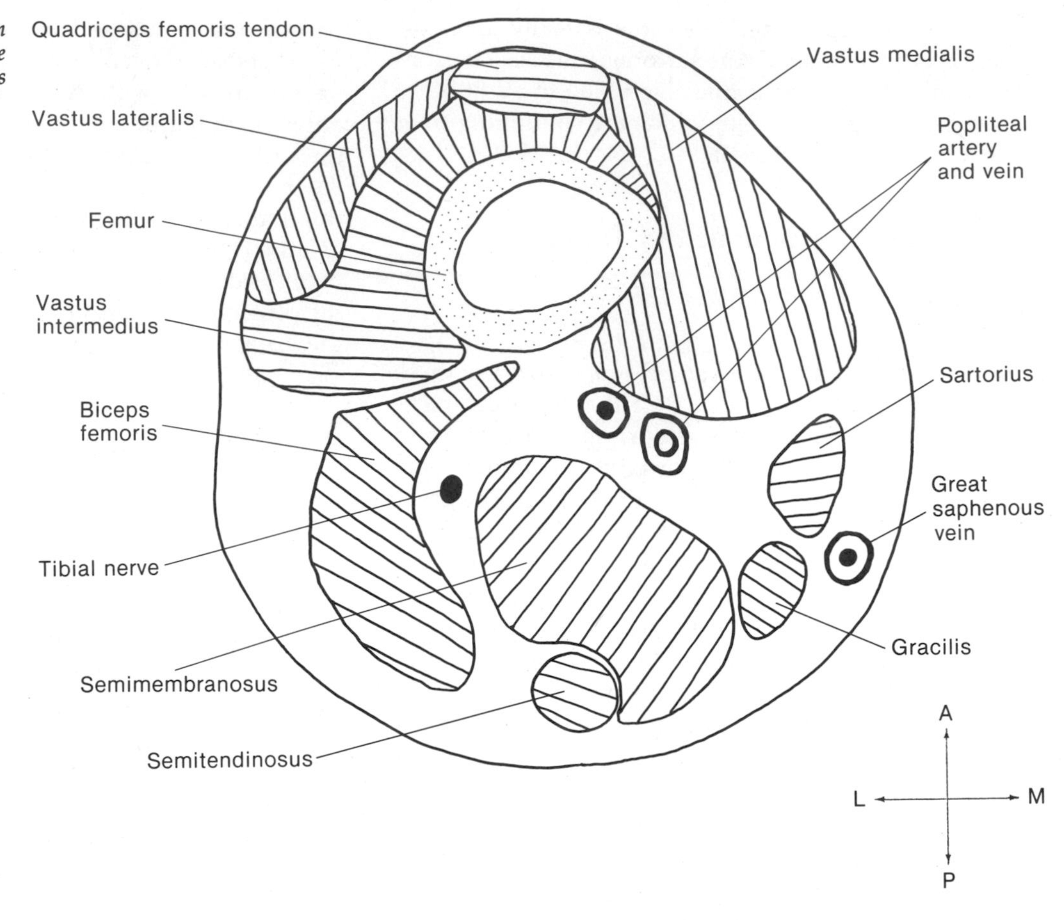

FIGURE 5-17. Transverse section through the distal portion of the femur. Note the great saphenous vein.

the ankle. The shafts of the tibia and fibula are connected by strong, oblique, connective tissue fibers that form an **interosseous membrane.**

MUSCLES LOCATED IN THE LEG

The muscles of the leg are divided into anterior, lateral, and posterior compartments by intermuscular septa, by the interosseous membrane, and by the tibia and fibula. A partition formed by the tibia, the fibula, and the interosseous membrane separates the anterior and posterior compartments. Intermuscular septa separate the lateral compartment from the anterior and posterior compartments. The muscles of the leg are summarized in Table 5-6.

ANTERIOR MUSCLE COMPARTMENT. Located anterior to the interosseous membrane, the muscles of the anterior compartment function primarily in dorsiflexion of the foot at the ankle joint and in extension of the toes. The four muscles in the anterior compartment are the **tibialis anterior,** the **extensor digitorum longus,** the **extensor hallucis longus,** and the **peroneus tertius.** Both the **tibialis anterior** and the **extensor digitorum longus** originate on the lateral condyle of the tibia and extend throughout the entire length of the leg. They are seen in transverse sections through the proximal leg with the tibialis anterior muscle medial to the extensor digitorum longus. The **extensor hallucis longus** and the **peroneus tertius** originate on the middle and distal portions of the fibula and interosseous membrane. In distal sections of the leg, the sequence of these muscles, from medial to lateral, is as follows: tibialis anterior, extensor hallucis longus, extensor digitorum longus, and peroneus tertius. The neurovascular bundle in the anterior compartment contains the **deep peroneal nerve** and the **anterior tibial artery,** with its associated veins.

LATERAL MUSCLE COMPARTMENT. The lateral compartment is separated from the other compartments by the anterior and posterior intermuscular septa and the lateral surface of the fibula. It extends from the head of the fibula to the lateral malleolus. The **peroneus longus** and **peroneus brevis** are the major muscles in the compartment. The peroneus longus is the more superficial of the two muscles. It originates on the proximal part of the fibula and on the lateral condyle of the tibia. Its tendon continues down the leg, passes behind and below the lateral malleolus, and then curves medially under the foot to attach to the first metatarsal. The peroneus brevis originates on the distal half of the fibula. Its tendon also curves around the lateral malleolus, but the peroneus brevis inserts on the fifth metatarsal. Both muscles evert the foot and are weak plantarflexors at the ankle. The muscles in the lateral compartment are innervated by the **superficial peroneal nerve,** a branch of the common peroneal nerve. The blood supply to the muscles is through branches of the **peroneal artery,** which arises from the posterior tibial artery.

TABLE 5-6. Muscles Located in the Leg

Muscle	Origin	Insertion	Action	Innervation
Anterior Compartment				
Tibialis anterior	Lateral condyle of tibia	First cuneiform and metatarsal	Dorsiflexes foot	Deep peroneal
Extensor digitorum longus	Lateral condyle of tibia	Phalanges of second to fifth toes	Extends toes	Deep peroneal
Extensor hallucis longus	Middle fibula	Distal phalanx of great toe	Extends great toe	Deep peroneal
Peroneus tertius	Distal fibula	Fifth metatarsal	Dorsiflexes foot	Deep peroneal
Lateral Compartment				
Peroneus longus	Lateral condyle of tibia; proximal fibula	First metatarsal	Everts foot and weak plantarflexor of ankle	Superficial peroneal
Peroneus brevis	Distal half of fibula	Fifth metatarsal	Everts foot and weak plantarflexor of ankle	Superficial peroneal
Posterior Compartment				
Superficial Layer				
Gastrocnemius	Lateral and medial condyles of femur	Posterior surface of calcaneus	Plantar flexes foot; flexes knee	Tibial
Soleus	Tibia and fibula	Posterior calcaneus	Plantar flexes foot	Tibial
Plantaris	Popliteal surface of femur	Posterior calcaneus or becomes part of Achilles tendon	Plantar flexes foot	Tibial
Deep Layer				
Tibialis posterior	Tibia and fibula	Tarsals and metatarsals	Plantarflex and invert foot	Tibial
Flexor digitorum longus	Middle tibia	Distal phalanges of lateral four toes	Flex lateral four toes; assist in plantarflexion of foot	Tibial
Flexor hallucis longus	Distal fibula	Distal phalanx of great toe	Flex great toe; assist in plantarflexion of foot	Tibial
Popliteus	Lateral condyle of femur	Tibia	Medially rotate leg	Tibial

POSTERIOR MUSCLE COMPARTMENT. The posterior compartment has the largest bulk of the three compartments as a result of the mass of the gastrocnemius and soleus muscles, which are located there. From medial to lateral, the anterior delineation of the compartment is as follows: the tibia, the interosseous membrane, the fibula, and the posterior intermuscular septum. The posterior compartment is subdivided into superficial and deep muscles by a transverse intermuscular septum.

Superficial Muscles. The principal superficial muscles of the posterior compartment are the **gastrocnemius** and the **soleus,** which form most of the contour of the calf of the leg. The gastrocnemius originates by two heads from the lateral and medial condyles of the femur. The two heads come together at the inferior margin of the popliteal fossa to form a single gaster of the muscle. From the gaster, a long, tough tendon descends to insert on the posterior surface of the calcaneus. The gastrocnemius muscle is a strong plantarflexor of the foot at the ankle. Because of its origin on the femur, it is also a weak flexor of the knee. The broad, flat, fleshy soleus muscle lies deep to the gastrocnemius, and it is attached to the posterior surface of the tibia and fibula. The tendons of the gastrocnemius and soleus muscles join, forming a single common tendon that inserts on the posterior surface of the calcaneus. This common tendon is the **tendo calcaneus,** or **Achilles tendon.** The soleus works with the gastrocnemius in plantar flexing the ankle joint, but it has no action on the knee.

Deep Muscles. From medial to lateral, the positioning of the deep muscles is as follows: the **flexor digitorum longus,** the **tibialis posterior,** and the **flexor hallucis longus.** The tendons of these muscles curve to the sole of the foot and insert on the tarsals, metatarsals, or phalanges. The flexor hallucis longus flexes the great toe (hallux), whereas the flexor digitorum longus flexes the lateral four toes. Both of these muscles also assist in plantar flexing the foot. The tibialis posterior plantar flexes and inverts the foot. A fourth deep muscle, the **popliteus,** is a thin muscle in the popliteal region. It forms the floor of the popliteal fossa and acts on the knee joint to medially rotate the leg. The deep muscles are innervated by the **tibial nerve,** and the blood supply is derived from branches of the **posterior tibial artery.**

Because the gastrocnemius and soleus together have three heads of origin but a single common insertion and because they function together, they are sometimes collectively called the **triceps surae muscle.**

The small **plantaris** is a weak muscle at the superior border of the lateral head of the gastrocnemius. It has a long, thin tendon, which may become part of the Achilles tendon or may insert directly on the calcaneus. The plantaris muscle varies in size and may be absent with no apparent effect. The weak tendon may rupture during violent ankle movements and cause severe pain in the calf of the leg. Since the muscle is of little practical use in movement of the knee and ankle, its long tendon is sometimes used as a graft in reconstructive hand surgery.

Blood Supply. The blood supply to the muscles in the posterior compartment of the leg is through the **posterior tibial artery** and the associated veins. Innervation is by the **tibial nerve.**

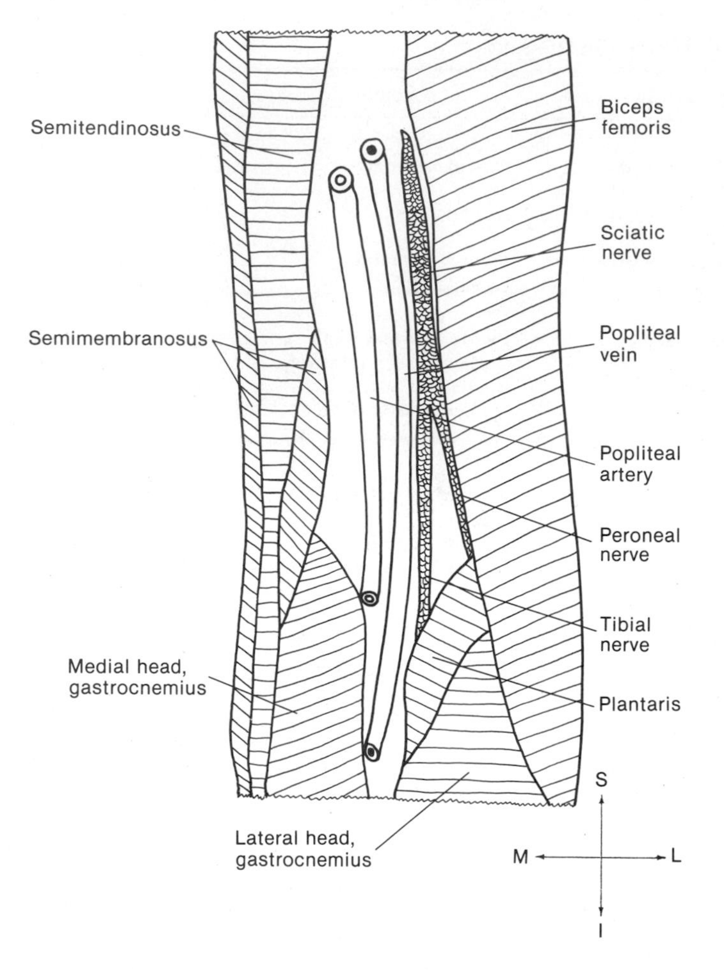

FIGURE 5–18. Surface view of popliteal fossa, showing boundaries and contents of the fossa. Note relationships of sciatic nerve and popliteal artery and vein.

POPLITEAL FOSSA

The **popliteal fossa** is the diamond-shaped, fat-filled region behind the knee joint. The vessels and nerves to the leg pass through this area. The upper two margins are formed by the diverging tendons of the **hamstring muscles.** The lateral and medial heads of the **gastrocnemius muscle** form the inferior two margins. The small **plantaris muscle** also contributes to the inferior lateral margin. The roof of the fossa consists of the deep fascia that envelops the muscles of the thigh and the leg. This is penetrated by the short saphenous vein and the posterior cutaneous nerve as they enter the superficial fascia. From superior to inferior, the floor of the fossa is formed by the popliteal surface of the femur, the capsule of the knee joint, and the popliteus muscle. As the femoral artery and vein enter the fossa, they become the **popliteal artery** and **vein.** Lymph nodes are usually in the region of the popliteal artery. The popliteal vein is superficial to the artery. The **tibial** and **common peroneal nerves,** which pass through the fossa, are terminal branches of the sciatic nerve. These two nerves are superficial to the popliteal vessels. The contents of the fossa are embedded in fat. The popliteal fossa is illustrated in Figure 5–18.

Transverse Sections of the Leg

TRANSVERSE SECTION THROUGH THE PROXIMAL LEG

The appearance of sections through the leg varies with the level because the muscles originate and insert in different regions. A transverse section through the proximal

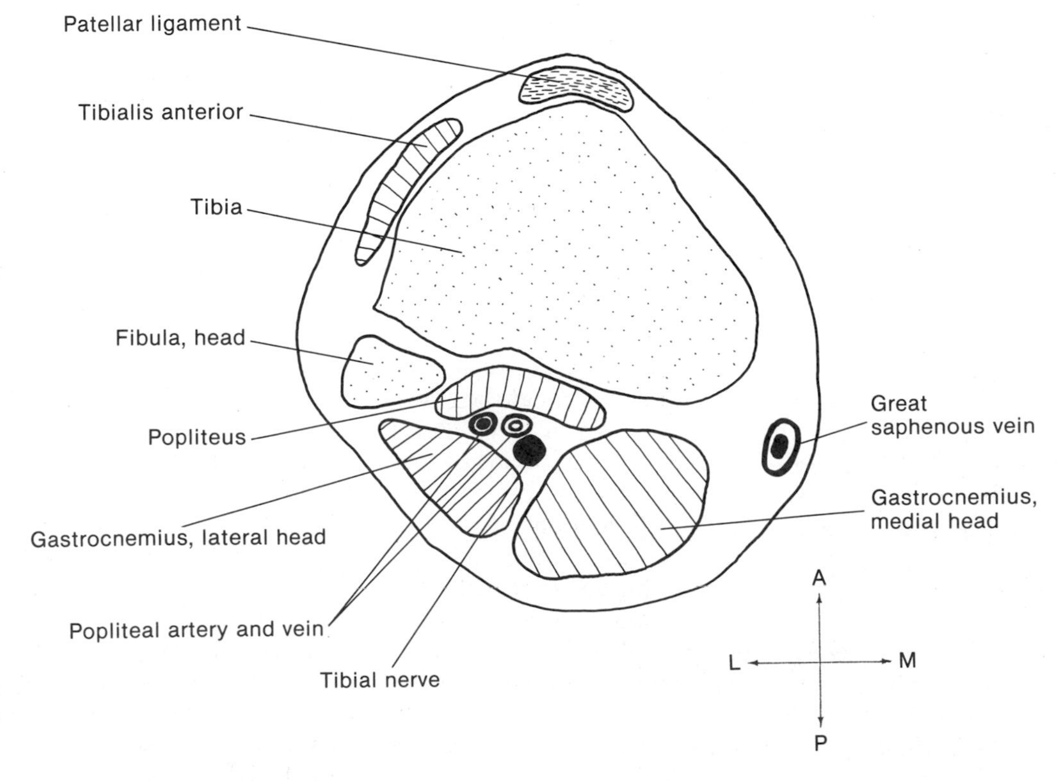

FIGURE 5–19. Transverse section through the proximal portion of the leg.

leg, near the lower region of the popliteal fossa, shows the **lateral** and **medial condyles** of the **tibia** and the **head of the fibula.** Muscle mass is minimal at this level, especially in the anterior and lateral compartments. The two heads of the **gastrocnemius** are evident in the posterior compartment. This is illustrated in Figure 5–19.

TRANSVERSE SECTION THROUGH THE MIDCALF REGION

At lower levels, through the shafts of the tibia and fibula in the midcalf region, more muscle mass is present and the intermuscular septa are more evident than in the proximal leg. The **tibialis anterior** and the **extensor digitorum longus muscles** are anterior to the **interosseous membrane.** The **peroneus longus** occupies most of the lateral compartment. The two heads of the **gastrocnemius** merge in the posterior compartment. The mass of the **soleus** is just beneath the gastrocnemius. The **tibialis posterior** is the most obvious deep muscle in this region. At more distal levels, near the ankle, the muscles become smaller and more tendinous. Here, the anterior tibial artery is near the surface along the anterior margin of the tibia. A transverse section through the midcalf region is illustrated in Figure 5–20.

ARTICULATIONS ASSOCIATED WITH THE LOWER EXTREMITY

Description of the Hip (Coxal) Joint

The hip (coxal) joint is formed by the head of the femur articulating in the acetabulum of the os coxa (hip bone). It is a multiaxial ball-and-socket joint, which allows flexion, extension, abduction, adduction, medial and lateral rotation, and circumduction. This joint bears the weight of the body; therefore, it is strong and stable. The strength and stability of the hip are enhanced by the shape and nature of the articular surfaces, the dense joint capsule, and the capsular ligaments.

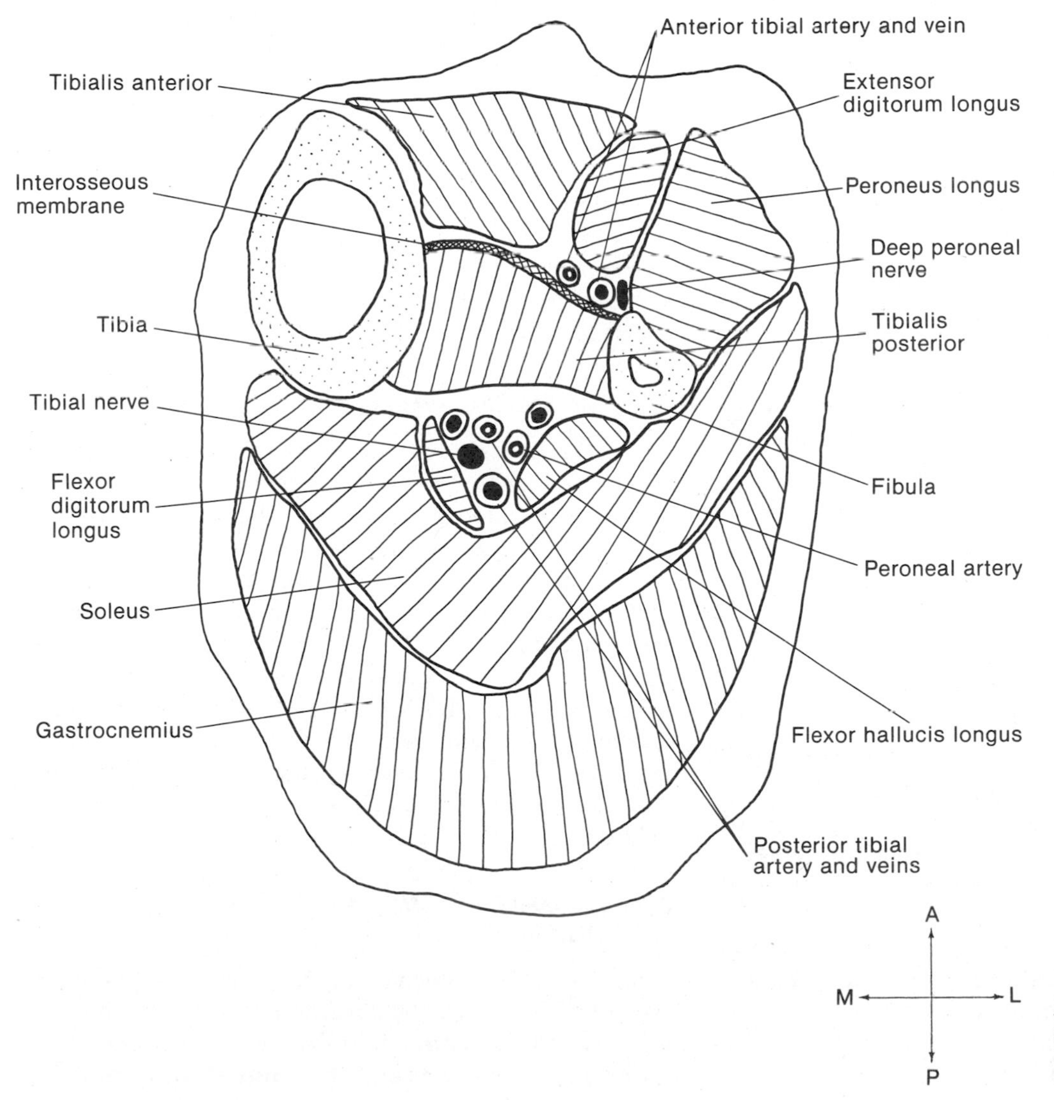

FIGURE 5–20. Transverse section through the midcalf region, showing the anterior, lateral, and posterior muscle groups.

ARTICULATING SURFACES

The articular surface of the femur is the head, which represents about two thirds of a sphere. The **femoral head** is covered by articular cartilage, except at a small depression, or pit, called the **fovea capitis femoris.** The **ligamentum teres femoris** (ligament of the head of the femur) attaches to the femoral head at the fovea. The femoral head fits into the **acetabulum,** a deep cup-shaped socket in the os coxa.

The rim of the acetabulum is incomplete at the inferior margin, leaving an **acetabular notch,** which is closed by the **transverse acetabular ligament.** The concavity of the acetabulum is further deepened by a fibrocartilaginous **acetabular labrum** that is attached to the bony margin of the socket. The articulating surface of the acetabulum consists of a C-shaped area covered by articular cartilage, which surrounds a centrally located, nonarticulating, **acetabular fossa.** The acetabular fossa contains a pad of fat that is covered by a synovial membrane and represents the coxal attachment of the ligamentum teres femoris. This ligament helps to hold the femoral head in the acetabulum.

The shaft of the femur is displaced about 2 in. from the head by the neck. This displacement moves the femoral shaft away from the pelvis to allow more freedom of movement.

JOINT CAPSULE AND ASSOCIATED MUSCLES

A dense fibrous joint capsule forms a cylinder that extends from the margin of the acetabulum to the neck of the femur near the trochanters. Synovial membrane lines the fibrous joint capsule. The capsule is thickened and reinforced by the **iliofemoral, pubofemoral,** and **ischiofemoral ligaments.**

Most of the muscles associated with the hip joint function to move the thigh as well as to stabilize the joint. All of these muscles originate on some part of the pelvic girdle, and most of them insert on some part of the femur. Exceptions to this are the gracilis and sartorius muscles, which extend down to the tibia and move the knee in addition to the hip. The muscles may be divided into anterior, posterior, and medial groups. The **iliopsoas** and the **sartorius** belong to the anterior group and flex the thigh. The posterolateral group is more extensive and contains the buttock muscles and the deep lateral rotators. The buttock muscles include the **gluteus maximus,** the **gluteus medius,** and the **gluteus minimus.** The deep rotators are located directly over the posterior portion of the joint. These muscles are the **piriformis,** the **gemelli,** the **obturators,** and the **quadratus femoris.** The medial muscle group is responsible for adduction of the thigh and includes the **pectineus** and the **gracilis** in addition to the **adductor longus,** the **adductor brevis,** and the **adductor magnus.** All of the muscles associated with the hip have been described in the sections dealing with the gluteal region and the thigh.

NEUROVASCULAR STRUCTURES

The principal neurovascular structures visible in sections of the hip are the **femoral artery,** the **femoral vein,** the **femoral nerve,** and the **sciatic nerve.** The femoral vessels and the femoral nerve are closely associated with, and superficial to, the iliopsoas muscle. The sciatic nerve, a branch of the sacral plexus, is the largest peripheral nerve in the body. It begins within the pelvis, anterior to the piriformis muscle. The sciatic nerve exits the pelvis through the greater sciatic notch, just inferior to the piriformis. It then descends, dorsal to the obturator and gemelli muscles but deep to the gluteus maximus.

Sectional Anatomy of the Hip Joint

TRANSVERSE SECTION THROUGH THE ACETABULUM

Figure 5–21 illustrates a transverse section through the upper portion of the hip. The **gluteal muscles** are clearly evident in the posterior compartment. The **piriformis muscle,** one of the deep rotators, is deep to the gluteus maximus. The piriformis is closely related to the **sciatic nerve,** as they both traverse the sciatic notch. The **obturator internus,** another of the deep lateral rotators, is medial to the bones of the pelvic girdle. In the anterior muscle compartment, the **sartorius** and **tensor fasciae latae** are located superficially, with the **iliopsoas** and **rectus femoris** just deep to the sartorius. The **pectineus** is medial to the iliopsoas.

Within the **acetabulum,** the humeral head articulates at the periphery of the socket, leaving a space, the **acetabular fossa,** in the center. The **ligamentum teres femoris** attaches the femoral head to the acetabular fossa.

SAGITTAL SECTION THROUGH THE ACETABULUM

Sagittal sections through the acetabulum, as illustrated in Figure 5–22, show the arrangement of the deep lateral rotators associated with the hip. The **piriformis,** at the margin of the **gluteus medius,** is the most superior of these muscles, followed by the **gemelli** and the quadratus femoris. The tendon for the **obturator internus** merges with the **superior** and **inferior gemelli muscles.** The **obturator externus** is deep to the **gluteus maximus** and inferior to the **piriformis.** Recall that the piriformis is a landmark for structures that are found deep to the gluteus maximus. The three layers of **adductor muscles** in the medial compartment, as well as the **pectineus,** are also evident.

CORONAL (FRONTAL) SECTION THROUGH THE ACETABULUM

Figure 5–23 illustrates a coronal section through the acetabulum. This shows the **obturator foramen** as an opening between the ilium and pubis, which is closed off by the **obturator internus** and **obturator externus** muscles.

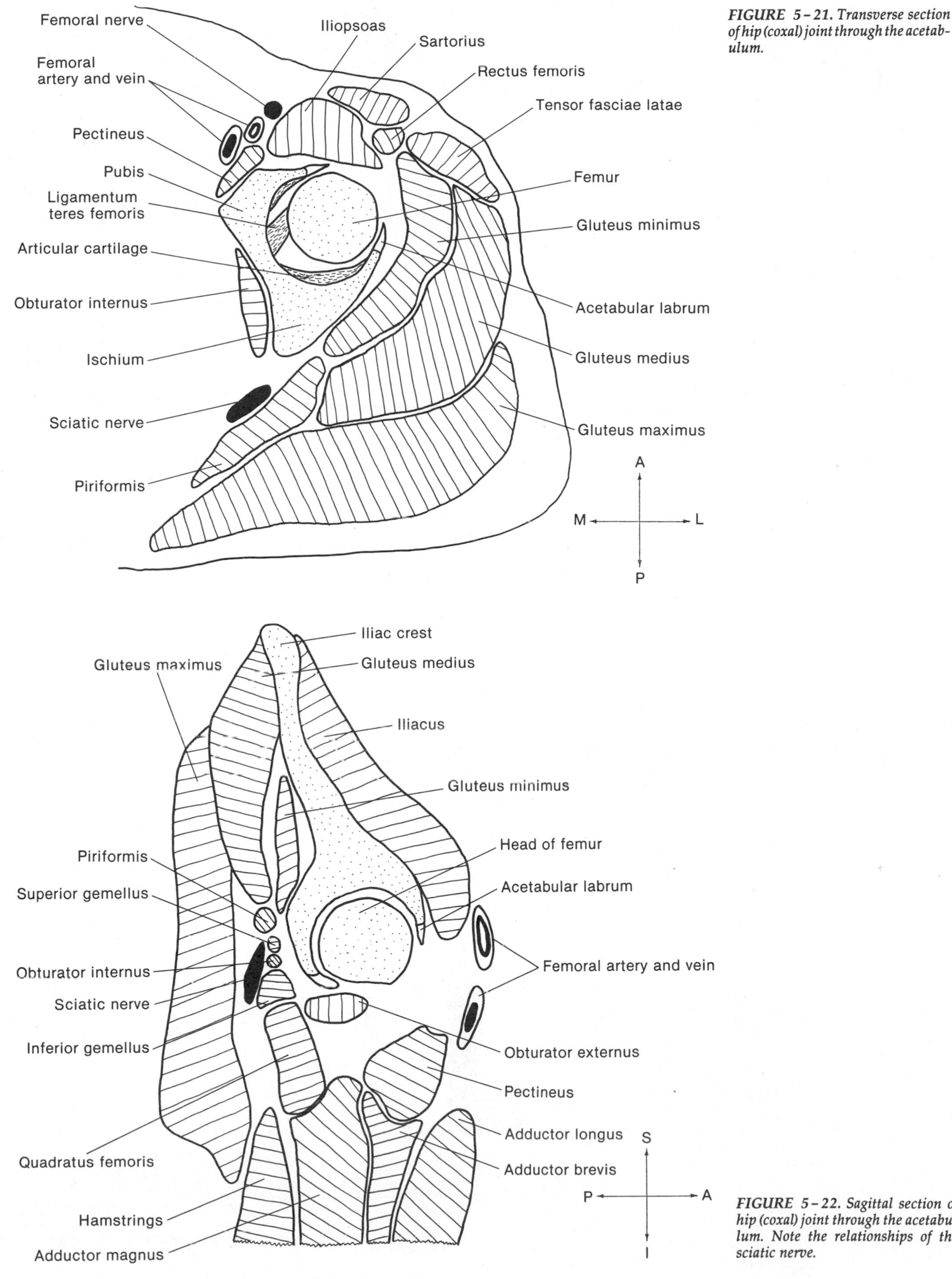

FIGURE 5-21. Transverse section of hip (coxal) joint through the acetabulum.

FIGURE 5-22. Sagittal section of hip (coxal) joint through the acetabulum. Note the relationships of the sciatic nerve.

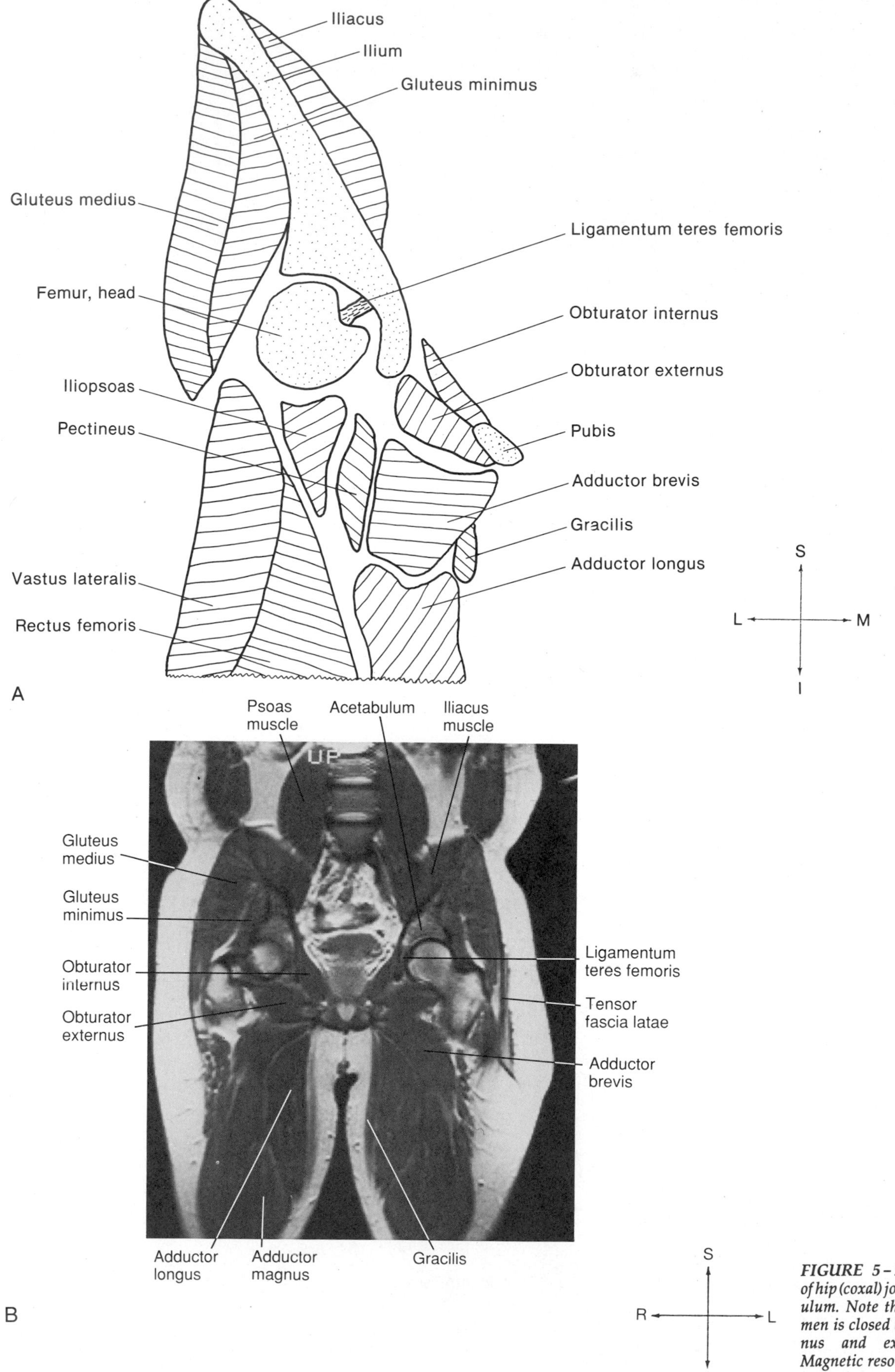

FIGURE 5-23. A, Coronal section of hip (coxal) joint through the acetabulum. Note that the obturator foramen is closed by the obturator internus and externus muscles. B, Magnetic resonance image of a similar region.

The lateral-to-medial sequence of the **iliopsoas, pectineus, adductor brevis,** and **gracilis** muscles is well represented.

Description of the Knee Joint

ARTICULAR SURFACES OF THE KNEE JOINT

The knee **(tibiofemoral)** joint is probably the most complex, yet most vulnerable, joint in the body. It is principally a hinge joint, but it does allow some gliding motion. The bony components are the **femur,** the **tibia,** and the **patella.** The large rounded condyles on the femur articulate with the rather flattened tibial plateaus. Anteriorly, the patella fits between the lateral and the medial condyles of the femur to form the patellofemoral articulation. The articulating surfaces are covered with hyaline articular cartilage.

LIGAMENTS ASSOCIATED WITH THE JOINT CAPSULE

A fibrous capsule forms a sleeve that encloses the joint. The capsule extends from a region on the femur just proximal to the lateral and medial condyles down to the tibia, just distal to the articulating surfaces. The fibrous capsule is lined with a synovial membrane, which secretes a lubricating synovial fluid into the joint cavity. Numerous ligaments and tendons add stability to the joint.

Concave, fibrocartilaginous pads called **menisci** (singular, meniscus) are interposed between the femoral condyles and the tibial plateaus. These lateral and medial menisci make shallow sockets for the rounded condyles. Anteriorly, the two menisci are joined together by a **transverse ligament.**

The fibrous capsule is strengthened by five ligaments. These are called external or extracapsular ligaments to distinguish them from those inside the capsule. Thickenings of the capsule on the lateral and medial side form **lateral (fibular)** and **medial (tibial) collateral ligaments.** Anteriorly, the continuation of the quadriceps femoris tendon forms the **patellar ligament,** which reinforces the fibrous capsule. Posterior to the joint, in the popliteal region, the capsule is strengthened by the **oblique popliteal** and **arcuate popliteal ligaments.** The oblique popliteal ligament is a broad expansion of the semimembranosus tendon. It arises from the posterior aspect of the medial tibial condyle and attaches near the center of the capsule. The Y-shaped arcuate popliteal ligament arises from the head of the fibula, then spreads out to attach to the lateral condyle of the femur and the intercondylar area of the tibia.

Intracapsular ligaments are inside the fibrous capsule rather than external to it. The anterior and posterior cruciate ligaments are intracapsular. They are located between the lateral and medial condyles of the femur, and they attach the femur to the tibia. The **anterior cruciate ligament** extends anteriorly from the lateral condyle of the femur to a point on the anterior surface of the tibia on the medial side of the intercondylar eminence. The **posterior cruciate ligament** extends between the medial condyle of the femur and the posterior surface of the tibia. Note that the anterior cruciate ligament attaches to the anterior tibial surface and that the posterior cruciate ligament attaches to the posterior surface of the tibia.

MUSCULAR SUPPORT FOR THE KNEE JOINT

Muscles and their tendons form a major support for the knee joint. Thus proper muscle conditioning can reduce the likelihood and severity of knee injuries. Conversely, weakened muscles contribute to instability of the knee joint. The primary muscles that contribute strength to the joint include the **quadriceps femoris,** the **hamstrings,** the **sartorius,** the **gracilis,** and the **gastrocnemius.** In addition to stabilizing the knee, these muscles move the leg by their action on the knee joint. See Table 5–5 and Table 5–6 for a review of these muscles and their actions.

NEUROVASCULAR STRUCTURES ASSOCIATED WITH THE KNEE JOINT

The principal neurovascular supply to the knee comes from the vessels and nerves in the popliteal fossa, which, from medial to lateral include the **popliteal artery,** the **popliteal vein,** the **tibial nerve,** and the **common peroneal nerve.** The popliteal artery is a continuation of the femoral artery. In the popliteal fossa, it gives off several genicular arteries, which course around the bones to supply the components of the joint. The tibial and common peroneal nerves are branches of the sciatic nerve.

Sectional Anatomy of the Knee Joint

SAGITTAL SECTION THROUGH THE LATERAL FEMUR AND THE TIBIA

Figure 5–24 illustrates a parasagittal section of the knee joint. The articulation of the **lateral condyle** of the femur with the **tibia** and the **lateral meniscus** between the two bones is shown. The **patella** is anterior to the femur. The **patellar ligament** extends from the patella down to the tibial tuberosity as an extension of the quadriceps femoris tendon. Three bursae associated with the patella, two subcutaneous and one deep, are illustrated.

CORONAL (FRONTAL) SECTION THROUGH THE KNEE JOINT

A coronal (frontal) section through the knee joint, as illustrated in Figure 5–25, shows both the **lateral** and the **medial menisci.** The joint capsule is reinforced at the sides by the **lateral** and the **medial collateral ligaments.** The intracapsular **anterior** and **posterior cruciate ligaments** are in the space between the two femoral condyles.

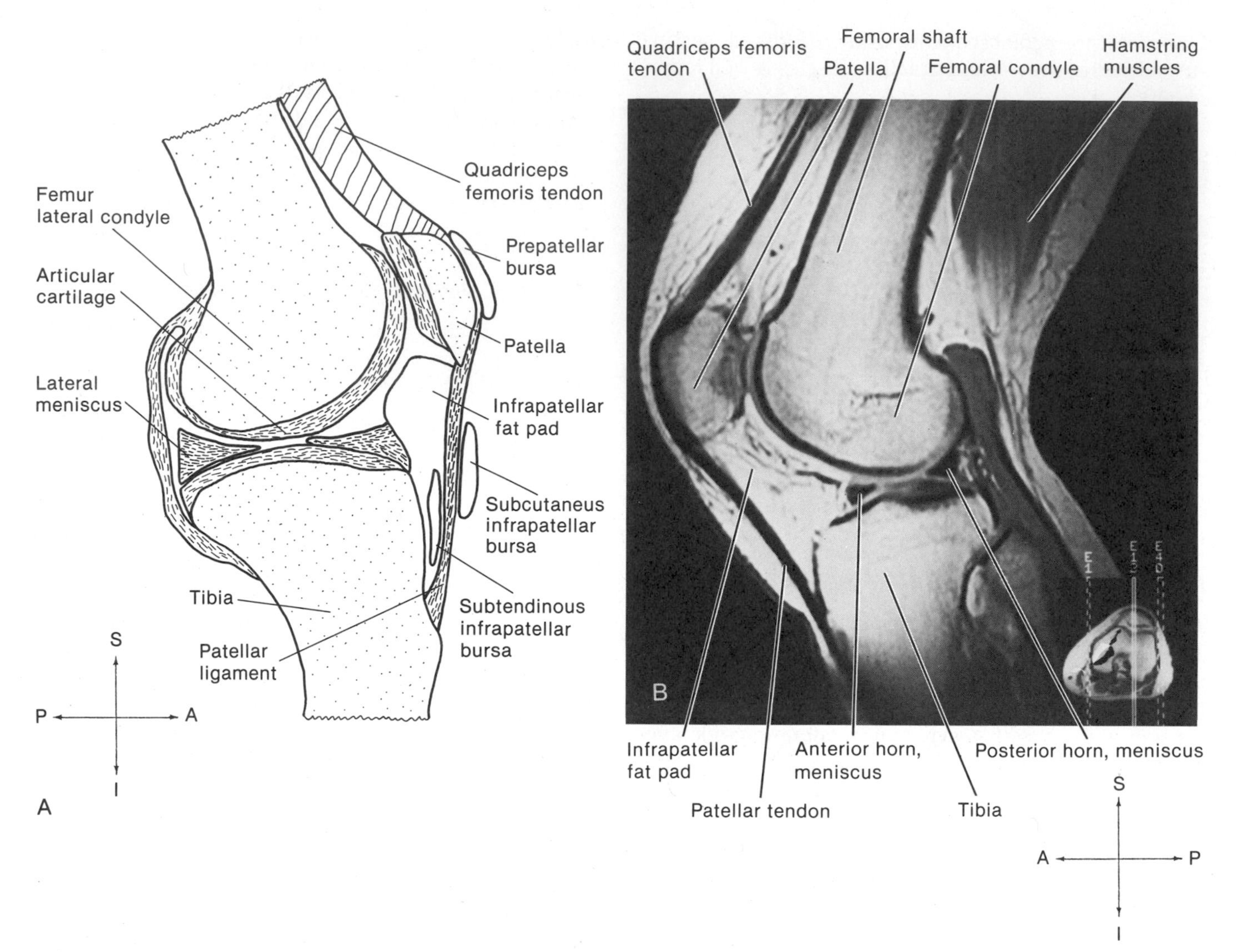

FIGURE 5-24. A, Sagittal section of knee joint through the lateral femur and tibia. Note the lateral meniscus, the patellar ligament, the subcutaneous and subtendinous bursae. B, Magnetic resonance image of a similar region.

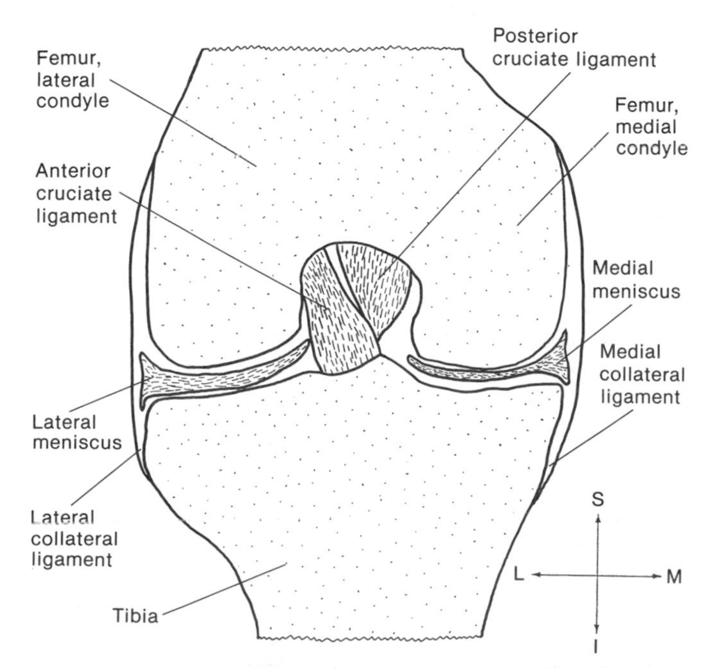

FIGURE 5-25. Coronal section through the knee joint. Note the lateral and medial menisci and the anterior and posterior cruciate ligaments.

Description of the Ankle Joint

The ankle, or **talocrural joint,** consists of the articulation between the distal **tibia** and **fibula** of the leg and the **talus** of the tarsal bones. In addition, some attention will be given to the relationships of the talus to the **calcaneus.** The articular surfaces of the talus are shown in Figure 5-26, which illustrates a coronal section through the ankle.

TALOCRURAL ARTICULATIONS

The talocrural joint is a hinge-type synovial joint, which permits dorsiflexion and plantarflexion of the foot. Medial and lateral movement is restricted by the medial malleolus of the tibia and the lateral malleolus of the fibula. The talus has three articular surfaces that contribute to the joint. The largest is the superior rounded surface, called the **trochlea,** which rests in the inferior concave surface of the distal tibia. Laterally and medially on the talus are facets for articulation with the malleoli of the fibula and tibia.

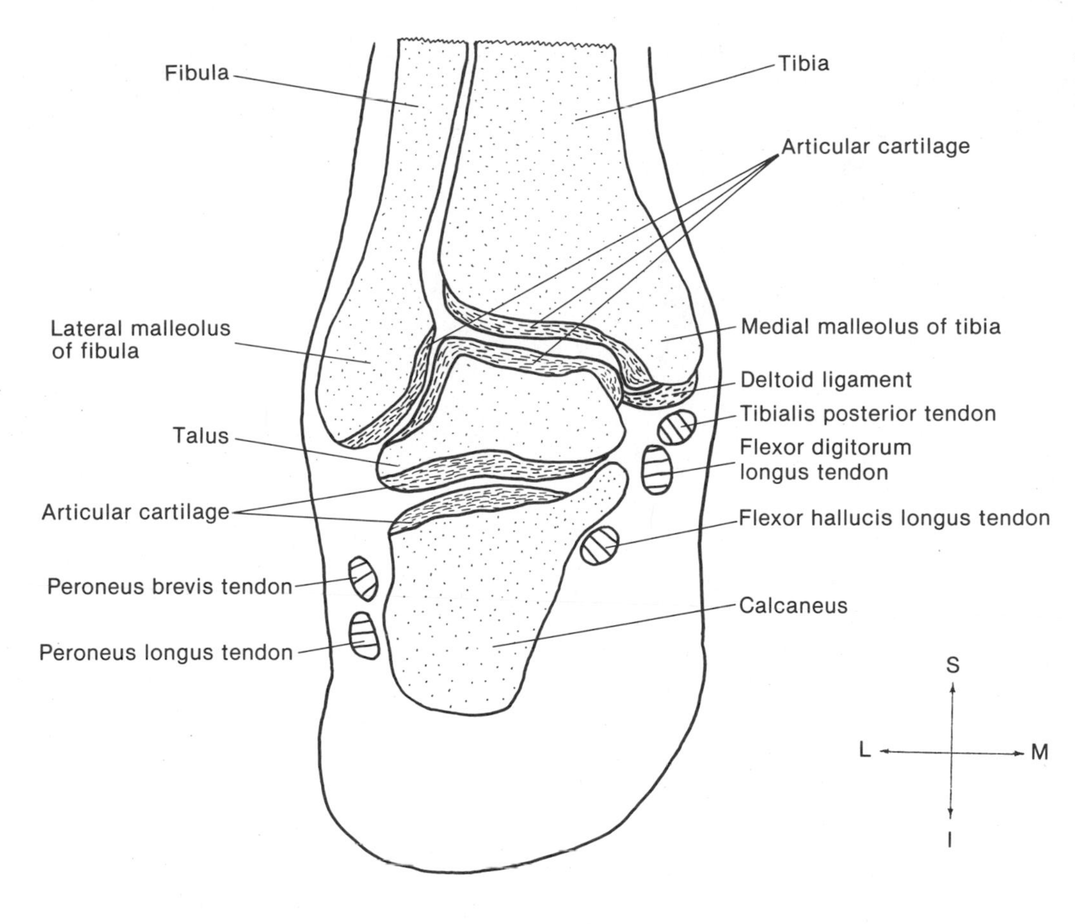

FIGURE 5-26. Coronal section through ankle. Note talocrural and talocalcanean articulations.

The talocrural joint is enclosed in a thin, fibrous capsule that is strengthened on the sides by ligaments. The **lateral (fibular) ligament** extends from the lateral malleolus to the talus and the calcaneus. Distally, the ligament is divided into three parts, according to the area of attachment. The anterior and posterior talofibular ligaments attach to the anterior and posterior regions of the talus. The calcaneofibular ligament forms a cord that extends from the lateral malleolus to the lateral surface of the calcaneus. The ligament on the medial side is much stronger than the lateral ligament. The **medial (deltoid** or **tibial) ligament** arises from the medial malleolus. The fibers of the ligament fan out to form a broad base of attachment on the anterior and posterior regions of the talus, the calcaneus, and the navicular. The four individual portions of the ligament are named according to their distal attachments as the anterior and the posterior tibiotalar, the tibiocalcanean, and the tibionavicular ligaments.

TALOCALCANEAN ARTICULATIONS

Inferiorly, in the subtalar region, three facets of the talus articulate with the calcaneus. This **subtalar** or **talocalcanean joint** has a fibrous capsule that is distinct from that of the ankle. The interosseous ligament, which extends from the midtalar region to the calcaneus, provides support for the joint. It is located in a space called the tarsal sinus. The talocalcanean joint is a gliding synovial joint that allows inversion and eversion of the foot.

MUSCULOTENDINOUS STRUCTURES ASSOCIATED WITH THE ANKLE JOINT

Much of the support for the ankle comes from the musculotendinous structures in the region. These may be divided into medial, lateral, anterior, and posterior groups. The relationships of the musculotendinous structures of the ankle are illustrated in Figure 5-27, which shows a transverse section just superior to the joint cavity.

MEDIAL MUSCLE GROUP. The medial group includes the **tibialis posterior**, the **flexor digitorum longus,** and the **flexor hallucis longus.** The tibialis posterior is the most anterior of the three and the flexor hallucis longus is the most posterior. The **posterior tibial artery** and **tibial nerve** are located between the flexor digitorum longus and the flexor hallucis longus. Posterior to the medial malleolus, the tibial nerve divides into the medial and the lateral plantar nerves.

LATERAL MUSCLE GROUP. The lateral group consists of the **peroneus longus** and the **peroneus brevis.** As the muscles descend the leg, the peroneus brevis is anterior to the peroneus longus. The two muscles follow a groove in the lateral malleolus, then curve forward under the malleolus, and extend anteriorly to the metatarsals. In the foot, the peroneus brevis is superior to the peroneus longus (see Fig. 5-26).

ANTERIOR MUSCLE GROUP. The anterior muscle group consists of the **tibialis anterior,** the **extensor hallucis longus,** the **extensor digitorum longus,** and the **peroneus tertius.** The tibialis anterior is the most promi-

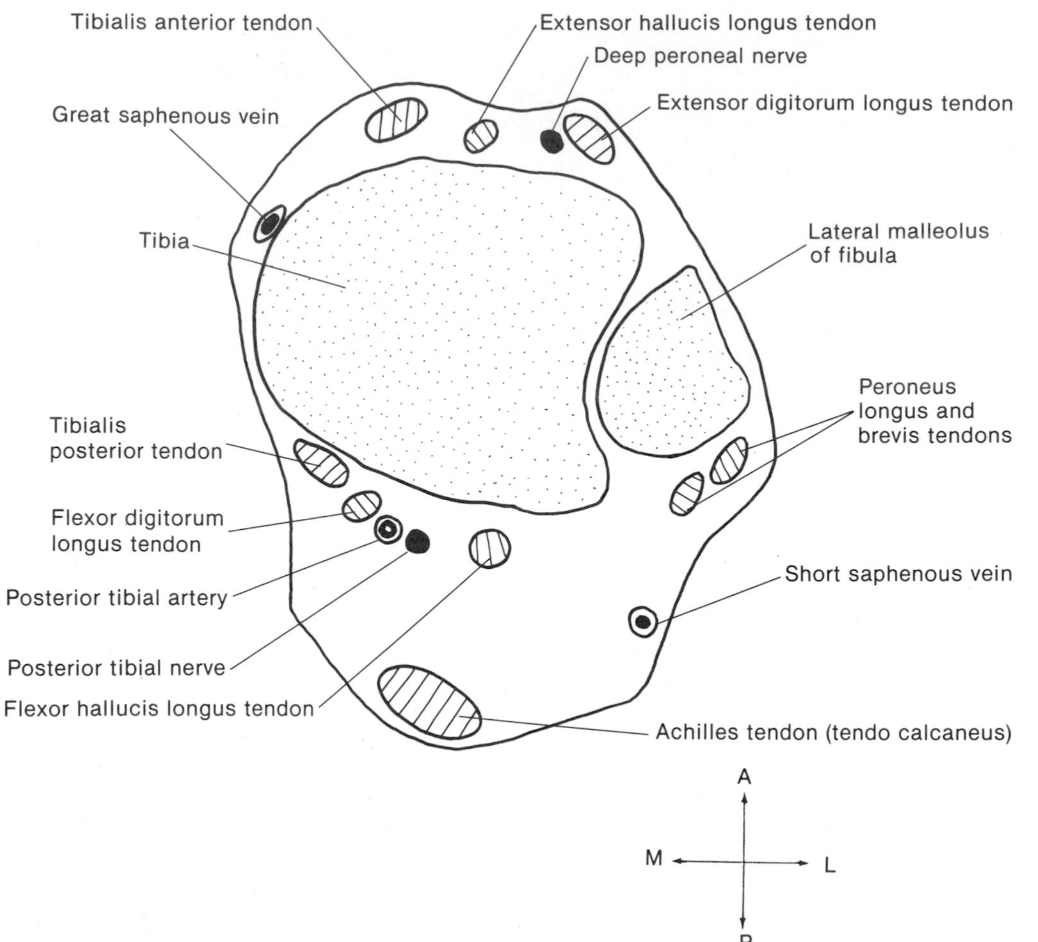

FIGURE 5-27. Transverse section through the lateral malleolus of the fibula, showing musculotendinous structures passing through the ankle joint.

nent and the most medial of these muscles. The extensor digitorum longus and peroneus tertius are difficult to separate and are the most lateral muscles of the group. The **anterior tibial artery** and **deep peroneal nerve** are deep to the extensor hallucis muscle.

POSTERIOR MUSCLE GROUP. The **tendo calcaneus,** or Achilles tendon, constitutes the only musculotendinous structure in the posterior group. This tendon arises from the gastrocnemius and soleus muscles and attaches to the posterior surface of the calcaneus. The Achilles tendon is the thickest and strongest tendon in the body.

REVIEW QUESTIONS

1. The two bones that form the pectoral girdle are the ______________ and ______________.
2. **True** or **False:** Muscles that extend between the scapula and humerus tend to also move the scapula.
3. Which one of the following is **NOT** a part of the boundaries of the axilla?
   a. intertubercular groove of the humerus
   b. pectoralis minor muscle
   c. subscapularis muscle
   d. brachial plexus
   e. ribs
4. Which one of the following statements about the components of the arm is **NOT** correct:
   a. The biceps brachii and the triceps brachii are located on opposite sides of the humerus.
   b. The only bone in the arm is the humerus.
   c. The cephalic and basilic arteries are superficial arteries in the arm and the brachial artery is a deep artery, which accompanies the brachial vein.
   d. The nerves in the arm are all branches of the brachial plexus.
   e. The coracobrachialis muscle located in the arm acts primarily on the shoulder joint.
5. The muscle that forms the lateral boundary of the cubital fossa is the ______________.
6. The ______________ nerve is located in the carpal tunnel.
7. The four muscles that form the rotator cuff of the shoulder are the ______________, ______________, ______________, and ______________.

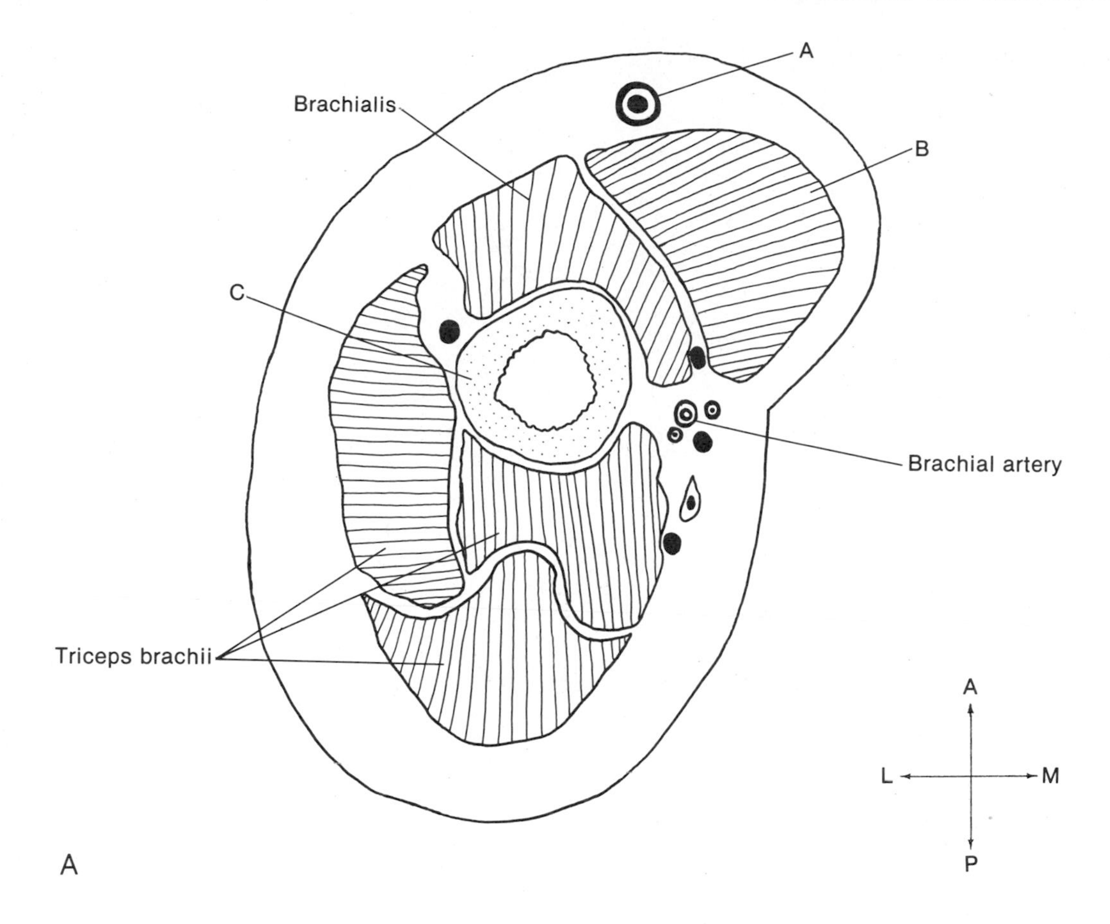

8. Which of the following statements about the forearm is **NOT** correct?
   a. The bones in the forearm are the radius and ulna.
   b. The muscles of the forearm are divided into anterior and posterior compartments by the radius, ulna, and interosseous membrane.
   c. Muscles in the forearm generally act on the wrist, hand, and fingers.
   d. The anterior muscles in the forearm are extensors and the posterior muscles are flexors.
   e. The brachioradialis muscle acts on the elbow rather than the wrist or hand.
9. **True** or **False:** The humeroradial articulation at the elbow involves the trochlea of the humerus and the radial notch of the radius.
10. Identify the structures indicated on Figure *A.*
11. Which of the following muscles is most superior?
    a. quadratus femoris
    b. piriformis
    c. gemelli
    d. gluteus minimus
    e. pectineus
12. The muscles in the posterior compartment of the thigh are collectively called the ______________.
13. **True** or **False:** From lateral to medial, the neurovascular structures in the femoral triangle are the femoral artery, the femoral nerve, and the femoral vein.
14. Which of the following is **NOT** a correct statement about the leg?
    a. The tibia is the lateral bone in the leg and the fibula is medial.
    b. The principal superficial muscles in the posterior compartment are the gastrocnemius and soleus.
    c. The muscles in the anterior compartment principally function to dorsiflex the foot and extend the toes.
    d. The blood supply to the posterior muscle compartment is by way of the posterior tibial artery and innervation is by the tibial nerve.
    e. The tendons of the gastrocnemius muscle extend over two joints; therefore the muscle affects both the knee and the ankle.
15. **True** or **False:** The popliteal fossa is basically diamond-shaped, with two sides formed by tendons of the hamstring muscles and two sides formed by tendons of the gastrocnemius.
16. The short ligament that extends from the fovea capitis femoris to the acetabular fossa is called the ______________.
17. The fibrocartilaginous pads that rest on the tibial plateaus and form shallow sockets for the femoral condyles are called ______________.
18. The superior rounded surface of the talus that contributes to the talocrural articulation is called the ______________.

19. Identify the structures indicated on Figure *B*.

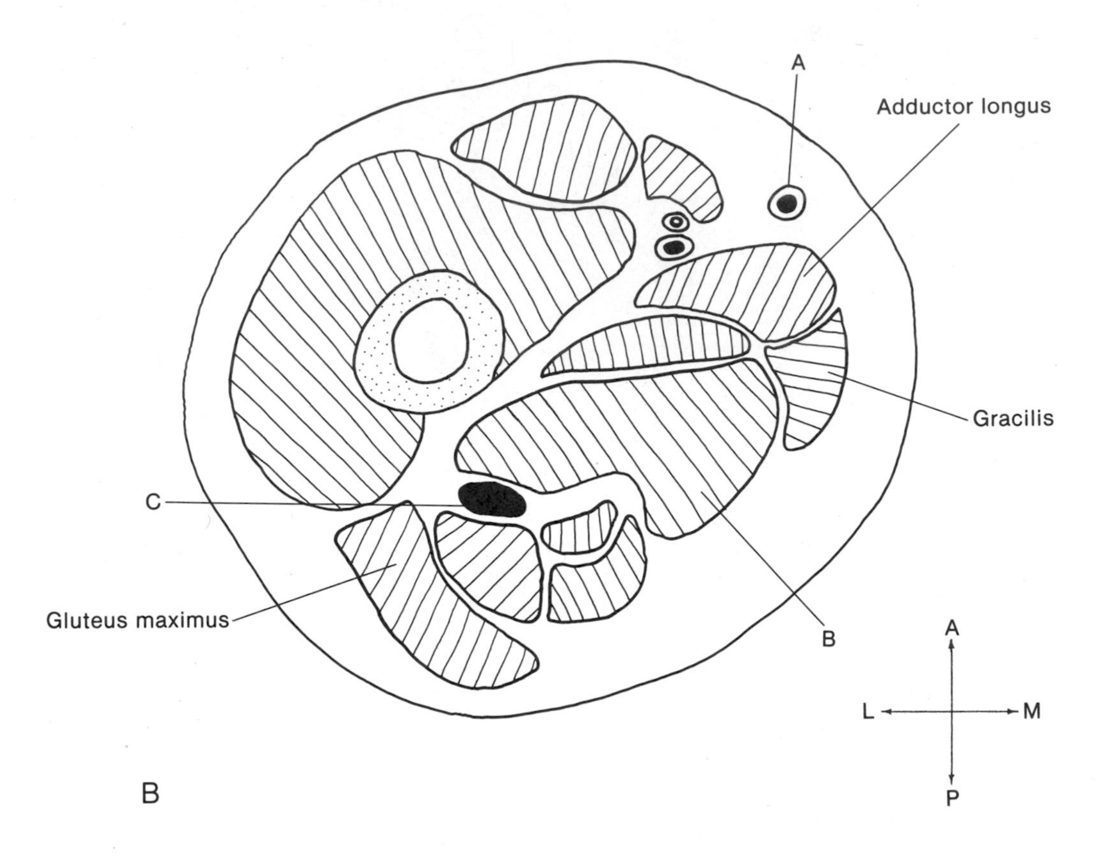

20. Identify the structures indicated on Figure *C*.

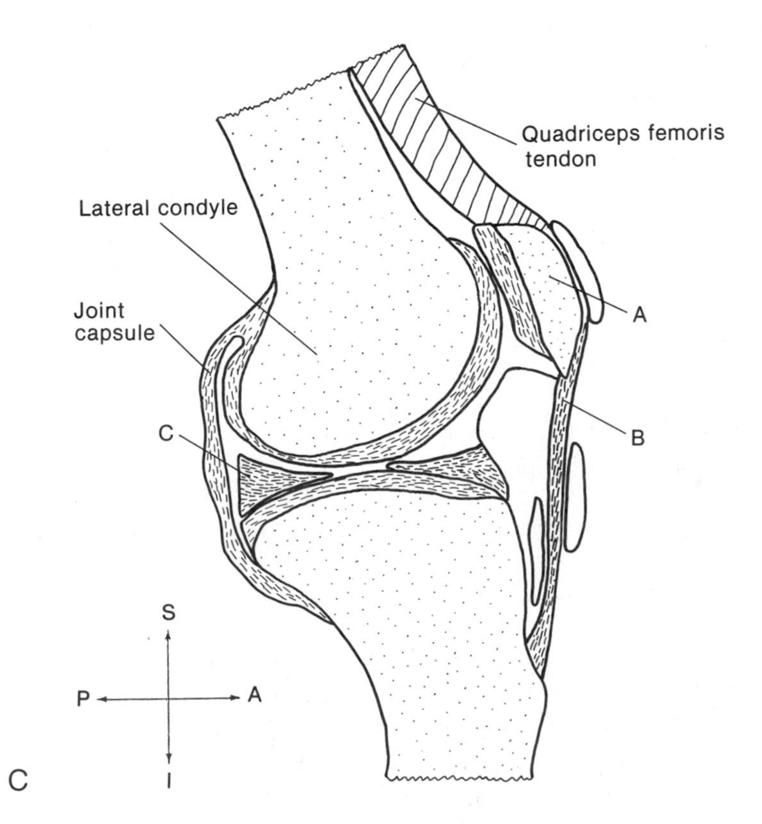

SUGGESTED READING

Anderson, J.E.: Grant's Atlas of Anatomy, 8th ed. Baltimore, Williams & Wilkins, 1983.

Berquist, T.H., Ehman, R.L., and May, G.R.: Pocket Atlas of *MRI* Body Anatomy. New York, Raven Press, 1987.

Bo, W.J., Meschan, I., and Krueger, W.A.: Basic Atlas of Cross-Sectional Anatomy. Philadelphia, W.B. Saunders Co., 1980.

Bo, W.J., Wolfman, N.T., Krueger, W.A., and Meschan, I.: Basic Atlas of Sectional Anatomy with Correlated Imaging, 2nd ed. Philadelphia, W.B. Saunders Co., 1990.

Bushong, S.C.: Magnetic Resonance Imaging: Physical and Biological Principles. St. Louis, C.V. Mosby Co., 1988.

Cahill, D.R. and Orland, M.J.: Atlas of Human Cross-Sectional Anatomy. Philadelphia, Lea & Febiger, 1984.

Carter, B., Morehead, J., Wolpert, S.M., Hammerschlag, S.B., Griffiths, H.J., and Kahn, P.C.: Cross-Sectional Anatomy. New York, Appleton-Century-Crofts, 1977.

Chiu, L.C., Lipcamon, J.D., and Yiu-Chiu, V.S.: Clinical Computed Tomography: Illustrated Procedural Guide. Rockville, Md., Aspen Systems Corporation, 1986.

Christoforidis, A.J.: Atlas of Axial, Sagittal, and Coronal Anatomy with CT & MRI. Philadelphia, W.B. Saunders Co., 1988.

El-Khoury, G.Y., Bergman, R.A., and Montgomery, W.J.: Sectional Anatomy by MRI/CT. New York, Churchill Livingstone, 1990.

Gosling, J.A., Harris, P.F., Humpherson, J.R., Whitmore, I., and Willan, P.L.T.: Atlas of Human Anatomy. New York, Gower Medical Publishing, 1985.

Hall-Craggs, E.C.B.: Anatomy as a Basis for Clinical Medicine, 2nd ed. Baltimore, Urban & Schwarzenberg, 1990.

Han, M.C. and Kim, C.W.: Sectional Human Anatomy. Seoul, South Korea, Ilchokak, 1985.

Haughton, V.M. and Daniels, D.L.: Pocket Atlas of Cranial Magnetic Resonance Imaging. New York, Raven Press, 1986.

Kieffer, S.A. and Heitzman, E.R.: An Atlas of Cross-Sectional Anatomy. Hagerstown, Md., Harper & Row, 1979.

Koritke, J.G., and Sick, H.: Atlas of Sectional Human Anatomy, 2nd ed. Baltimore, Urban & Schwarzenberg, 1988.

Ledley, R.S., Huang, H.K., and Mazziotta, J.C.: Cross-Sectional Anatomy. Baltimore, Williams & Wilkins, 1977.

McGrath, P. and Mills, P.: Atlas of Sectional Anatomy, Head, Neck and Trunk. Basel, Switzerland, Karger, 1984.

McMinn, R.M.H. and Hutchings, R.T.: Color Atlas of Human Anatomy, 2nd ed. Chicago, Year Book Medical Publishers, 1988.

Middleton, W.D. and Lawson, T.L. (eds.): Anatomy and MRI of the Joints: A Multiplaner Atlas. New York, Raven Press, 1989.

Moore, K.L.: Clinically Oriented Anatomy, 2nd ed. Baltimore, Williams & Wilkins, 1985.

Netter, F.H.: Atlas of Human Anatomy. Summit, N.J., Ciba-Geigy Corporation, 1989.

Peterson, R.R.: A Cross-Sectional Approach to Anatomy. Chicago, Year Book Medical Publishers, 1980.

Rifkin, M.D., and Waldroup, L.: Pocket Atlas of Normal Ultrasound Anatomy. New York, Raven Press, 1985.

Rohen, J.W. and Yokochi, C.: Color Atlas of Anatomy. New York, Igaku-Shoin, 1984.

Schnitzlein, H.N. and Murtagh, F.R.: Imaging Anatomy of the Head and Spine. Baltimore, Urban & Schwarzenberg, 1985.

Snell, R.S.: Clinical Anatomy for Medical Students, 2nd ed. Boston, Little, Brown and Co., 1981.

Stewart, J.V.: Clinical Anatomy & Physiology for the Frustrated and Angry Health Professional. Miami, MedMaster, Inc., 1986.

Tortora, G.J. and Anagnostakos, N.P.: Principles of Anatomy and Physiology, 6th ed. New York, Harper & Row, 1990.

Van De Graaff, K.M. and Fox, S.I.: Concepts of Human Anatomy & Physiology, 2nd ed. Dubuque, Ia., William C. Brown Publishers, 1989.

Von Hagens, G., Romrell, L.J., Ross, M.H., and Tiedemann, K.: The Visible Human Body: An Atlas of Sectional Anatomy. Philadelphia, Lea & Febiger, 1990.

Warwick, R. and Williams, P.L. (eds.): Gray's Anatomy, 36th British ed. Philadelphia, W.B. Saunders Co., 1980.

Weinstein, J.B., Lee, J.K.T., and Sage, S.S.: Pocket Atlas of Normal CT Anatomy. New York, Raven Press, 1985.

Wyman, A.C., Lawson, T.L., and Goodman, L.R.: Transverse Anatomy of the Human Thorax, Abdomen, and Pelvis. Boston, Little, Brown and Company, 1978.

GLOSSARY

Abdomen. Area between the thorax and the pelvis; the belly
Abdominal cavity. Superior portion of the abdominopelvic cavity; contains the stomach, liver, spleen, pancreas, gallbladder, and most of the small and large intestines
Abdominopelvic cavity. Inferior portion of the ventral body cavity; can be subdivided into an upper abdominal cavity and a lower pelvic cavity
Abducens nerve. Cranial nerve VI
Abduction. Movement of a body part away from the axis or midline of the body or one of its parts
Abscess. Localized accumulation of pus
Absorption. The taking up of substances by the skin or other tissues of the body
Acetabular labrum. A rim of fibrocartilage around the margin of the acetabulum
Acetabulum. The cup-shaped depression on the lateral surface of the os coxa (hipbone) in which the head of the femur fits
Achilles tendon. Calcaneal tendon
Acidity. The state of being acidic; the acid content of a fluid; opposite of alkalinity
Acromion process. Flattened portion of bone at the lateral end of the spine of the scapula; most prominent point of the shoulder
Acute. A disease state with sudden onset or of short duration; opposite of chronic
Adduction. Movement of a body part toward the axis or midline of the body or one of its parts
Adductor brevis. One of the muscles that adducts the thigh; lies deep to the adductor longus but anterior to the adductor magnus
Adductor longus. One of the muscles that adducts the thigh; the most anterior muscle of the adductor group
Adductor magnus. One of the muscles that adducts the thigh; the largest and the most posterior muscle of the adductor group
Adenoids. Pharyngeal tonsils; enlargement of the pharyngeal tonsils, as a result of chronic inflammation
Adhere. To stick, to bind, or to hold fast
Adipose. A type of connective tissue containing large quantities of fat; fatty tissue
Adjoin. To be in contact with or to lie next to
Adrenal. A gland located on top of each kidney; also called the suprarenal gland
Afferent. Carrying a nerve impulse or fluid toward an area or an organ
Aggregation. An accumulation or clump
Ala (pl. **alae**). A winglike structure
Alkalinity. A state of being alkaline or basic; the alkaline content of a fluid; opposite of acidity
Alveolus. A small, hollow area or cavity, e.g., the socket for a tooth or an air sac in the lungs
Amylase. An enzyme that splits starches into smaller molecules called disaccharides
Anastomose. Opening one into another; generally pertains to blood vessels or lymphatics
Anastomosis. The connection, or union, between tubular structures, e.g., in blood vessels or lymphatics
Anatomy. A study of the structure of the body and the relation of the parts to each other
Anesthesia. A partial or total loss of sensation
Aneurysm. A saclike bulge in a blood vessel, usually an artery, as a result of a weakening of the vessel wall
Annular. Relating to a ring-shaped or circular structure
Antagonistic. Working in opposition to each other; frequently pertains to muscle action or drugs
Anteflexion. A displacement that is characterized by a bending forward of the top part of an organ
Anterior. To the front of the body; ventral
Anterolateral. In front and to one side of the body
Anteroposterior. Directed from the front to the back
Anterosuperior. To the front and the superior (above) part of the body
Anteversion. The leaning forward (tilting) of an organ as a whole, without bending
Antrum. A nearly closed cavity or chamber
Anulus fibrosus. The fibrous outer part of the intervertebral disc
Aorta. The main trunk of the systemic arterial circulation
Aperture. A hole or opening
Apex. The pointed end of a conical structure
Aponeurosis. A white fibrous sheet composed of closely packed collagenous fibers; serves as a connection between a muscle and its attachment; a broad, flat sheet of tendon
Appendage. A structure attached to the body
Appendicular. Relating to an appendix or an appendage
Aqueduct. A canal or passageway, especially used for the conduction of fluid
Aqueduct of Sylvius. A channel through the midbrain that connects the third and fourth ventricles; also called the cerebral aqueduct
Aqueous humor. A watery, clear solution that fills the anterior cavity of the eye
Arachnoid. The thin cobweb-appearing layer of the meninges, located between the dura and pia mater
Arbor vitae cerebelli. The treelike branching arrangement of white matter in the cerebellum
Arcuate artery. The artery between the interlobar and the interlobular arteries in the kidney; forms an arch between the cortex and the medulla of the kidney
Areola. The pigmented area of skin around the nipple of the breast
Artery. A blood vessel that carries blood away from the heart
Articular. Referring to an articulation or a joint
Articulate. To join or to connect by means of a joint
Articulation. A joint; a place of contact between bones
Aryepiglottic fold. The membranous tissue in the larynx that

attaches the arytenoid cartilage to the border of the epiglottis

Arytenoid cartilages. A pair of small pyramidal cartilages located in the posterior region of the larynx

Ascites. An accumulation of serous fluid in the peritoneal cavity

Atrial. Relating to an atrium, especially in the heart

Atrioventricular. Relating to both an atrium and a ventricle of the heart

Atrium (pl. **atria**). The upper chambers of the heart, which receive blood from the veins

Auditory. Relating to the sense of hearing

Auricle. The external portion of the ear; also, the pouchlike projections from the atria of the heart

Autonomic nervous system. Division of the nervous system that transmits impulses from the brain and spinal cord to the visceral effectors, such as smooth muscles, glands, or cardiac muscle

Axial. Forming an axis; relating to the head, the neck, and the trunk of the body, as distinguished from the extremities

Axilla (pl. **axillae**). The region where the arm meets the chest; commonly called the armpit

Axillary artery. Artery that arises from the subclavian artery and passes through the axilla

Axon. The process of a neuron that carries the nerve impulse away from the cell body; the efferent process of a neuron

Azygos vein. An unpaired vein that drains the thoracic wall and empties into the superior vena cava

Baroreceptor. Sensory nerve ending that responds to changes in blood pressure; also called pressoreceptor

Bartholin's glands. A pair of glands, one on either side of the vaginal orifice, which open by a duct into the vestibule; also called the greater vestibular glands

Basal ganglia. Clusters of gray matter located within the white matter of each cerebral hemisphere; also called cerebral nuclei

Basilar artery. The artery that receives blood from the vertebral arteries and carries this blood into the circle of Willis at the base of the brain; located in the pontine cistern

Basilic vein. Superficial vein located on the medial side of the arm

Biceps brachii. A muscle of the arm, with two heads of origin, which flexes the forearm

Bicipital groove. A groove on the proximal humerus for the tendon of the long head of the biceps brachii

Biconvex. Having a protruding surface on both sides

Bicuspid valve. The left atrioventricular valve that has two cusps or flaps; also called the mitral valve

Bifid. Divided into two parts, e.g., the spinous processes of the cervical vertebrae

Bifurcate. To divide or separate into two parts

Bile. A secretion of the liver, stored in the gallbladder, that aids in the digestion of fats

Bipolar. Having two poles; relating to both ends of a cell

Brachial. Relating to the arm

Brachialis muscle. A muscle, located on the arm, that flexes the forearm

Brachiocephalic artery. The most anterior branch of the aortic arch; provides blood to the arm and the head on the right side

Brachioradialis muscle. A muscle, located on the arm, that flexes the forearm

Brachium (pl. **brachia**). The arm, especially above the elbow

Brainstem. Portion of the brain that connects the cerebrum and spinal cord; consists of the midbrain, the pons, and the medulla oblongata

Bronchial. Relating to the bronchi

Bronchopulmonary. Relating to the bronchi and the lungs

Bronchus (pl. **bronchi**). Either of the two branches of the trachea; further subdivides into secondary and tertiary branches and conveys air to and from the lungs

Buccal. Relating to the cheek or the mouth

Buccinator muscle. A muscle in the cheek

Bulbospongiosus muscle. One of the muscles of the perineum

Bulbourethral glands. Small, paired accessory glands of the male reproductive system; also called Cowper's glands

Bulbus oculi. The eyeball

Bursa (pl. **bursae**). A sac, or pouch, of synovial fluid located at friction points, especially in the region of the joints

Buttock. One of two fleshy masses on the posterior surface of the lower trunk, formed by the gluteal muscles

Calcaneofibular. Relating to the calcaneus and the fibula

Calcaneus. One of the tarsal bones, commonly called the heel bone

Calvaria. The upper part of the skull

Canaliculus (pl. **canaliculi**). A small channel or canal

Capitulum. A small rounded eminence or articular extremity of a bone; rounded surface on distal humerus that articulates with the radius; also called capitellum

Carbohydrates. A group of organic compounds composed of carbon, hydrogen, and oxygen with a 2:1 ratio of hydrogen to both carbon and oxygen; sugars, starches, cellulose

Carcinoma. A malignant tumor originating in epithelial cells

Cardiac. Relating to the heart; relating to the esophageal entrance to the stomach

Carina. Ridge formed at the bifurcation of the trachea into the two mainstem bronchi

Carotid. Relating to the major arteries of the neck

Carpal. Relating to the bones in the wrist

Cartilage. A tough, nonvascular type of connective tissue

Cartilaginous. Consisting of cartilage

Caudal. Relating to the tail

Caudate nucleus. One of several masses of gray matter located in the cerebrum

Caval. Relating to the vena cava, either superior or inferior

Cavernous sinus. A space, or cavity, on either side of the body and dorsum sellae of the sphenoid bone

Cecum. The blind pouch, or cul-de-sac, at the beginning of the large intestine, where the ileum enters

Celiac trunk. First major branch from the abdominal aorta, divides into the common hepatic, the left gastric, and the splenic arteries

Cephalic. Relating to the head; superior in position

Cerebellar. Relating to the cerebellum portion of the brain

Cerebellomedullary cistern. Subarachnoid cistern between the cerebellum and the medulla oblongata; also called the cisterna magna

Cerebellum. Portion of the brain that is below and posterior to the cerebrum and that is concerned with the coordination of movements

Cerebral aqueduct. A channel between the third and fourth ventricles in the brain; passes through the midbrain and contains cerebrospinal fluid; also called the aqueduct of Sylvius

Cerebrospinal fluid. A fluid that is produced by the choroid plexus of the ventricles and that circulates through the ventricles of the brain and in the subarachnoid space around the brain

Cerebrum. The largest part of the brain, consisting of two hemispheres

Cervical. Relating to the neck

Cervix. A neck-shaped structure; often used to denote the narrow, neck region of the uterus

Chemoreceptor. A nerve ending or sense organ that is sensitive to chemical stimuli

Chiasma. An X-shaped crossing; the optic chiasma is the X-shaped crossing of the optic nerves
Chiasmatic cistern. The subarachnoid space that contains cerebrospinal fluid in the region of the optic chiasma
Choana (pl. **choanae**). A funnel-shaped opening, especially the opening of the nasal cavity into the nasopharynx; also called the internal nares
Chordae tendineae. String or chordlike structures that extend between the papillary muscles and the flaps of the atrioventricular valves in the ventricles of the heart
Choroid. The middle or vascular layer of the eyeball
Choroid plexus. Specialized vascular structures in the ventricles of the brain that produce cerebrospinal fluid
Chronic. Pertains to a slowly progressing disease that persists over a long period of time; opposite of acute
Cilium (pl. **cilia**). A microscopic hairlike projection from the cell surface for cell locomotion or to move substances across the cell surface
Circulus arteriosus cerebri. Circle of anastomosing arteries at the base of the brain; also called the circle of Willis
Circumcision. Surgical removal of the prepuce or the foreskin from the glans penis
Circumduction. Circular movement of a part in which the distal end of the bone moves in a circle, whereas the proximal end remains relatively stable
Circumflex. Relating to arched structures
Circumflex artery. A branch of the left coronary artery that is located in the left atrioventricular sulcus
Cistern. An enclosed space or reservoir for body fluids, especially areas of the subarachnoid space that act as reservoirs for cerebrospinal fluid
Cisterna ambiens. The subarachnoid space at the posterior end of the corpus callosum; also called the superior cistern or the cistern of the great cerebral vein
Cisterna chyli. The dilatation at the beginning of the thoracic duct in the abdomen
Cisterna magna. Subarachnoid cistern located between the cerebellum and the medulla oblongata; also called the cerebellomedullary cistern
Claustrum. A thin layer of gray matter on the lateral margin of the external capsule of the brain
Clavicle. Collarbone; a bone that extends from the sternum to the acromion process of the scapula and that forms the anterior portion of the pectoral girdle
Cleft palate. Condition in which there is a fissure in the palate, or roof of the mouth, because the palatine processes of the two maxillae did not unite before birth
Clitoris. A small, erectile organ at the anterior end of the vulva in the female; homologous to the penis in the male
Coccygeus muscle. The smaller of the two muscles in the anal region of the pelvic floor
Coccyx. Three or four fused bones at the distal end of the vertebral column, commonly called the tailbone
Cochlea. The spiral cavity in the inner ear, which contains the essential organ of hearing
Coitus. Sexual intercourse or copulation
Colon. Portion of the large intestine that extends between the cecum and the rectum; divided into ascending, transverse, descending, and sigmoid regions
Collateral circulation. A secondary or alternate path of blood flow through anastomosing vessels
Colliculus (pl. **colliculi**). A small round elevation, as in the midbrain
Concave. Characterized by having a hollow or depressed shape
Concha (pl. **conchae**). A scroll or shell-shaped bone found in the nasal cavity; also called the turbinate
Condyle. A rounded prominence at the end of a bone for articulation with another bone
Confluence of sinuses. The junction of the venous sinuses in the dura mater, in the region of the internal occipital protuberance
Congenital. Present at birth
Conjunctiva. The delicate mucous membrane that lines the eyelids and that covers the exposed surface of the outer layer of the eyeball
Connective tissue. The most abundant tissue of the four basic tissue types; serves to bind and to support
Constrict. To make narrow or to reduce in size
Contraction. The shortening of a muscle or an increase in tension in a muscle
Conus medullaris. The tapered distal end of the spinal cord
Convex. Having a rounded or bulging surface
Convoluted. Twisted, coiled, or rolled
Cooper's ligament. Suspensory ligament of the breast
Copulation. Sexual intercourse, coitus
Coracobrachialis. A muscle that flexes and adducts the arm
Coracoid process. A thick, curved process at the superior border of the scapula
Cornea. The transparent anterior portion of the outer layer of the eyeball
Corniculate cartilage. A pair of small cartilages in the larynx
Coronal plane. Plane parallel to the long axis of the body that divides the body into anterior and posterior portions; also called the frontal plane
Coronary arteries (left and right). First branches from the ascending aorta; located at the level of the aortic semilunar valve; supply blood to the heart wall
Coronoid process. Certain processes of bones such as the coronoid process of the mandible
Corpora quadrigemina. Four bodies that form the dorsal part of the midbrain
Corpus (pl. **corpora**). Body, or main portion, of a structure
Corpus callosum. A large band of white fibers that connects the two cerebral hemispheres
Corpus spongiosum. Ventral column of erectile tissue that surrounds the urethra in the penis
Corpus cavernosum. Either of the two dorsal columns of erectile tissue in the penis
Cortex. The outer layer of an organ; the outer layer of gray matter of the cerebrum
Costal. Pertaining to the ribs
Cowper's gland. An accessory gland of the male reproductive system; also called bulbourethral gland
Coxa (pl. **coxae**). The hipbone, or the os coxa
Cranial cavity. One of the divisions of the dorsal body cavity; contains the brain
Cranium. The bones of the skull that enclose the brain
Cremaster muscle. An extension of the internal oblique muscle that is found in the spermatic cord and that contracts to elevate the testes
Crest. A bony ridge
Cribriform plate. Flat region on either side of the crista galli of the ethmoid bone, perforated by many small holes called olfactory foramina
Cricoid cartilage. Most inferior cartilage of the larynx
Cruciate. Overlapping, or crossing, such as the cruciate ligaments of the knee
Crus (pl. **crura**). Derived from Latin for leg; slender, tapered portion of a muscle or organ, such as the crus of the diaphragm or the crura of the penis
Cubital. Relating to the forearm; cubitus refers to the elbow joint and the cubital fossa is the depression on the anterior surface of the elbow joint
Cul-de-sac. A blind-ended pouch or sac
Cuneiform. Wedge-shaped; three of the tarsal bones

Cusp. A triangular piece of an atrioventricular valve in the heart; an elevation or projection on a tooth
Cutaneous. Pertaining to the skin
Cystic duct. Duct from the gallbladder

Dartos. Contractile tissue in the subcutaneous tissue of the scrotum
Decussation. Crossing over to the other side, especially in nerve tracts; if bilateral, then forms the shape of the letter X
Degeneration. Deterioration or breakdown of tissues
Delineate. To mark boundaries, to delimit
Deltoid. Superficial muscle over the shoulder; abducts the arm
Demarcation. The marking of boundaries; delimitation
Detrusor muscle. Smooth muscle in the wall of the urinary bladder
Diaphragma sella. Extension of dura mater over the sella turcica of the sphenoid bone
Diaphragmatic. Relating to the diaphragm
Diencephalon. Portion of the brain that is surrounded by the cerebral hemispheres and that encloses the third ventricle; principal components are the thalamus and hypothalamus
Digastric. A muscle in the floor of the oral cavity
Dilatation. An enlarged or swollen area, usually used in reference to a tubular structure, opening, or cavity
Dilate. To expand, to enlarge, or to swell
Dislocation. Displacement of a part from normal position, usually used in reference to bones in joints
Dissect. To cut apart or to separate tissues or organs in the study of anatomy
Distal. Directional term for something located away from a point of reference, such as the center, the midline, the point of attachment, or the point of origin; opposite of proximal
Distensible. Capable of being stretched
Diverge. To spread apart from a common point
Diverticulum (pl. **diverticula**). A pouch, or sac, protruding from the wall of a tubular organ
Dopamine. A compound produced in the caudate nucleus and putamen of the brain, diminished in some disease states such as parkinsonism
Dorsal. Directional term relating to the back or posterior surface of a structure
Dorsum. The back or posterior surface of a structure
Duct. A tube or channel, usually for carrying secretions
Duct of Wirsung. Pancreatic duct
Ductus arteriosus. A channel or vessel between the pulmonary artery and the aorta in the fetus that allows blood to bypass the lungs in fetal circulation
Ductus deferens. A tube or channel that conveys sperm from the epididymis to the ejaculatory duct in the male
Duodenum. The first part of the small intestine
Dura mater. The outer, tough covering around the brain and spinal cord; the outer layer of the meninges

Echogenic. Tissues that produce echos
Ectopic. Located outside the normal place
Edema. Swelling caused by an abnormal accumulation of interstitial fluid
Efferent. Carrying a nerve impulse or fluid away from an area or an organ
Ejaculatory duct. The tube that transports sperm from the vas deferens to the prostatic urethra
Embryo. The developing human, from the time of conception to the end of the eighth week of development
Eminence. An elevated area or prominence, especially used in reference to regions on bones
Encircle. To surround
Endocardium. The innermost layer of the heart, composed of endothelium
Endocrine. A gland that secretes its hormones into the bloodstream; a ductless gland
Endometrium. The mucous membrane that makes up the inner layer of the uterine wall
Endosteum. Membranous lining of bone cavities
Endothelium. A thin layer of epithelial cells that lines the heart, blood vessels, and lymphatics
Enunciation. The formation of words
Enzyme. A protein, secreted by the body, that acts as a catalyst to speed up chemical reactions in the body
Epicardium. The visceral or innermost layer of the serous pericardium, which is in contact with the heart and forms the outermost layer of the heart wall
Epicondyle. Projection above or upon a condyle
Epicranium. The scalp; structures that cover the skull, including the muscle, the aponeurosis, and the skin
Epididymis. A comma-shaped structure on the posterior surface of the testis; site of sperm maturation
Epidural. Located above or upon the dura mater
Epigastric region. The upper, middle abdominal region, directly superior to the umbilical region
Epiglottic. Relating to the epiglottis
Epiglottis. Leaf-shaped piece of cartilage that covers the trachea during swallowing to prevent food from blocking the airway
Epiploic. Relating to the omentum
Epiploic appendages. Bodies of fat along the taeniae coli of the large intestine
Epiploic foramen. The opening along the right margin of the lesser omentum that allows access to the omental bursa; also called the foramen of Winslow
Epithalamus. A small area of the diencephalon above the thalamus
Epithelium. The nonvascular layer of cells that lines body cavities and covers the exterior surface of the body; one of the four main types of tissue in the body
Equilibrium. A state of balance
Erectile tissue. Tissue that becomes rigid when filled with blood
Erector. A muscle that can raise or cause a structure to become erect
Esophagus. Muscular passageway that takes food from the pharynx to the stomach
Estrogens. Term for female sex hormones that maintain secondary sex characteristics
Ethmoid. A bone resembling a sieve because of many foramina, located behind the nose
Eustachian tube. Tube that connects the middle ear to the nasopharynx; auditory tube
Eversion. The movement of turning a joint outward, e.g., eversion of the foot
Evert. To turn outward
Excretory. Relating to excretion
Exocrine. Glands that secrete their products to a surface via ducts
Expel. To force outward
Extension. The act of increasing the angle at a joint; opposite of flexion
Extensor. A muscle that increases the angle between bones at a joint
Extensor carpi ulnaris. Muscle of the forearm that extends the hand
Extensor carpi radialis. Muscle of the forearm that extends the hand
Extracapsular. Located outside of a capsule
Extradural. Located outside of the dura mater
Extrapelvic. Located outside of the pelvic cavity
Extremity. A limb; an arm or a leg

Extrinsic. Located on the outside of the structure that is being acted upon

Facet. A very smooth bone surface for articulation with another structure
Facial. Relating to the face; cranial nerve VII
Falciform ligament. Fold of peritoneum extending from the diaphragm and the anterior abdominal wall to the surface of the liver, between the two major lobes
Fallopian tube. Duct that carries the ova from the ovary to the uterus; also called the uterine tube or the oviduct
Falx cerebelli. The triangular extension of the cranial dura mater that is located between the two lobes of the cerebellum
Falx cerebri. The fold of cranial dura mater that extends into the longitudinal fissure between the two cerebral hemispheres
Fascia. Loose connective tissue, located under the skin, or a fibrous membrane that covers and separates muscles
Fauces. The opening from the oral cavity into the oropharynx
Femoral. Relating to the femur or the thigh
Femur. Large bone of the thigh
Fertility. The ability to start or to support conception
Fertilization. The union of a spermatozoan with an ovum
Fetal. Relating to the fetus
Fetus. The developing offspring in the uterus from the beginning of the third month of development until birth
Fibrocartilage. Cartilage that contains a large number of collagenous fibers
Fibromuscular. Denotes a tissue that is both muscular and fibrous
Fibroserous. Denotes a tissue that is both serous and fibrous
Fibrous. Composed of fibers of connective tissue
Fibula. The smaller and more lateral bone of the lower leg
Filtration. Movement of a liquid through a filter as a result of a pressure difference
Fimbriae. Fingerlike projections that surround the openings of the uterine tubes
Fissure. A slit, cleft, or groove
Flaccid. Muscles without tone, flabby
Flexion. The act of decreasing the angle at a joint; opposite of extension
Flexor. A muscle that decreases the angle between bones at a joint
Flexor carpi ulnaris. A muscle of the forearm that flexes the hand
Flexor carpi radialis. A muscle of the forearm that flexes the hand
Flexure. A bend or turn
Foliated. Resembling a leaf
Follicle. A mass of cells containing a cavity; a small depression in the skin from which the hair emerges
Foramen (pl. **foramina**). A hole or opening
Foramen of Magendie. The medial opening in the roof of the fourth ventricle through which cerebrospinal fluid enters the subarachnoid space; also called median aperture
Foramen of Monro. Opening from the lateral ventricle into the third ventricle; also called the interventricular foramen
Foramen magnum. Large hole located in the occipital bone through which the spinal cord passes
Foramina of Luschka. Two lateral openings in the roof of the fourth ventricle through which cerebrospinal fluid enters the subarachnoid space; also called lateral apertures
Fornix (pl. **fornices**). An arched structure, or the space caused by such a structure, e.g., the fornix of the vagina
Fossa (pl. **fossae**). A pit or shallow depression
Fovea. A small depression
Frenulum. A small fold of mucous membrane that connects two parts and limits movement, e.g., the lingual frenulum on the underside of the tongue
Frontal. Relating to the forehead
Frontal plane. Plane parallel to the long axis of the body that divides the body into anterior and posterior portions; also called the coronal plane
Frontalis muscle. Muscle that wrinkles the forehead
Fructose. A simple sugar or monosaccharide
Fundus. The part of a hollow organ that is the farthest from, above, or opposite of its opening

Galea aponeurotica. The broad, flat tendon that connects the frontalis muscle with the occipitalis muscle; a part of the scalp; also called the epicranial aponeurosis
Gallbladder. An oblong sac that is located on the underside of the liver and stores bile
Ganglion (pl. **ganglia**). A group of nerve cell bodies located outside the brain and spinal cord
Gaster. Refers to the belly
Gastric. Relating to the stomach
Gastrocnemius. Muscle of the calf that plantar flexes the foot
Gastrocolic. Relating to the stomach and colon or to the large intestine
Gastroduodenal. Relating to the stomach and the duodenum
Gastroepiploic. Relating to the stomach and the omentum
Gastroesophageal. Relating to the stomach and the esophagus
Gastrohepatic. Relating to the stomach and the liver
Gastrointestinal. Relating to the stomach and the intestines (gastroenteric)
Gastrosplenic. Relating to the stomach and the spleen
Gemellus (pl. **gemelli**). Means twins; two muscles deep to the gluteal muscles that laterally rotate and abduct the thigh
Geniohyoid. One of the muscles that acts on the hyoid bone
Genital. Relating to reproduction
Genitalia. Reproductive organs
Genitourinary. Relating to the genital system and the urinary system; also called urogenital
Genu. Any structure that resembles a flexed knee, e.g., genu of the corpus callosum
Gingiva. The gum; mucous membrane that surrounds the alveolar process and the neck of the tooth
Glans penis. The caplike extension of the corpus spongiosum at the distal tip of the penis
Glenohumeral. Relating to the glenoid fossa of the scapula and to the humerus
Glenoid. Relating to an articular depression or socket of a joint, e.g., glenoid fossa of the scapula
Globus pallidus. The medial region of gray matter in the lentiform nucleus
Glossopharyngeal. Relating to the tongue and the pharynx
Glottis. Vocal apparatus in the larynx, consisting of the vocal folds and the rima glottidis between them
Gluteal. Relating to the buttocks
Gluteus maximus. The large, fleshy muscle that forms the prominent portion of the buttocks
Gluteus medius. The broad, thick muscle situated on the outer surface of the pelvis
Gluteus minimus. The smallest and deepest muscle of the three gluteal muscles
Gonad. Primary reproductive organ, the ovary in the female and the testes in the male
Gonadal. Refers to the gonad
Graafian follicle. A fluid-filled follicle that contains an immature ovum and its surrounding, estrogen-secreting cells; also called a vesicular ovarian follicle
Gracilis. Straplike muscle that is located on the medial surface of the thigh and that functions in adduction of the thigh
Gyrus (pl. **gyri**). Rounded elevations on the surface of the brain

Hallucis muscles. Muscles that work on the big toe, e.g., the adductor hallucis

Hallux. The big toe
Hamate. One of the bones of the wrist (carpals)
Hemidiaphragm. Half or one dome of the diaphragm
Hemorrhage. Profuse bleeding
Hepatic. Relating to the liver
Hepatoduodenal. Relating to the liver and the duodenum
Hepatogastric. Relating to the liver and the stomach
Hepatorenal. Relating to the liver and the kidney
Hernia. A protrusion of an organ through an abnormal opening or weakening of the wall that usually contains it
Hiatus. An opening or an aperture
Hilum. The region where vessels and nerves enter and leave an organ
Homologous. Organs that are alike in structure or origin
Hormones. Substances secreted by endocrine glands that are released into the blood and cause a response in the target organ
Humeral. Relating to the humerus
Humeroradial. Relating to the humerus and the radius
Humeroscapular. Relating to the humerus and the scapula
Humeroulnar. Relating to the humerus and the ulna
Humerus. Bone of the upper arm
Hyaline cartilage. Gelatinous material with a glassy appearance found in many extracellular areas
Hyoid. A U-shaped bone found in the neck, which serves as an anchor for the tongue
Hypertrophy. An increase in bulk or in size of tissue without cell division
Hypochondriac region. Abdominal region located superiorly and laterally, near the ribs, on either side of the epigastric region
Hypogastric region. The lower, middle abdominal region located directly inferior to the umbilical region
Hypoglossal. Located beneath the tongue
Hypopharynx. Below the pharynx
Hypophysis. Endocrine gland located under the brain; also called the pituitary gland
Hypothalamus. A portion of the diencephalon located inferior to the thalamus; also forms the floor of the third ventricle

Ileocecal. Relating to the ileum and the cecum
Ileum. The third part of the small intestine, which is connected to the colon
Iliac. Relating to the ilium of the os coxa
Iliacus. Muscle that covers the ilium of the os coxa
Iliofemoral. Relating to the ilium and the femur
Iliopsoas. Muscle that results from the joining of the iliacus and the psoas muscles, located in the upper thigh
Iliotibial. Relating to the ilium and the tibia
Ilium. The superior, broad portion of the os coxa
Implantation. The attaching of the fertilized ovum into the uterine wall
Incisura angularis. A notch, or indentation, between the body and the pylorus of the stomach
Incontinence. The inability to control the passage of urine or feces as a result of loss of sphincter control
Incus. One of three ossicles of the ear, small anvil-shaped bone of the middle ear, located between the malleus and the stapes
Inferiorly. Located below or inferior to another structure
Infraglenoid. Located below the glenoid fossa of the scapula bone
Infraglottic. Located below the glottis
Infrahyoid. Located below the hyoid bone
Infraorbital. Located below the orbit of the eye
Infrapubic. Located below the pubis
Infraspinatus muscle. Muscle located below the spine of the scapula
Infraspinous. Located below the spinous process, e.g., the infraspinous fossa located under the spine of the scapula
Infrasternal. Located below the sternum
Infundibulum. Funnel-shaped structure, e.g., the infundibulum of the pituitary gland or the infundibulum of the uterine tube
Inguinal. Refers to the groin
Innervation. Nerve or nerves that supply an area
Innominate artery. The first and most anterior arterial branch off the aorta; also called the brachiocephalic artery
Inspiration. The act of breathing in, inhalation
Insula. The hidden lobe of the brain, deep to the temporal lobe; also called the island of Reil
Interatrial. Located between the two atria
Interclavicular. Located between the two clavicles
Intercondylar. Located between two condyles
Intercostal. Located between adjacent ribs
Intercristal. Located between two crests
Interiliac. Located between the two ilia
Intermuscular. Located between muscles
Interosseous. Connecting or lying between bones
Interpeduncular. Located between the two cerebral peduncles
Interposed. Located between two structures
Intertubercular. Located between two tubercles
Interventricular. Located between the two ventricles of the heart
Interventricular foramen. Opening located between a lateral ventricle and the third ventricle; also called the foramen of Monro
Intervertebral. Located between two vertebrae
Intestine. The portion of the gastrointestinal tract that extends from the stomach to the anus
Intracapsular. Located within a capsule
Intracranial. Located within the skull
Intramuscular. Located within a muscle
Intrathoracic. Located within the thorax
Inversion. Turning inside out or reversing the normal relationship between organs; turning the sole of the foot medially; the opposite of eversion
Iris. The colored portion of the eye, a circle of smooth muscles that controls the size of the pupil
Ischiocavernosus. Muscles of the perineum
Ischiofemoral. Relating to the ischium and the femur
Ischiopubic. Relating to the ischium and the pubis
Ischium. The lowest and posterior portion of the ox coxa
Island of Reil. The hidden lobe of the brain, deep to the temporal lobe; also called the insula
Isthmus. A narrow band of tissue that connects two larger parts or a narrow passageway between two larger cavities

Jejunum. The middle portion of the small intestine, located between the duodenum and the ileum
Jugular. Relating to the neck; structures located in the neck

Labia majora. Two large folds of fat-filled tissue that are lateral to the labia minora
Labia minora. The two narrow folds of tissue that enclose or delineate the vestibule in the female
Labrum. A lip or an edge
Laceration. A jagged tear of tissues
Lacerum. Foramen resembling a jagged tear
Lacrimal. Relating to tears
Lactiferous. Secreting or carrying milk
Lamina (pl. **laminae**). A thin layer or flat plate
Laryngeal. Relating to the larynx
Laryngopharynx. The lowest part of the pharynx, located posterior to the larynx, that leads into the esophagus
Larynx. Voicebox, organ of voice production that contains the vocal folds

Lateral. Located on the side away from the middle
Latissimus dorsi. Broad superficial muscle of the back
Lenticular nucleus, lentiform nucleus. A mass of gray matter, lateral to the internal capsule, that is a part of the basal ganglia
Leptomeninges. Collective term relating to the arachnoid and pia mater
Levator ani muscle. Muscle of the pelvic floor
Levator costarum muscle. Muscle that raises the ribs
Lienorenal. Relating to the spleen and the kidney; also called splenorenal
Ligament. A band of fibrous tissue that connects bones together
Ligamentum arteriosum. Band of tissue that represents the remnant of the ductus arteriosus from the fetal circulation
Linea alba. White line, a narrow band of the anterior aponeurosis, between the two rectus abdominis muscles, in the middle of the abdomen, from the xiphoid process to the pubic symphysis
Linea aspera. A long ridge on the posterior surface of the femur
Lingual. Referring to the tongue
Lobule. A small lobe
Longitudinal fissure. Fissure of the brain that separates the cerebrum into two cerebral hemispheres
Lordosis. Abnormally exaggerated lumbar curve, swayback
Lumbar region. Abdominal region on either side of the umbilical region; related to the loins
Lumbosacral. Relating to the lumbar region of the spine and the sacrum
Lumen. Open space in the interior of a tubular structure
Lymph. Clear or yellowish fluid that is derived from the interstitial fluid and that is contained in the lymph vessels
Lymphatic. Relating to lymph, lymph nodes, or lymph vessels
Lymphoid. Referring to or resembling lymph or lymphatic tissue

Malignant. Refers to a disease that resists treatment and tends to be fatal; frequently pertains to tumors that have the properties of uncontrolled growth and dissemination
Malleolus. Projections on either side of the ankle; one is located on the tibia, the other on the fibula
Malleus. One of the three ear ossicles in the middle ear; the ossicle located next to the tympanic membrane and shaped like a hammer or mallet
Mammary. Referring to the breasts
Mammillary. Structures that are breast-shaped and are located at the base of the hypothalamus
Mandible. Bone of the lower jaw
Manubrium. Means handle; refers to the superior part of the sternum
Masseter. Muscle that acts to close the jaw, a muscle of mastication
Mastication. The process of chewing
Mastoid. The projection on the temporal bone located posterior to the ear
Maxilla (pl. **maxillae**). Bones that form the upper jaw
Maxillary. Relating to the upper jaw
Meatus. A channel or opening
Medial. Toward the middle or median plane of a body or organ
Median. Centrally located
Mediastinal. Relating to the mediastinum
Mediastinum (pl. **mediastina**). The central region of the thoracic cavity
Medulla. The inner part of an organ
Medulla oblongata. Inferiormost part of the brainstem; extends from the pons to the spinal cord and is continuous with the spinal cord at the foramen magnum
Melanin. Dark pigment found in the skin, the hair, and the iris of the eye
Meninges. Membranes that cover the brain and the spinal cord
Meningitis. An inflammation of the meninges
Meniscus (pl. **menisci**). A crescent-shaped structure; frequently refers to the pad of fibrocartilage found in certain joints, e.g., the knee
Menopause. The normal termination of the menstrual cycle
Mesenteric. Relating to the mesentery
Mesentery. Double layer of peritoneum that is attached to the intestines and to the body wall and that contains blood vessels and nerves
Mesocolon. The fold of peritoneum that attaches the colon to the posterior wall of the abdomen
Mesovarium. The fold of peritoneum that attaches the ovary to the posterior layer of the broad ligament
Metacarpals. Bones located in the hand, between the carpals and the phalanges
Metastasis. The transfer of disease from its starting point to a distant point
Metatarsals. Bones of the foot, between the tarsals and phalanges
Midbrain. The part of the brain that is located between the pons and the diencephalon; also called the mesencephalon
Midsagittal. The plane that divides the body into equal right and left sides
Mitral valve. Valve located between the left atrium and the left ventricle; also called the bicuspid valve
Mnemonic. Relating to or assisting the memory
Molar. A posterior tooth that grinds food
Mons pubis. The prominence caused by a pad of fat located over the symphysis pubis in the female
Morphology. The study of the form or the structure of living organisms
Motility. The ability to move
Mucoid. Resembles mucus
Mucosa. The mucous membrane that lines a cavity opening to the exterior; consists of epithelium and lamina propria
Mucous. Relating to, consisting of, or producing mucus
Mucus. Viscous secretions produced by specialized membranes (mucous membranes)
Multiaxial. Having many axes
Musculature. The system of muscles in the body or in a body part
Myelin. The fatty substance that surrounds and insulates the axon of some nerve cells
Mylohyoid. One of the muscles associated with the hyoid bone, located in the neck
Myocardium. The muscular layer of the heart
Myometrium. The muscular layer of the uterus

Naris (pl. **nares**). One of the external openings of the nose; also called nostril
Nasal. Relating to the nose
Nasolacrimal. Relating to the nose and the lacrimal bones
Nasopharynx. The superior part of the pharynx, located above the level of the soft palate, posterior to the nasal cavity
Navel. Umbilicus; depressed area where the umbilical cord was attached
Navicular. Anterior bone of the ankle or tarsus
Neural. Relating to the nervous system
Neurons. Nerve cells; the basic functional unit of the nervous system
Neurovascular. Relating to the nervous and the vascular systems
Neutralize. To render ineffective or to make neutral
Nodules. Small nodes
Nuchal. Relating to the back or the nape of the neck
Nucleus pulposus. Inner, soft core of an intervertebral disc

Oblique. A slanting or sloping direction; deviating from the perpendicular or horizontal
Obturator externus. One of the muscles that closes the obturator foramen; one of the lateral rotators of the thigh
Obturator internus. One of the muscles that closes the obturator foramen and is part of the lateral pelvic wall; one of the lateral rotators of the thigh
Occipital. Relating to the back of the head
Ocular. Relating to the eye
Oculomotor. Cranial nerve III, innervates muscles that affect eye movement
Odontoid. Shaped like a tooth; process on the second cervical vertebra, also called the dens
Olecranon. Curved process on the ulna that forms the point of the elbow
Olfactory. Relating to the sense of smell; cranial nerve I
Omentum (pl. **omenta**). A fold of peritoneum in the abdominal cavity, usually associated with the stomach
Omohyoid. One of the infrahyoid muscles
Oocyte. An immature cell in the ovary that, after undergoing meiosis, produces an ovum
Ophthalmic. Pertaining to the eye
Optic. Cranial nerve II; refers to the eye, vision, or properties of light
Optimum. Most suitable or favorable conditions
Orbicularis. A circular muscle
Orbicularis oculi. Circular muscle around the eye
Orbicularis oris. A circular muscle around the mouth
Orbit. Cavity in the skull that contains the eyeball; called the eye socket
Orbital. Relating to the orbit
Organ of Corti. Sensory receptors for hearing, located in the inner ear; also called the spiral organ of Corti
Orifice. An opening or aperture
Oropharynx. Middle portion of the pharynx, directly posterior to the oral cavity; extends from the soft palate to the hyoid bone
Os. Mouth or opening, such as the os of the cervix; may also refer to bone, such as os coxa
Osseous. Consisting of bone
Ossicle. A small bone; auditory ossicles are three small bones (malleus, incus, stapes) in the middle ear (see individual definitions for each ossicle)
Ossification. Formation of bone
Ovarian. Relating to the ovary
Ovary. Female gonad that produces ova, estrogens, and progesterone
Oviduct. Slender tube that extends from the uterus to the region of the ovary; also called the uterine tube or the fallopian tube
Ovulation. Rupture of a mature follicle in the ovary with the release of a secondary oocyte into the pelvic cavity
Ovum (pl. **ova**). Female gamete or germ cell; egg cell
Oxytocin. Hormone produced in the hypothalamus and stored in the posterior pituitary, stimulates smooth muscle contractions in the pregnant uterus and ejection of milk from the breasts

Pacemaker. A structure that establishes a basic rhythmic pattern; sinoatrial node in the heart
Palate. Roof of the mouth, separates the oral and the nasal cavities
Palatine. Relating to the roof of the mouth
Palmar. Relating to the palm of the hand
Palpate. To examine by touch or by feel
Palpebra (pl. **palpebrae**). Eyelid
Pampiniform plexus. In the male, a plexus of veins from the testicle and epididymis that is included in the spermatic cord; in the female, a plexus of ovarian veins in the broad ligament
Pancreatic. Relating to the pancreas
Papillary muscles. Conical projections of myocardium on the inner surface of the ventricles
Papilledema. Swelling of the optic nerve as a result of increased intracranial pressure
Paranasal. Located adjacent to or near the nose
Parasagittal plane. A vertical plane that divides the body into unequal right and left portions; does not pass through the midline
Parasympathetic. Refers to the part of the autonomic nervous system that is concerned with conserving and restoring energy
Parathyroid gland. One of four small glands embedded on the posterior surface of the thyroid gland
Paraurethral glands. A pair of small, mucus-secreting glands associated with the orifice of the urethra in the female; also called Skene's glands
Parenchyma. The characteristic or functional tissue of an organ or gland
Parietal. Pertaining to the outer wall of a body cavity
Parotid. Largest of the salivary glands
Parturition. Act of giving birth; childbirth; delivery
Patella. Flattened bone in front of the knee; also called the kneecap
Pathogen. Microorganism capable of causing disease
Pectinate. Comb-shaped; region of the atria of the heart with comb-shaped muscular ridges
Pectineus muscle. One of the muscles that adducts and flexes the thigh
Pectoral. Relating to the chest or thorax
Pedicle. A stem or stalk; a short process that connects the body with the lamina of a vertebra
Peduncle. A stalk or ropelike mass of nerve fibers that connects one part of the brain to another
Pelvic. Relating to the pelvis
Pelvis. A basinlike skeletal structure that acts as an attachment for the lower extremity and contains viscera; a funnel-shaped region, such as the pelvis of the kidney
Pendulous. Drooping or sagging
Penile. Relating to the penis
Penis. Male copulatory organ and organ of urinary excretion
Pericardium. Membranous sac that encloses the heart
Perineal. Relating to the perineum
Perineum. Region bounded by the pubis, coccyx, and thighs; clinical perineum is the region between the anus and the external genitalia
Periosteum. Connective tissue membrane that covers bone
Periphery. Area away from the center; outer surface of the body
Peritoneal. Relating to the peritoneum
Peritoneum. Serous membrane of the abdominal cavity
Peroneal. Relating to the fibula or lateral portion of the leg
Peroneus muscles. A group of muscles that originate on the fibula and function to evert the foot
Petrosal. Relating to the petrous portion of the temporal bone
Peyer's patches. Aggregated lymphatic follicles in the wall of the small intestine, especially in the ileum
Phalanx (pl. **phalanges**). A bone of the finger or toe
Pharyngeal. Relating to the pharynx
Pharynx. A musculomembranous cavity or tube posterior to the nasal, oral, and laryngeal cavities
Phonation. Production of sound
Phrenic. Relating to the diaphragm
Pia mater. A delicate membrane, the innermost layer of meninges around the brain
Piriform. Pear-shaped

Piriformis muscle. One of the muscles that laterally rotates the thigh, located in the greater sciatic notch
Plantar. Relating to the sole of the foot
Plantar flexion. Flexion at the ankle that bends the foot downward toward the sole of the foot
Plantaris muscle. One of the leg muscles that plantar flexes the foot
Platysma. A thin muscle in the superficial fascia of the anterior neck region that contracts to depress the lower jaw and to wrinkle the skin of the neck
Pleura. Serous membrane that covers the lungs and lines the walls of the chest cavity
Pleural. Relating to the pleura
Plexus. A network of blood vessels, lymphatic vessels, or nerves
Pons. Central part of the brainstem, located between the cerebral peduncles and the medulla oblongata
Pontine. Relating to the pons
Popliteal. Relating to the back of the knee
Popliteus muscle. A small muscle on the medial side of the back of the knee that flexes and medially rotates the leg
Porta. Entrance or door; the region where blood vessels, nerves, lymphatic vessels, and ducts enter and leave an organ
Porta hepatis. Fissure on the visceral surface of the liver through which the hepatic portal vein, the hepatic artery, and the hepatic ducts pass
Posterior. Nearer to or toward the back of the body; dorsal
Prepuce. Loose fold of skin that covers the glans penis; also called the foreskin
Progesterone. A hormone that is produced by the corpus luteum in the ovary; it functions to prepare the lining of the uterus for implantation of a fertilized ovum
Prolactin. A hormone that is produced in the anterior pituitary gland and that stimulates the production of milk
Prolapse. To fall or to slip down; the inferior displacement of an organ or part of an organ from its normal position
Promontory. A projection or elevation
Pronate. To place in a face-down position; to turn the hand so that the palm is turned downward or backward
Proprioception. The sense of body position and movement as a result of information received from sense receptors in the muscles and tendons
Prostate. An accessory gland in the male reproductive system, located inferior to the urinary bladder
Proximal. Located nearest the center, midline, point of attachment, or point of origin; opposite of distal
Psoas muscle. Large muscle mass, located on either side of the lumbar vertebrae, that flexes and medially rotates the thigh
Pterygoid. Wing-shaped; refers to the pterygoid processes on the sphenoid bone and the pterygoid muscles
Ptosis. Drooping or sagging
Pubic. Relating to the pubis
Pubis. The pubic bone or the region over the pubic bone
Pudendal. Relating to the genital area
Pudendum. External genital organs, especially in the female; also called the vulva
Pulmonary. Relating to the lungs
Purkinje fibers. Conduction myofibers of the heart; specialized cells that are part of the conduction system of the heart
Pus. The liquid product of inflammation that contains cellular debris, dead leukocytes, and dead bacteria
Putamen. The lateral portion of the lentiform nucleus, a region of gray matter in the cerebrum
Pyloric. Relating to the pylorus
Pylorus. The distal portion of the stomach; opening between the stomach and the duodenum
Pyramidal. Relating to or having the shape of a pyramid

Quadrant. One of four sections; used to designate regions of the abdomen
Quadrate. Having four sides, such as the quadrate lobe of the liver
Quadratus lumborum. One of the muscles of the posterior abdominal wall, lateral to the psoas muscles
Quadriceps femoris. Muscle mass of the anterior thigh; consists of four muscles

Radial. Relating to the radius, a bone in the forearm; diverging in various directions from a central point
Radiograph. The processed photographic film used in radiography
Radius. The lateral bone in the forearm
Ramus (pl. **rami**). Branch; a branch of an artery, vein, or nerve
Raphe. A ridge or line that marks the union of two similar structures
Receptor. A sensory end-organ; it receives a stimulus and converts it into a nerve impulse
Rectouterine pouch. Peritoneal space between the rectum and uterus; also called the cul-de-sac or pouch of Douglas
Rectovesical pouch. Peritoneal space between the rectum and the urinary bladder
Rectum. Terminal portion of the intestinal tract; section of intestine that extends from the sigmoid colon to the anus
Rectus. Straight; used to describe some muscles that run a straight course, such as the rectus abdominis muscle and the rectus muscles of the eye
Reflux. Backward flow
Renal. Pertaining to the kidney
Respiration. The physical and chemical processes by which an individual acquires oxygen and releases carbon dioxide
Respiratory. Pertaining to respiration
Retina. The innermost layer of the eyeball; part of the eye that contains the visual receptors
Retinaculum. A bandlike ligament found in the wrist and in the ankle
Retinal. Relating to the retina
Retromammary. Located behind or deep to the mammary gland
Retroperitoneal. Located behind the peritoneum
Retropharyngeal. Located behind or dorsal to the pharynx
Retroversion. Backward tilt of an organ
Rima. A slitlike opening
Rima glottidis. The slitlike opening between the true vocal folds
Rima vestibuli. The slitlike opening between the vestibular or false vocal folds
Rotation. Circular motion around an axis
Rotator. A muscle that rotates a part
Ruga (pl. **rugae**). A fold or wrinkle; gastric rugae are folds in the lining of the stomach
Rupture. To break or to tear apart; hernia

Sac. An anatomic structure that resembles a bag or pouch
Saccule. A small sac
Sacral. Relating to the sacrum
Sacroiliac. Relating to both the sacrum and the ilium, e.g., the sacroiliac joint
Sacrum. Curved, triangular bone made up of five fused vertebrae that are inferior to the lumbar vertebrae and are wedged between the two hip bones
Sagittal. A vertical plane that divides the body into right and left portions
Salivary. Relating to saliva; salivary glands produce saliva
Saphenous. Relating to either of two long, superficial veins in the leg, the great (long) saphenous vein and the small

(short) saphenous vein; the great (long) saphenous vein is the longest vein in the body
Sartorius. A long, straplike muscle that courses obliquely across the anterior thigh and flexes the thigh and leg
Scalene muscles. A group of muscles associated with the first and second ribs
Scalp. The skin that covers the cranium
Scapula (pl. **scapulae**). Part of the pectoral girdle that provides the attachment for the upper extremity; also called the shoulder blade
Scapular. Relating to the scapula
Sciatic. Relating to the hip or ischium, such as the sciatic nerve
Sclera. The outermost layer of the eyeball
Scrotum. The pouch or sac that encloses the testes and the lower part of the spermatic cord
Sebaceous. Relating to a fatty or oily substance called sebum
Sella turcica. A depression on the upper surface of the sphenoid bone that marks the location of the pituitary gland
Semimembranosus. One of the hamstring muscles located on the thigh
Seminal. Relating to the semen; seminal vesicles are paired glands, posterior and inferior to the bladder in the male, that secrete a component of semen into the ejaculatory duct
Seminiferous tubules. Tightly coiled ducts, located in the testes, where spermatozoa are produced
Semitendinosus. One of the hamstring muscles located on the thigh
Septal. Relating to a septum
Septum (pl. **septa**). A wall or partition
Septum pellucidum. A thin partition between the two lateral ventricles in the brain
Sequela (pl. **sequelae**). An abnormal condition, which follows and is caused by, another disease
Sigmoid. Crooked, or having the shape of the letter S; the sigmoid colon is the S-shaped portion of the colon, between the descending colon and the rectum
Sinoatrial node. The portion of the conduction system of the heart that initiates and establishes the rhythm of the heart beat; also called the pacemaker or the SA node
Sinus. An air-filled cavity in a cranial bone; a dilated channel for the passage of blood or lymph
Sinusoid. Like a sinus; venous channels in some organs, such as the liver and spleen
Skeletal. Relating to the skeleton or bone
Skullcap. The top portion of the cranium; also called the calvaria
Soleus. Muscle in the calf of the leg, deep to the gastrocnemius muscle; plantarflexes of the foot
Somatic. Pertaining to the body or the soma
Sperm. Spermatozoa; mature male gametes or germ cells
Spermatic cord. Supporting structure of the male reproductive system that contains the vas deferens, the blood vessels, the lymphatics, the cremaster muscle, and the connective tissue
Spermatogenesis. The formation of spermatozoa
Spermatozoon (pl. **spermatozoa**). Mature male gamete or germ cell; sperm
Sphenoethmoidal. Relating to both the sphenoid bone and the ethmoid bone
Sphenoidal. Relating to the sphenoid bone
Sphincter. A circular muscle that closes an opening when it contracts
Spinal. Relating to a spine; relating to the vertebral column
Spine. A short projection of bone, a spinous process; the vertebral column
Splenic. Relating to the spleen
Squamous. Flat or scaly
Squama (pl. **squamae**). Thin, flat portion of the temporal bone; squamous region of the temporal bone
Stapes. One of three ear ossicles in the middle ear; attached to the oval window
Stasis. A stoppage or halt in the normal flow of fluids
Stenson's duct. The duct from the parotid gland that empties into the oral cavity; also called the parotid duct
Sternal. Relating to the sternum
Sternum. A long, flat bone that forms the anterior portion of the thoracic cage; also called the breast bone
Stratified. Consisting of, or arranged in, many layers
Stroma (pl. **stromata**). The supporting or connective tissue framework of an organ, as opposed to its functional part, the parenchyma
Styloid process. A long, pointed process on the temporal bone
Substantia nigra. A region of deeply pigmented cells in the cerebral peduncles
Sulcus (pl. **sulci**). A groove or furrow
Superficial. On, or near, the surface; shallow, such as a superficial wound
Superior. Located higher, or toward the head; the upper surface of an organ
Supination. Act of lying on the back; rotation of the forearm so that the palm of the hand is forward or upward
Supinator. A muscle that supinates the forearm
Supine. Lying on the back
Sura. Calf of the leg
Sutures. Immovable joints that unite bones of the skull
Sympathetic. One of the two divisions of the autonomic nervous system
Symphysis. A slightly movable joint that has a pad of fibrocartilage between the opposing bones, such as the symphysis pubis
Synapse. The junction or region of communication between nerve cells
Syndrome. A set of signs and symptoms that are characteristic of a particular abnormal condition
Synovial joints. Freely movable articulations characterized by a membrane that secretes a fluid for lubrication of the joint
Systemic. Relating to, or affecting, the entire body

Tailbone. The coccyx
Talus. One of the tarsal bones that articulates with the tibia and fibula to form the ankle joint
Tarsal. Relating to the bones that form the heel, ankle, and instep of the foot
Tarsus. A collective term for the seven bones that form the heel, ankle, and instep of the foot
Temporal. Relating to and the bone that forms the temple or side of the head
Temporalis. Muscle of mastication that extends from the temporal bone to the mandible
Tendinous. Having the nature of or relating to a tendon
Tendon. A cord of dense, fibrous connective tissue that attaches muscle to bone
Tensor fasciae latae. Superficial muscle of the lateral thigh; muscle that flexes and abducts the thigh
Tentorium cerebelli. An extension of dura mater that acts as a partition between the cerebrum and the cerebellum
Teres major and minor. Muscles associated with movement of the shoulder joint
Tertiary. Relating to the number three
Testicular. Relating to the testes
Testis (pl. **testes**). Male gonad that produces spermatozoa and the male hormone testosterone; also called the testicle
Thalamic. Relating to the thalamus
Thalamus (pl. **thalami**). An oval mass of gray matter, located

on either side of the third ventricle in the cerebrum; primarily functions as a relay center for sensory impulses
Thigh. The proximal portion of the lower extremity, located between the hip and the knee
Thoracic. Relating to the thorax or chest
Thorax. The upper part of the body, between the neck and the diaphragm; the chest
Thrombus. A blood clot in an unbroken blood vessel, located at its point of formation, usually within a vein
Thymus. A lymphoid structure, located in the mediastinum posterior to the sternum; plays a role in the development of the immune system
Thyroid. A bilobed endocrine gland, located on either side of the trachea, that functions in the regulation of the body's metabolism
Tibia. Long bone in the thigh
Tibial. Relating to the tibia
Tibialis anterior. Muscle along the anterior surface of the tibia
Tonsil. A small mass of lymphoid tissue embedded in mucous membrane
Tortuous. Twisted
Trabecula (pl. **trabeculae**). A supporting cord of connective tissue
Trabeculae carneae. Muscular bands or ridges of myocardium on the inner surface of the ventricles of the heart
Trachea. Passageway for air between the larynx and the bronchi, also called the windpipe
Tracheal. Relating to the trachea
Transected. Cut across
Transverse. Directional term indicating horizontal or crosswise direction
Trapezius. Superficial muscle of the upper back
Trauma. An injury
Traverse. To navigate, cross, bridge, or span
Triceps brachii. Large muscle with three heads, located on the posterior surface of the arm that functions to extend the arm at the elbow
Tricuspid. Having three cusps, the valve between the right atrium and the right ventricle
Trigeminal. Cranial nerve V
Trigone. Triangular area on the floor of the bladder marked by the openings for the two ureters and the urethra
Trochanter. One of two large prominences on the proximal femur
Trochlea. Any pulleylike structure or surface
Trochlear. Cranial nerve IV
Tubercle. A rounded elevation on a bone
Tuberosity. A rounded projection from a bone
Tunic. A layer or coating
Turbinate. Shaped like a scroll; a bone that is shaped like a scroll or conch shell; one of the nasal conchae

Ulna. The medial bone of the forearm
Ulnar. Relating to the ulna
Umbilical. Relating to the umbilicus or navel
Umbilicus. Region on the abdomen that marks the former attachment of the umbilical cord; also called the navel
Uncinate. Shaped like a hook; an extension or region of the pancreas
Undifferentiated. Not specialized, usually in reference to cells
Uniaxial. Allows movement along one axis, such as a hinge joint
Unmyelinated. Does not possess myelin, a fatty sheath; refers to nerve axons
Ureter. Tubular structure that conveys urine from the renal pelvis to the urinary bladder
Urethra. Tubular structure that conveys urine from the urinary bladder to the exterior
Urogenital. Relating to both the urinary (renal) system and the genital (reproductive) system; also called genitourinary
Uterus. The hollow, muscular organ in the female pelvis that is the site of menstruation, implantation of the fertilized egg, and the development of the embryo and fetus; also called the womb
Utricle. A part of the membranous labyrinth, located in the vestibule of the inner ear, that contains sense receptors for static equilibrium
Uvula. The fleshy projection at the posterior end of the soft palate

Vagina. The muscular tube that extends from the vestibule to the cervix of the uterus in the female
Vaginal. Relating to the vagina
Vagus. Cranial nerve X
Vallecula. A shallow groove; the groove between the epiglottis and the root of the tongue
Vas (pl. **vasa**). A channel or duct, which conveys a liquid
Vascular. Pertaining to, or containing, vessels
Vasculature. The blood vessels of an organ
Vastus muscles. Part of the quadriceps femoris group of muscles on the anterior thigh: vastus lateralis, vastus intermedius, and vastus medialis
Vein. A vessel that carries blood toward the heart
Vena (pl. **venae**). Latin for vein
Venous. Relating to veins
Ventral. Pertaining to the front or anterior side of the body; opposite of dorsal
Ventricle. A chamber or cavity, especially in the heart or brain
Ventricular. Pertaining to a ventricle
Vermiform. Shaped like a worm, such as the vermiform appendix
Vermis. Latin for worm; the central part of the cerebellum that connects the two cerebellar hemispheres
Vertebra (pl. **vertebrae**). One of the bones that form the spinal column or backbone
Vertebral. Relating to the vertebrae
Vesicle. A small pouch or sac that contains liquid
Vestibule. A chamber or space, such as the vestibule of the larynx or vestibule of the inner ear
Vestibulocochlear. Cranial nerve VIII; functions in the sense of hearing and equilibrium
Villus (pl. **villi**). A small, hairlike projection from the surface of a membrane
Viscera. The organs inside the ventral body cavity
Visceral. Relating to the viscera
Vitreous body. The soft, gelatinous substance that fills the posterior portion of the eyeball, between the lens and the retina
Vomer. Bone of the face that forms the lower portion of the nasal septum
Vulva. Collective term for the external genitalia of the female; also called the pudendum

Wharton's duct. Duct of the submandibular salivary gland; also called submandibular duct
Wharton's jelly. Homogeneous intercellular substance of the umbilical cord; also called mucous connective tissue

Xiphoid. Shaped like a sword; the distal portion of the sternum

Zygomatic. Relating to the cheekbone or the zygomatic bone, such as the zygomatic process or the zygomatic arch
Zygomatic bone. Cheekbone; also called zygoma
Zygote. Fertilized egg; single cell, resulting from the union of male and female gametes

COMBINING FORMS

Many anatomic and medical terms are made up of combinations of word roots, or combining forms, together with prefixes or suffixes. Some of the more commonly used word roots and combining forms and their definitions are listed here, along with examples of their uses in a word and their definitions.

Ab- away from; *abduct*—to take away from
Abdomino- abdomen; *abdominopelvic cavity*—the portion of the ventral body cavity that includes the abdomen and the pelvis
Acro- extremity; *acromion process*—the extremity or extreme end of the spine of the scapula
Ad- toward; *adduct*—to move toward the axis of the body
Alb- white; *linea alba*—vertical white line in the center of the abdomen
Alveol- cavity, socket; *alveolus*—air cavity in the lung
Ante- before; *antecubital*—region in front of the elbow
Antero- in front, ventral; *anterolateral*—located in front and to one side
Atrio- relating to an atrium of the heart; *atrioventricular valve*—valve between an atrium and a ventricle

Bi- two; *biceps brachii*—muscle in the arm with two heads
Brachi- arm; brachial artery—the principal artery in the arm
Bronch- relating to the bronchi; *bronchoscopy*—direct visual examination of the bronchi

Capit- relating to the head; *capitulum*—a head-shaped eminence on a bone
Cardi- heart; *cardiology*—study of the heart and its diseases
Cephal- head; *cephalad*—toward the head
Cerebello- relating to the cerebellum; *cerebellomedullary*—relating to the cerebellum and the medulla oblongata
Cerebro- brain; *cerebrospinal fluid*—fluid that circulates around the brain and spinal cord
Chondr- cartilage; *hypochondriac region*—abdominal region below the cartilage of the ribs
Coraco- relating to the coracoid process of the scapula; *coracoacromial ligament*—ligament between the coracoid process and the acromion process
Cor-, Coron- heart; *coronary arteries*—arteries that supply blood to the heart muscle
Cost- rib; *costal cartilage*—the cartilage that connects the ribs to the sternum
Crani- skull; *craniotomy*—surgical opening of the skull
Cysti- sac or bladder; *cystic duct*—duct from the gallbladder

Dorsi- denotes a relationship to the dorsal or the posterior surface; *dorsilateral*—on the posterior surface and to one side
Duodeno- denotes a relationship to the first part of the small intestine; *duodenojejunal flexure*—turn at the junction of the duodenum and the jejunum
Dura- hard or tough; *dura mater*—outer, tough membrane that covers the brain and the spinal cord

Ecto- outside; *ectopic pregnancy*—pregnancy or gestation outside the uterus
Endo- inside; *endocardium*—membrane that lines the inside of the heart wall
Epi- above or upon; *epidural*—above the dura mater
Ex-, Exo-, Extra- out or away from; *extrinsic muscles*—muscles outside the structure being acted upon

Fibro- denotes fibers; *fibrocartilage*—cartilage with an abundance of white fibers in the matrix

Gastr- stomach; *gastritis*—inflammation of the stomach
Gingiv- gum; *gingivitis*—inflammation of the gums
Glosso- tongue; *glossopharyngeal nerve*—nerve that provides innervation for the tongue

Hepato- liver; *hepatitis*—inflammation of the liver
Hyper- beyond or excessive; *hypertrophy*—excessive growth of a tissue
Hypo- under, below, deficient; *hypoglossal*—below the tongue
Hyster- uterus; *hysterectomy*—surgical removal of the uterus

Ileo- ileum; *ileocecal valve*—junction of the ileum and the cecum
Ilio- ilium; *iliosacral*—pertaining to both the ilium and the sacrum
Infra- beneath; *infraorbital foramen*—foramen below the orbit of the eye
Inter- among or between; *intercostal muscles*—the muscles between the ribs
Intra- within or inside; *intracranial pressure*—pressure inside the cranium

Labi- lip; *glenoid labrum*—rim or lip of fibrocartilage around the margin of the glenoid fossa
Laryngo- relates to the larynx; *laryngopharynx*—the portion of the pharynx in the region of the larynx
Lieno- relates to the spleen; *lienorenal ligament*—peritoneal ligament between the spleen and the kidney
Lingua- tongue; *lingual frenulum*—fold of tissue that anchors the tongue to the floor of the mouth
Lumbo- lower back, loin; *lumbar vertebrae*—vertebrae in the lower back region

Meningo- membrane; *meningitis*—inflammation of the membranes around the brain and spinal cord
Meta- after or beyond; *metacarpals*—bones of the hand, beyond the wrist

Metr- uterus; *endometrium*—lining of the uterus
Mid- located in the middle; *midsagittal*—sagittal section in the middle, divides into equal right and left portions
Musculo- muscular system; *musculoskeletal*—pertains to both the muscular system and the skeletal system
Myo- muscle; *myocardium*—muscle in the heart wall

Naso- referring to the nose; *nasolacrimal duct*—duct that drains tears into the nasal cavity
Nephro- kidney; *nephron*—basic functional unit of the kidney
Neuro- nerve; *neuromuscular*—pertaining to both nerves and muscles

Oculo- eye; *oculomotor*—nerve that transmits impulses to the muscles of the eye
Odont- tooth; *odontoid process*—toothlike process on the second cervical vertebra
Oo- egg; *oocyte*—egg cell
Ophthalm- eye; *ophthalmology*—study of the eye and its diseases
Orchid- testicle; *cryptorchidism*—failure of the testicle to descend into the scrotum
Oro- mouth; *oropharynx*—region of the pharynx posterior to the cavity of the mouth
Oss-, osseo-, osteo- bone; *osseous*—containing bone

Palpebr- eyelid; *levator palpebrae superioris*—muscle that raises the upper eyelid
Para- adjacent, alongside, or deviation from normal; *parasagittal section*—a sagittal section that is not in the midline
Parieto- relates to the wall of a cavity; *parietal peritoneum*—layer of peritoneum that lines the wall of the abdominal cavity
Patho- disease; *pathology*—study of disease mechanisms
Peri- around; *pericardium*—membrane around the heart
Phon- voice, sound; *phonation*—the production of sound
Phren- diaphragm; *phrenic nerve*—nerve that stimulates the diaphragm
Postero- a direction toward the back of the body; *posterolateral*—toward the sides on the back of the body
Pre- before, either in time or in space; *prenatal*—before birth
Pulmon- lung; *pulmonary*—relating to the lungs

Quadr-, quadri- of or relating to the number four; *quadriceps femoris*—muscle group that contains four muscles

Radio- relating to radiation; *radiograph*—picture formed by radiation
Radio- relating to the radius; *radioulnar joint*—articulation between the radius and the ulna
Recto- relating to the rectum; *rectouterine pouch*—pouch or cul-de-sac between the rectum and uterus
Retro- situated behind or backward; *retroperitoneal*—located behind the peritoneum

Sacr-, sacri- relating to the sacrum; *sacroiliac*—relating to both the sacrum and the ilium
Semi- denoting half or partial; *semicircular canals*—canals that are shaped like half of a circle
Spheno- relating to the sphenoid bone; *sphenoethmoidal*—relates to both the sphenoid bone and the ethmoid bone
Sterno- relating to the sternum; *sternopericardial ligament*—membranous ligament located between the sternum and the pericardium
Stylo- relating to the styloid process of the temporal bone; *stylomastoid foramen*—foramen located between the styloid process and the mastoid
Sub- below or less than; *submandibular*—below the mandible
Supero- relating to superior; *superomedial*—in a direction that is superior and medial to the reference
Super-, supra- meaning above or too much; *supraspinatus muscle*—muscle above the spine of the scapula

Trans- across, through, or beyond; *transected*—cut through
Tri- three; *trigone*—a triangular region, marked by three openings, on the floor of the urinary bladder

Uro- relating to urine; *urogenital organs*—organs that pertain to the urinary system and the genital system
Utero- relating to the uterus; *uterovesicle pouch*—peritoneal pouch between the uterus and the urinary bladder

Vas- vessel or duct; *cerebrovascular*—pertaining to the blood vessels of the brain
Vertebro- relating to the vertebrae; *vertebrosternal ribs*—ribs that are attached to the vertebrae and the sternum
Vesico- relating to the urinary bladder; *vesicoprostatic*—relating to the urinary bladder and the prostate gland
Viscer- organ; *visceral peritoneum*—peritoneum that covers the abdominal organs

EPONYMS

Eponyms are names of diseases, structures, or procedures that include the name of an individual. The current trend is to avoid, or at least to discourage, the use of eponyms because they are nondescriptive and often vague. Eponyms are still widely used, however, especially in clinical practice. This list has been prepared to help you relate common eponyms to the preferred terminology.

Eponym	Preferred Terminology
Achilles tendon	calcaneal tendon
Adam's apple	thyroid cartilage
ampulla of Vater	hepatopancreatic ampulla
antrum of Highmore	maxillary sinus
aqueduct of Sylvius	cerebral aqueduct
Bartholin's duct	major sublingual duct
Bartholin's gland	greater vestibular gland
bundle of His	atrioventricular (AV) bundle
canal of Schlemm	scleral venous sinus
circle of Willis	cerebral arterial circle
Cooper's ligament	suspensory ligament of the breast
Cowper's gland	bulbourethral gland
crypt of Lieberkuhn	intestinal gland
duct of Rivinus	lesser sublingual duct
duct of Wirsung	pancreatic duct
Eustachian tube	auditory tube
Fallopian tube	uterine tube
fissure of Rolando	central fissure (sulcus)
fissure of Sylvius	lateral cerebral fissure
foramen of Luschka	lateral aperture
foramen of Magendie	median aperture
foramen of Monro	interventricular foramen
foramen of Winslow	epiploic foramen
graafian follicle	vesicular ovarian follicle
island of Reil	insula
ligament of Treitz	suspensory muscle of the duodenum
organ of Corti	spiral organ
Peyer's patches	aggregated lymphatic follicles
pouch of Douglas	rectouterine pouch
Purkinje fibers	conduction myofibers
Skene's gland	paraurethral gland
Stensen's duct	parotid duct
Wharton's duct	submandibular duct
Wharton's jelly	mucous connective tissue

INDEX

Note: Page numbers in *italics* refer to illustrations;
page numbers followed by (t) refer to tables.

J

K

L

M

T

U

V